D0026218

Programmable Controllers
& Designing
Sequential Logic

The Saunders College Publishing Series in Electronics Technology

Bennett
Advanced Circuit Analysis
ISBN: 0-15-510843-4

Brey
The Motorola Microprocessor Family: 68000, 68010, 68020, 68030, and 68040; Programming and Interfacing with Applications
ISBN: 0-03-026403-5

Carr
Elements of Microwave Technology
ISBN: 0-15-522102-7

Carr
Integrated Electronics: Operational Amplifiers and Linear IC's with Applications
ISBN: 0-15-541360-0

Driscoll
Data Communications
ISBN: 0-03-026637-8

Filer/Leinonen
Programmable Controllers and Designing Sequential Logic
ISBN: 0-03-032322-3

Garrod/Borns
Digital Logic: Analysis, Application and Design
ISBN: 0-03-023099-3

Greenfield
The 68HC11 Microcontroller
ISBN: 0-03-0515882

Grant
Understanding Lightwave Transmission: Applications of Fiber Optics
ISBN: 0-15-592874-0

Harrison
Transform Methods in Circuit Analysis
ISBN: 0-03-020724-X

Hazen
Exploring Electronic Devices
ISBN: 0-03-028533-X

Hazen
Fundamentals of DC and AC Circuits
ISBN: 0-03-028538-0

Hazen
Experiencing Electricity and Electronics
Conventional Current Version
ISBN: 0-03-007747-8
Electron Flow Version
ISBN: 0-03-03427-2

Ismail/Rooney
Digital Concepts and Applications
ISBN: 0-03-026628-9

Laverghetta
Analog Communications for Technology
ISBN: 0-03-029403-7

Leach
Discrete and Integrated Circuit Electronics
ISBN: 0-03-020844-0

Ludeman
Introduction to Electronic Devices and Circuits
ISBN: 0-03-009538-7

McBride
Computer Troubleshooting and Maintenance
ISBN: 0-15-512663-6

Oppenheimer
Survey of Electronics
ISBN: 0-03-020842-4

Prestopnik
Digital Electronics: Concepts and Applications for Digital Design
ISBN: 0-03-026757-9

Seeger
Introduction to Microprocessors with the Intel 8085
ISBN: 0-15-543527-2

Spiteri
Robotics Technology
ISBN: 0-03-020858-0

Wagner
Digital Electronics
ISBN: 0-15-517636-6

Winzer
Linear Integrated Circuits
ISBN: 0-03-03246-8

Yatsko/Hata
Circuits: Principles, Analysis, and Simulation
ISBN: 0-03-00933-2

two-year associate degrees to graduate engineering students. George is responsible for training people from industry through Allen-Bradley's training programs, and his classes range through an equally wide spectrum. This indicates that training is needed at all levels so programmable controllers can be understood and used effectively.

Philosophy

This text is designed for a hands-on approach for learning how to use a PLC. Our ultimate goal is to take the student to a point where they could independently write a properly functioning control program on a PLC. When introducing concepts, we use simple and easy to understand examples so ideas are clear. The students' exercises help crystalize these concepts by guiding them through problems which start out easy but become progressively more challenging. Programmable controllers initially evolved because relay ladder logic control had become too expensive to design and change. This book is unique because it explains how this evolution took place and thoroughly explains relay ladder logic in the beginning chapters. Having this background makes it much easier to understand the instructions unique to PLC's. The manuscript was tested by using it for courses at Michigan Tech. The proof that the text was working was in seeing the students' projects at the end of the second term. They could work independently with little supervision using the PLC effectively to control a process.

Content Overview

The first seven chapters are intended to be background information for the students so that they know the basics necessary to use any programmable controller. Some of these chapters can be skipped depending on what the students have had before they start the course. Almost all students, taking the course I teach, will already have had digital and numbering systems. Chapters 8 through 14 introduce the Allen-Bradley PLC-5 instructions. Chapters 9 and 10 introduce the basic primary instruction necessary to write most programs and Chapters 11 through 14 introduce the more complex specialized instruction. Chapters 15 through 17 put it all together by teaching the students how to write programs in a structured disciplined way, how to communicate with other devices, and introduce advanced programming techniques. Chapter 15 is unique in that it shows the students how to write programs in a structured manner. One criticism of ladder logic is that it is generally done by trial and error; this chapter teaches a structured approach. State, Petri, and Sequential Function charts are studied in depth and example control programs are given for each technique. Finally, Chapter 18 deals pragmatically with the difficult task of selecting a PLC.

Course Length/Prerequisites

This text is designed to be taught over two terms. The first term would be introductory and used to teach students the basics (using Chapters 1 through 9) of how

Preface

The introduction of programmable controllers and their acceptance in industry made it imperative that a course be developed at the institution where I was teaching, Michigan Tech. The first task was deciding what such a course should cover and what objectives needed to be met. The second task was to find a book that matched this content and those objectives. Initially, there were very few books to choose from other than manufacturer's instruction manuals. Although effective, these manuals tended to be too terse, with almost no practical examples or practice problems for the students use. We used these manuals for several years waiting for books to be published which contained the normal pedagogy necessary for teaching college students. During this period, I began developing my own notes and labs to use for class. The books then being published either lacked the content I wanted or had nothing about the specific programmable controller I was using for class. I find it very ineffective to use books which use the generic approach.

Finally, I decided that the only way I was going to get what I wanted was to write a book myself. My one nagging concern however, was how to get up-to-date, state-of-the-art information, given that programmable controllers were changing so fast. I needed someone in industry who was knowledgeable. I asked George Leinonen, a former teaching associate then working for Allen-Bradley, to write the book with me, and to my great pleasure he agreed. George and I had co-taught courses together at Michigan Tech. before he went to work for Allen-Bradley, and we knew we could work together.

This book is a result of that collaboration between George and me, and it is written around Allen-Bradley's flagship programmer controller the PLC-5. The micro controller version of the PLC-5 is the SLC-500, and it also uses the PLC-5 instruction set. Allen-Bradley is the most popular controller sold in the USA and therefore was the logical choice for this book.

Audience

The book is targeted for teaching technicians and engineers how to use and program programmable controllers. This would include engineers in a baccalaureate program, technicians in associate or in a baccalaureate technology program, and industrial personnel. The mix in my classes at Michigan Tech has ranged from

To my wife, Barbara, and my children,
Mark, Kristain, Anne, and Sarah

To my wife, Charlotte, and my children,
Jodi and Cory

Text Typeface: Times Roman
Compositor: BookMasters, Inc.
Acquisitions Editor: Barbara Gingery
Assistant Editor: Laura Shur
Managing Editor: Carol Field
Project Editor: Michael Bass & Associates
Copy Editor: Elliot Simon
Manager of Art and Design: Carol Bleistine
Art Director: Doris Bruey
Art and Design Coordinator: Caroline McGowan
Text Designer: Rebecca Lemna
Cover Designer: Laurence Didona
Text Artwork: Rolin Graphics
Director of EDP: Tim Frelick
Production Manager: Charlene Squibb

Printed in the United States of America

Programmable Controllers and Designing Sequential Logic (TITLE)

0-03-032322-3 (ISBN)

Library of Congress Catalog Card Number:

0123 0016 987654321

THIS BOOK IS PRINTED ON **ACID-FREE, RECYCLED** PAPER

Programmable Controllers & Designing Sequential Logic

Robert Filer
Michigan Technological University

George Leinonen
Allen-Bradley Company

SAUNDERS COLLEGE PUBLISHING
A Harcourt Brace Jovanovich College Publisher

Ft. Worth Philadelphia San Diego Chicago San Francisco
New York Orlando Austin San Antonio Montreal Toronto
London Sydney Tokyo

to use the PLC-5; and structured programming through State diagrams is covered in Chapter 15. This course requires a minimum of prerequisites. The student should have had AC and DC circuit theory and college algebra; digital circuits is suggested. An advance project course would be taught in the second term. The text would be used to teach the more complex uses and instruction dealt with in Chapters 11 through 18 and would also be used as a reference as the students work on a project involving a fairly complex control. The second-term Project course is more a matter of maturity and motivation. Because of this, it is better to teach this at the sophomore level or above.

Pedagogical Features

Each chapter includes the following design features to make it easier for the student to follow and use the text:

Chapter outline
Chapter objectives
Chapter introduction or preview
Figures to support text
Easy to follow worked example
Chapter summary
Exercises

End of text Material

Glossary
JIC standards Graphic Symbols for Electrical Control Diagrams
Example projects for Programmable Controllers
Answers for odd numbered exercises

Ancillaries

Instructors Solution Manual to end of chapter problems

Acknowledgments

There are many people who have helped us along the way. Special thanks to the reviewers who were very helpful and gave constructive criticism:

William Mack, Harrisburg Area Community College
Sohil Anwar, Luzerne County Community College
Richard Polanin, Illinois Central College
Daniel Cronauer, Luzerne County Community College
Russell Jerd, Sinclaire Community College

The Saunders Staff was the catalyst and provided essential support:

Barbara Gingery, Senior Acquisitions Editor

Laura Shur, Assistant Editor

Charlene Squibb, Production Manager

Michael Bass & Associates

Michigan Technological University provided essential environment and support needed for writing a text. Phyllis Tapani and Pammi Kuivanen provided secretarial help.

My wife, Barbara, (even though she has the responsibility of a full-time job and taking care of a household with demands from our five-year-old daughter, Sarah) gave suggestions, provided proofreading, and exercised patience while I was writing. Peter Tampas provided the support I needed and obtained two programmable controllers through John Rodkey of Allen-Bradley. These PLC-5's were essential for testing the control circuit employed in this book. My students were most helpful in testing the book's usefulness and in providing feedback for required changes. Two in particular, John Suomi and David Loch, helped test the control circuits in the chapters by entering the control into the PLC-5 and checking that these programs worked properly. This they did with little supervision from me, which drew on their patience and fortitude and demanded a lot of hard work as they pored over instruction manuals to figure out how to do things. Mark Kilpela was continually called on to add auxiliary equipment to programmable controllers and to keep equipment repaired and running properly. I also pestered my good friend, Dale Walivaara, to look for technical problems in controls I came up with, or to advise on writing. I am sure I have overlooked many others who have helped me, and even though they aren't specifically mentioned, I wish to thank them.

Robert Filer
January 1992

Contents

1 Sequential Control

OUTLINE

OBJECTIVES

Upon completion of this chapter, the student will be able to:

- Describe how sequential control has evolved and what equipment has been used as this type of control modernized.
- Define *sequential control,* and explain the difference between it and automatic control.
- Describe how ladder logic has evolved and why it is still the preferred language for programming programmable controllers.
- Explain the role programmable controllers play in replacing relay logic.
- Give the advantages and disadvantages of relay logic and programmable controllers.
- Cite some of the computer languages presently used in programmable controls.
- Describe the factors influencing the development of sequential control, and give a pragmatic view of decision making in industry.
- Discuss how industry changes as new equipment evolves.

THIS chapter will give a brief history of the evolution of sequential control, including the equipment that has been used as this type of control modernized. The reasons for changing or not changing to various types of control will be discussed. Sequential control will be defined, and the difference between it and automatic control will be explained. Included will be how ladder logic has evolved and why it is still the preferred language for programming programmable controllers. The present and future roles of other languages will also be discussed. This chapter will help the novice technician or engineer understand what factors are influencing the development of sequential control and give a pragmatic view of decision making in industry. It is intended to help the reader gain some insight into how industry changes as new equipment evolves.

1-1 History

An essential part of the development of the present-day programmable controller was the introduction of the integrated circuit (IC) in 1959. This made solid-state control in small, low-cost packages possible. It took about ten years for IC's to become sophisticated, readily available, and inexpensive. Figure 1-1 shows a typical integrated circuit.

The first programmable controller was designed by the Hydra-Matic Division of General Motors in 1968. Prior to that year, electromechanical relays had been used that required expensive rewiring and reworking with each yearly model change. General Motors was looking for a way to save money on this costly process. The new programmable controllers were employed on a limited basis, and it took some time to work out the bugs and make this type of control reliable and cost effective.

A real boost came with Intel's introduction of the microprocessor in 1971. The microprocessor put a central processing unit (CPU) on a single chip and made it possible to put a computer on a single printed circuit board by adding memory and input/output (I/O) ports. High-level programming languages and development systems that would allow easy development of programming for these CPU's made the microprocessor a natural for the brains of a programmable controller. Microprocessors have continually become more powerful through an increase in the number of digits of data they can process at one time. The first ones intro-

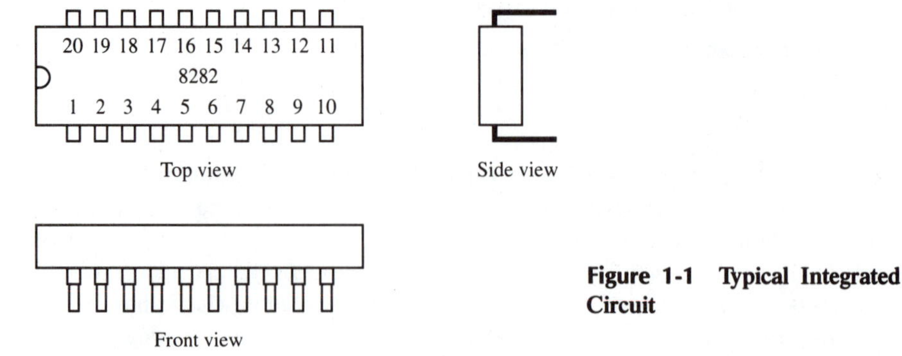

Figure 1-1 Typical Integrated Circuit

duced could only handle four digits, or bits. Today, 32-bit processors are both readily available and inexpensive.

The memory that is required for computer power and speed also made great strides during this period. The first memory-IC's had a capacity of 1 kilobyte. Today, 1-megabyte memory chips are common. These two things together make it possible to put powerful computers in small packages. Such computers are approaching the large, mainframe capabilities of just a few years back. Single-board computers are the backbone of the modern programmable controller (see Figure 1-2).

The development of the modern programmable controller pretty much parallels that of the single-board computer. There was a time lag of a few years until the newly introduced computer components became readily available and inexpensive. The introduction of very small and inexpensive PLC's, commonly called *micro PLC's,* has made it practical to use programmable controllers for control on a very small scale. A relay control that required half a dozen relays can be replaced economically by one of these micros. Such PLC's are priced low, and a few hundred dollars can buy a flexible, sophisticated, versatile control.

Figure 1-2 Modern Programmable Controller
(Courtesy of Allen-Bradley, a Rockwell International Company)

1-2 Definitions of Sequential Control, Programmable Controller, and Relay

An automobile plant provides a classic example of the need for sequential control. The processes needed to build a car must follow one another in a particular order. You can't put the tires on if the axles haven't been installed, for instance. Most of these processes need to have some equipment turned on for a period of time and then turned off. Controlling when a device is to be on or off is accomplished via sensing devices such as limit switches. These switches can also generate discrete types of information in a form such as off/on. Although today's programmable controllers can handle analog feedback that is continuously changing and adjusting, we will deal with control that uses discrete, or binary, information. The definitions to be used in this book follow.

Sequential Control is a process that dictates the correct order of events and ensures that one event will occur only after the completion of another. We will also stick to **discrete control,** in which inputs and outputs are binary in form, such as off and on.

Automatic Control is a process in which an analog feedback is used to continuously change and adjust the output. Once feedback is involved, special precautions are required to ensure that the system is stable. For further details, the reader is referred to a book specifically on the topic of automatic control.

Programmable Controller A **programmable controller** is a computer that has been hardened to work in an industrial environment and is equipped with special I/O and a control programming language. The common abbreviation used in industry for these devices, PC, can be confusing because that is also the abbreviation for *personal computer.* Therefore, some manufacturers refer to their programmable controller as a *PLC,* which stand for *programmable logic controller.* We will use PLC in this book to avoid confusion.

Relay The term *relay* has different meanings, depending on the industry involved. In the power industry it covers devices that range from a simple, general-purpose relay with an actuating coil and contacts (see Figure 1-3a) to complicated electromechanical and solid-state devices (see Figure 1-3b) that communicate over microwave links to exchange information in order to make decisions. The sophisticated relays do not resemble the simple electromechanical relays used in sequential control. In this book, relay will refer to the simple general-purpose or industrial control relay with an actuating coil and contacts (see Figures 1-3a and c). The industrial relay is a heavy-duty version of the general-purpose relay.

1-3 The Roles of Training and Equipment Reliability

That present-day programmable controllers use relay ladder logic is basically due to the role the relay played in the development of sequential logic. The relay is a very simple electrical device that has been around almost as long as people have been using electricity. It was first employed as a repeater for telegrams in the

Figure 1-3a Simple, General-purpose Relay *(Courtesy of Allen-Bradley)*

1840s. Its simplicity and reliability have kept it the predominant device in control for well over a hundred years. Solid-state devices did not successfully start replacing relays in control circuits until the microprocessor chip came along, bringing with it significant advantages.

One can observe, without any special equipment, whether a relay is operating properly. It can easily be made tough and capable of withstanding heavy power surges and harsh environmental conditions. The power that operates the relay can readily be isolated from the contacts on the relay so that the two don't interfere with each other. As long as the control isn't too complicated, very little training is required to troubleshoot and maintain this type of logic. These advantages took a while for solid-state devices to overcome.

The advantages of solid-state devices over relays were irrelevant until the solid-state devices could be proven reliable in harsh environments. Also, the

Figure 1-3b Sophisticated Relay *(Courtesy of Upper Peninsula Power Co.)*

amount of training and the complexity of equipment needed to troubleshoot solid-state devices initially was a problem. When solid-state devices were first introduced in the electrical power area, the manufacturers promised long life, because solid-state devices have no moving parts. Some of the problems with these devices did not become apparent until they were put into environments that were truly hostile. Many new installations failed because of power surges nearby that could couple with these devices either electrostatically or magnetically. Such surges didn't affect relays but were devastating to solid-state devices. When solid-state relays were first introduced in the electrical power industry, failures would occur for various reasons. Here are some examples.

EXAMPLE 1-1

A conductor would be used to feed an input signal to a solid-state relay, and this conductor would be bundled in the same harness as a conductor feed power to an

Figure 1-3c Industrial Relay *(Courtesy of Allen-Bradley)*

alarm bell to indicate low oil pressure in a generator. When the alarm bell went off, the bell would be rung by making and then breaking the electrical connection to a coil that operated the hammer that strikes the bell housing. This put high-voltage sparks on the circuit feeding the bell due to the inductive kick when the bell's coil was interrupted. This high-voltage surge could be coupled to the input of the solid-state relay through conductors that shared the same harness and ran side by side for relatively long distances.

Whereas modern solid-state relays have surge protection on their inputs, the ones first introduced did not. The many failures initially encountered made those who tried these devices back away from their use.

EXAMPLE 1-2

Solid-state equipment was damaged by the operation of lightning arrestors during a normal lightning storm. A control with solid-state devices could be mounted on a power pole to control an automatic circuit interrupter mounted at the top of the pole. Any lightning strike would cause a ground wire to carry heavy current, which would be taken to the ground by a conductor running down the side of the pole. On that same pole would be the control cable going from the control near the ground up to the device it was controlling (see Figure 1-4). A lightning arrestor gave the lightning a low-impedance path to ground, thereby limiting equipment

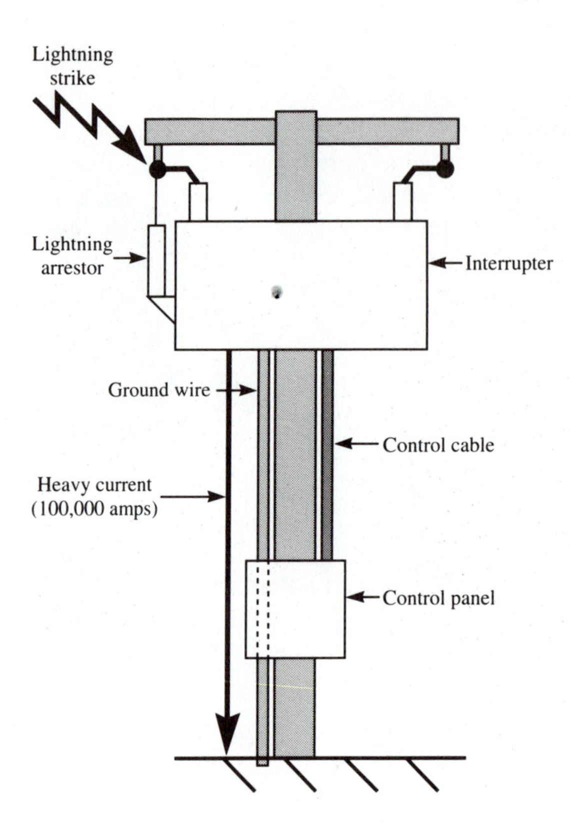

Figure 1-4 Example of a High-current-surge Problem at a Power Pole

damage. The current that must be conducted to ground can be easily 100,000 amps. So much current coming down the pole would set up a strong magnetic field that would couple to the control cable between the solid-state control and the device it was controlling.

Such surges initially did extensive damage to the solid-state controls. Then the industry learned that a grounded coaxial cable and surge protection were needed to protect the controls.

E X A M P L E 1 - 3

Arc welding produces surges on a power system, thereby causing failures. If solid-state control shares the same power source as arc welders on a production line, a malfunction occurs because of the power surges created by the welding. By simple resistive coupling, these surges would be transferred to any equipment sharing the same power supply. (*Resistive coupling* occurs when two processes share the same circuit or have conductors or conduction paths with some or no resistance in common.)

EXAMPLE 1-4

Failures were caused by extreme industrial environments, such as warm sunlight shining on a control box during the day followed by extreme cold at night. Such sudden and drastic changes sometimes caused solid-state devices to fail. Certain environments, such as foundries, are dirty and hot, and sometimes the same process produces a corrosive atmosphere. These environments cause solid-state controls to fail.

EXAMPLE 1-5

Solid-state devices operate on low voltage and cannot be mixed with the power voltages. If 120-V AC comes in contact with the low-voltage feeding, the solid-state equipment will be damaged and will fail. It is imperative that this isolation be maintained, because any carelessness could result in a failure. Michigan Tech. has a relay panel it has been using for about ten years (see Figure 1-5), and the mechanical relays have had perhaps one failure due to constant use and abuse by students. However, the solid-state relays on these panels are regularly damaged by students and need to be replaced. Putting 120-V AC in a 5–30-V DC input does the damage. The mechanical relays are not hurt by putting the DC into the AC.

Figure 1-5 Training Panels *(Courtesy of Michigan Technical University)*

In these examples, there was little or no effect on mechanical relays, which operated reliably. Any environmental problems were easily rectified by putting the mechanical relays in simple enclosures, and ventilation often wasn't required because such relays can tolerate a wide range of temperatures. However, for solid-state devices it took a while for these problems to be discovered and rectified. The initial introduction of solid-state devices was made without sufficiently taking these kinds of problems into account, and this caused equipment failures and unhappy customers.

Troubleshooting when these solid-state control devices failed was also a problem. Technicians required more training and needed more sophisticated equipment. Mechanical relays were simple, visual, reliable, and very easy to troubleshoot.

Consider the impact of a manufacturer trying a control that caused a plant assembly line to be shut down because the control failed. Let's say it took several hours to find the failed components and that the unit had to be replaced. To further complicate matters, let's say it failed again shortly after being replaced. The failed unit eventually would be removed, to be replaced by something more reliable and proven. It would be a long time before that particular plant would use that type of control again, and the word would also be passed around to other professionals.

Due to such problems, solid-state devices were slow to succeed in power-control areas. But since solid-state devices were reliable in low-power applications such as communication, these devices started to evolve rapidly with the introduction of the transistor and, later, integrated circuits. Before solid-state devices substantially penetrated the control for power, the computer had evolved. The lessons learned with solid-state devices did not have to be relearned when applied to the computer, for computer manufacturers already knew that the computer needed to be protected from harsh environments encountered in the power industry. Out of this came a computer designed to operate in such environments and to interface with power equipment. This device is the programmable controller. In other words, a programmable controller is really a computer with inputs and outputs (I/O) that has been designed to operate in electrically and mechanically harsh environments.

Your home computer could be made into a programmable controller if you provided some way for the computer to receive information from devices such as push buttons or switches. You would also need a program to process the inputs and decide the means of turning off and on devices such as motors. Having a way for your computer to react and work with external noncomputer devices is known as *interfacing*. A normal computer doesn't have I/O that can work with power voltages such as 120-V AC, but a programmable controller does. You would have to come up with some way of turning off and on 120-V devices with the digital signals your computer puts out, which are usually 0–5 V. You also must make sure that your computer is isolated from the 120-V system or it could damage the computer's circuits.

Again, a programmable controller is a computer with special I/O and hardening to survive in harsh environments. When the programmable controller was first introduced to manufacturers, calling it a computer presented a problem. Many of the control engineers in the late 1960s had been educated and trained before small computers were available, and they were not interested in learning the complicated procedures required to run these early computers. It seemed from their experience with mainframe computers that you practically had to have a degree in computer science to operate the machines of the 1960s and early 1970s. It wasn't until the late 1970s and early 1980s that the personal computer started to become user-friendly.

In order to get the first programmable controllers accepted by the sequential-control industry, they needed to be called by a name other than *computer,* so the term *programmable controller* was devised. A language had to be developed that didn't look like a computer language and that was user-friendly. That language was *relay ladder logic,* and it was designed so people could use it immediately without special training. It is tough to sell a new device that will solve all your problems if you have to learn a special computer language to use it. The industry was not ready to accept a new device that required training to operate and that hadn't yet proved itself. If it wanted to break into this area, the PLC industry had little choice but to create a special language and a keyboard with symbols familiar to people using control. This shouldn't be too surprising, because many new computer languages have evolved for special applications in industry, for example, Fortran for engineering and COBOL for business.

Once PLC's with an easily used and understood language were available, acceptance began. Considering that the language was developed specifically for this field, it will probably predominate for some time. Other languages may evolve, but it is unlikely that a language developed for another field will take over a language developed specifically for this field. New technicians and engineers familiar with other languages but unfamiliar with relay ladder logic might want to use other languages for this type of control. However, it will take a while before newly trained technicians and engineers reach the decision-making level of management.

One language that has become popular due to the introduction of personal computers is BASIC, which nowadays is introduced at the high school level and sometimes even in grade school. It is fairly easy to use, and many people operating personal computers know it. Most programmable controllers manufactured today have available for programming both relay ladder logic and BASIC. Since the younger generation is more comfortable with BASIC than with relay ladder logic, BASIC may well become more popular. Also, a language designed specifically for control that lends itself to a structured approach to design may develop.

Relay ladder logic tends to be done in a nonstructured fashion. A lot of control has been designed basically by trial and error: Make a control for something; if it doesn't work, modify it until it does. This was true in early control when prototype control was created. Computers, which are powerful tools for a structured approach to design, were unavailable when plants were originally automating production. The trial-and-error approach is actually a powerful method when systems are not too complicated; however, once a system get complicated, such an approach can be inefficient and costly. It is much harder to troubleshoot if a control scheme lacks a structured layout. A programmable controller makes a structured approach to design easy. We will introduce a method in Chapter 15 for making relay logic control schemes that lets the control engineer or technician use relay ladder logic, and thus not have to learn a new programming language, but at the same time design control in a very structured manner.

Programmable controllers cannot replace all relay applications, for example, motor starters, where the size of the current is out of the range of current and voltages that the normal interface devices on a PLC can handle. Also, for some applications a PLC is both too expensive and an overkill. When only a few relays are needed and no changes are anticipated, a relay control may be appropriate.

Because these powerful controllers are revolutionizing the control industry, the demand for PLC's has resulted in over 100 manufacturers of programmable controllers and a billion-dollar industry by the early 1990s. What does the PLC have to offer that is making it so successful? Here are some of its advantages.

Advantages of the PLC

Cost Effectiveness If an application has more than a half-dozen relays, it probably will be less expensive to install a PLC. One problem with relay control panels is the expense of making controls for special applications that are one of a kind. This automatically means there can be no cost reduction by making a production run of multiple units of one particular type of control. But PLC's are mass produced and thus inexpensive. At the same time, the cost advantage of mass production is realized; timers, counters, sequencers, and other devices can be simulated, and the number of such devices is determined by how much memory is available. Cost savings are enormous because only a few bytes of memory costing only a few cents are required to simulate each device, whereas the actual devices would cost several dollars. The number of such devices one can have is basically a function of the amount of memory in the PLC. Simulating a hundred relays, timers, and counters is not a problem even on small PLC's. The amount of inexpensive memory available for computers has constantly increased. Machines with more memory can do more tasks faster, and a wide range of memory capabilities has evolved in programmable controllers.

Flexibility The PLC can be moved from one application to another simply by reprogramming it, whereas a hard-wired relay control must be physically rewired, which is prohibitively expensive. It has been so expensive that often a relay control has been scrapped and a new one built rather than to modify an old one. The advantage of the programmable controller is that it can easily be adapted and changed to fit one application or another.

Powerful Computer Capabilities Since the heart of the PLC is actually a computer, you really get the capabilities of a computer with the control, which means you can collect and process data. The computer can diagnose and indicate problems to aid in troubleshooting. One PLC can communicate with another PLC or, for that matter, with another computer. Controls can be easily displayed and documented. A printout of the control scheme can be made.

Ease in Troubleshooting Control A monitor or display can show the control schematic, to reveal what is happening in the control in real time. This lets a person review the control and make changes simply by entering a command at a keyboard. Because of this, programmable controllers have evolved that can do far more than the relay controls they are replacing. Modular design has simplified troubleshooting. The PLC runs diagnostic tests and indicates when a unit is bad. The usual way of fixing the problem is simply to replace a particular module.

Reliability Today's PLC is a reliable and powerful means of implementing sequential control as well as automatic control with feedback. It is accepted as reliable, and is continually overtaking control, whereas relays used to be the mainstay.

Readily Available Training Training in PLC operation is offered by the industries that produce the PLC's as well as by colleges and universities with courses for both technicians and engineers. Training can be justified because of the numerous advantages. The benefits of PLC's are so great that it is essential that engineers and technicians have some training in this area.

It would be a serious mistake not to consider a programmable controller in any application that involves more than a few relays. Any engineer or technician who lacks training in this area is unprepared to work with most manufacturers or industries that use sequential control. Electrical engineers and technicians will obviously work with these devices, but mechanical, chemical, civil, metallurgical, and other engineers should learn about this device too. If a process is sequential, a programmable controller can probably do it better and less expensively than other available controls. Manufacturers hiring engineers or technicians expect new graduates to be familiar with programmable controllers and to know how to use them.

The key features and advantages of PLC's are summarized in Table 1-1.

In the chapters that follow, we will introduce the basic concepts that will enable the reader to understand the programmable controller as well as the relay logic that has evolved as the predominant language for programming controllers. The first chapters will deal with relays, how they work, and how to use them to design control. A great number of these controls still exist, and learning about them makes it easy to understand how to program controllers with relay ladder

Table 1-1 Summary of Advantages and Key Features of the PLC

Feature	Advantage
Cost effectiveness	Less expensive than relay control
Flexibility	Can be moved from one application to another by reprogramming
Powerful computer capabilities	Data collection and processing easily done
Ease of troubleshooting	Reduces downtime
Reliability	Requires minimum maintenance
Training availability	Minimizes personnel training time and expense

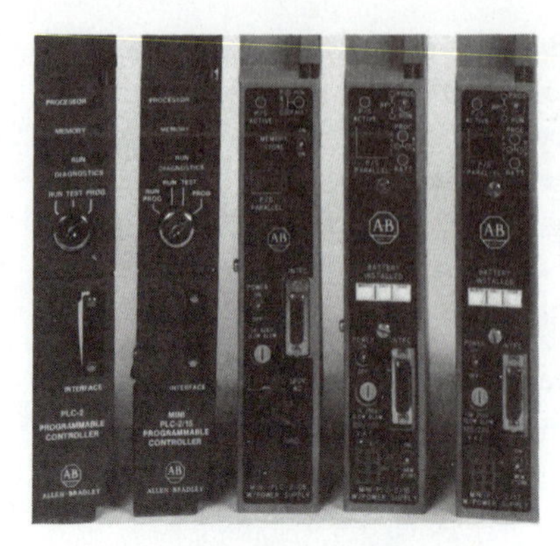

Figure 1-6 Programmable Controllers *(Courtesy of Allen-Bradley)*

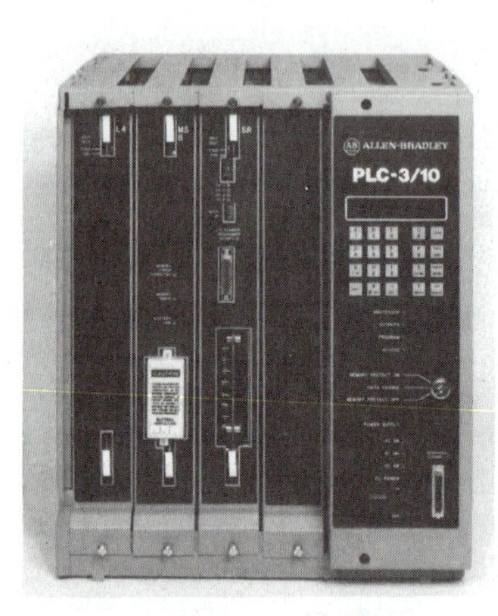

logic. Many of the control schemes that were designed with relays have proven themselves through years of reliable performance and thus should be examined for their benefits and advantages. This information can help us design new controls. Relay logic is explained so this can be done.

1-4 Equipment Available

The equipment available from various manufacturers is an indication of the strength of the market for PLC's. There are many manufacturers, and the range of what is available is considerable. Small PLC's are available at a competitive price to replace a 4–10 relay control. Add-on modules for interfacing have become extensive. These include analog-to-digital and digital-to-analog modules; special communication modules; AC- or DC-discrete on/off control at various voltage levels; TTL-compatible digital modules; intelligent modules for proportional, integral and derivative control; language modules; and low-level-input modules for connecting directly to transducers. The equipment available now ranges from small PLC's to large, versatile PLC's (Figure 1-6). It is difficult to categorize

PLC's because the small units now have the capability of much bigger units, and what is available is constantly changing. For a while, the amount of I/O and memory available were used to distinguish among small, medium, and large PLC's; but this has become inadequate to describe modern PLC's. (Chapter 18 will cover in detail how to select a PLC for a particular application.)

Summary

Sequential control can be implemented by three different technologies: Relay control was the first, and is still being used. The second technology was solid-state, or digital, control which is a minor player because it introduced the technology to build inexpensive electronic computers. The third, and latest, technology is the microcomputer, which has many important advantages over the other two. Microcomputer technology has been responsible for the programmable controller, which is now the device of choice for both discrete sequential control and automatic control. All three types of control will be encountered by technicians and engineers as they go out into industry. A well-rounded professional should be able to work with and understand each of these methods of implementing control. A knowledge of relay logic and digital logic will help the engineer understand the present-day programmable controller. Learning how to use and work with a programmable controller is paramount for the modern technician or engineer.

The dominant player in the programmable controller arena is Allen-Bradley, which is the IBM of progammable controllers. We will use the Allen-Bradley PLC-5, a state-of-the-art machine, for teaching the PLC.

Exercises

1-1. List some advantages of a programmable controller over relays for implementing a control.

1-2. List five programming languages used for either computers or programmable controllers.

1-3. What are some advantages and disadvantages of relay logic?

1-4. Why was the relay logic language made available on PLC's?

1-5. When a new product is introduced, what are some major factors that will determine its acceptance?

1-6. Explain how relay ladder logic influenced the development of programming for PLC's.

1-7. What is the difference between a personal computer and a programmable controller?

1-8. What is the difference between discrete sequential control and automatic control?

1-9. Name a manufacturing process in which you would consider employing a programmable controller. Give an example of how it would be used.

1-10. What steps would be required to move a programmable controller from one application to another?

1-11. Discuss some of the problems the solid-state industry had when it introduced new solid-state control into control where high-voltage power was used.

1-12. What determines how many devices such as relay coils, contacts, counters, and timers a programmable controller can simulate?

2

Relays

OUTLINE

OBJECTIVES

Upon completion of this chapter, the student will be able to:

- Explain how a relay is constructed and how it operates.
- Cite different types of relays, list their advantages and disadvantages, and explain why they are needed.
- Discuss standard ratings and their purpose.
- Give examples of simple control circuits, such as motor control, and explain how devices in such controls function.
- Explain why the relay is still the device of choice in controls for certain applications.
- Explain the function of normally opened vs. normally closed contacts and how these are shown in controls.
- Describe the effect of DC vs. AC power sources on contact ratings.
- Select the correct type of relay, properly rated, for an application.

THIS chapter will explain how a control relay is constructed and operates. The advantages and disadvantages of relays will be discussed. It will briefly give an overview of some special-purpose relays. Examples of simple control circuits, such as motor control, will illustrate the function of these devices. The role of the relay in these examples will show that it is still the device of choice for certain applications. The state of relay shown in control will lead to an explanation of normally opened vs. normally closed contacts. DC vs. AC power sources and their effects on contact ratings will be covered.

2-1 Relay Construction

The relay is the workhorse of relay logic control, and it can be made in several configurations. We will use the balanced beam relay to explain how relays operate. Figure 2-1 shows the basic parts of this type of relay, in its deenergized state. *Deenergized* means no power is being applied to the coil. The coil is made by winding many turns of wire around a core of a magnetic material such as iron. In addition to being insulated, the wire is wound on a nonconducting form that is fitted over the iron core to isolate the coil from the rest of the relay. A spring pulls the beam down on the right side; consequently the beam goes up on the left side. Contacts are attached to the beam on the right, and the beam itself is made of a metal that will conduct electricity. A conductor is taken from the beam to a terminal so that an electrical connection can be made to the beam. The beam has silver-plated contact buttons attached to the top and bottom. Two other contact buttons are placed above and below the beam. These also are connected via a conductor to terminals so that outside connections can be made. The bottom contact button is held in a position so that when the spring is pulling down on the beam, the bottom contact button and the beam make contact. This pair forms what is called a *normally closed contact*. When the relay's coil is deenergized, this contact is closed. The top contact button and the upper contact at this time are opened; these two contacts are referred to as *normally opened contacts.*

Figure 2-2 shows the relay coil energized. Current flow through the coil magnetizes the iron core within the coil. This magnetizes the iron thereby attracting the iron on the beam and pulling the beam down on the left-hand side. This force is enough to overcome the counterforce on the spring, and the beam moves up on the right side. The upward motion puts the top contact button on the beam in touch with the upper contact button, making electrical contact. The normally

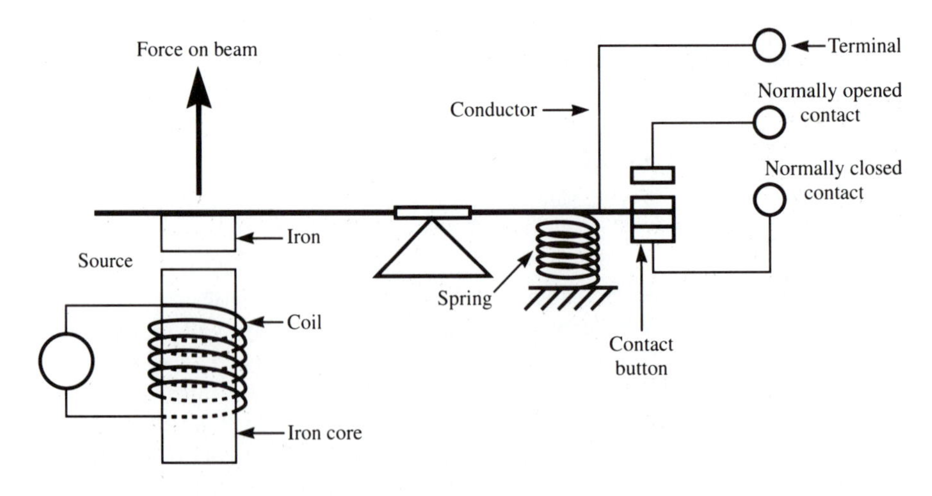

Figure 2-1 Balanced Beam Relay, Deenergized

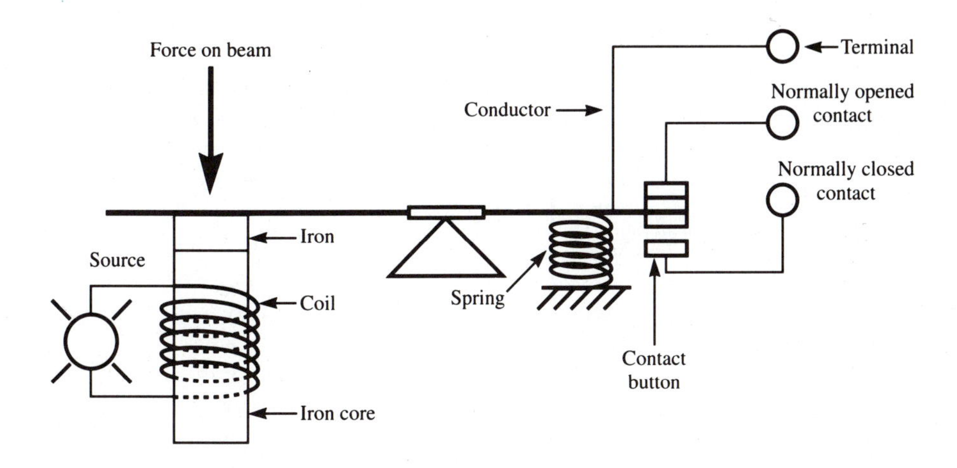

Figure 2-2 Balanced Beam Relay, Energized

opened contact now is closed. The lower contact button on the beam and the lower contact button move apart, and this normally closed contact opens.

Energizing the coil forces both pairs of contacts to change states. Thus, we can use these contacts to turn devices off and on. Figures 2-3 and 2-4 show how this relay is depicted schematically. By standard convention, relay control shows the contacts as they would be with the power off. This ensures that two different people looking at the same control scheme will interpret the control in the same way. A definition of a control relay follows.

Control Relays

A **control relay** is a low-power electromechanical device with a wound coil that can be activated and deactivated by applying voltage to the coil's inputs. Associated with this coil are sets of normally opened and normally closed contacts that change with the activating and deactivating of the coil.

Control relays are constructed in many different ways, but the operation is basically the same as just described. Relays are available with more than one normally opened and one normally closed contact. We could modify our example so that the coil pulled down on three beams at once and thus have three times the

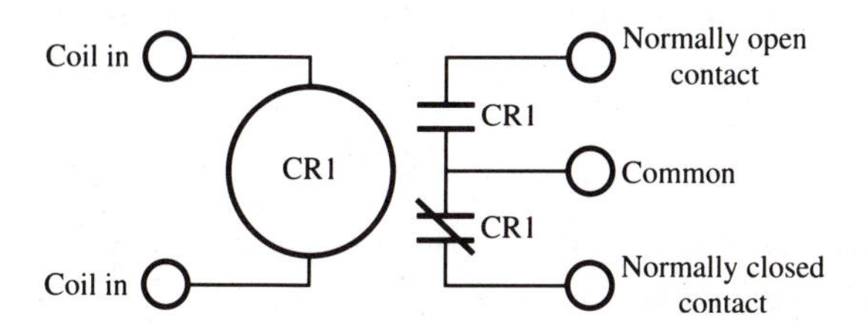

Figure 2-3 Relay, Shown Schematically

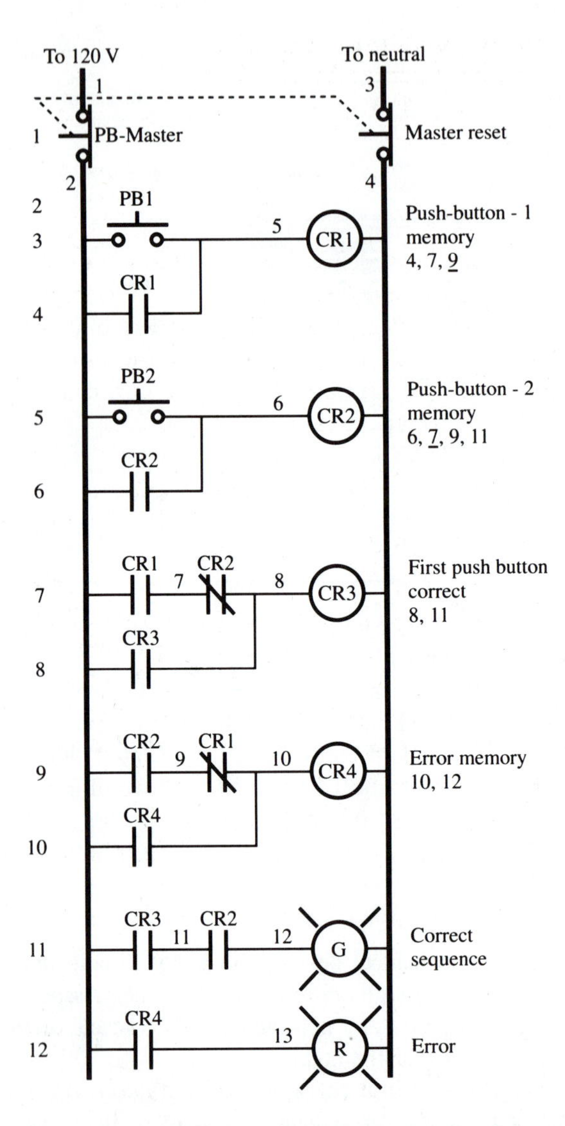

Figure 2-4 Relay Schematic

number of contacts. Control relays generally do not need to carry heavy currents or high voltages. The contacts are rated usually between 5 and 10 amps, with the most common rating for the coil voltage being 120-V AC. Other voltage ratings are sometimes used, and both AC and DC are included. The actual ratings depend somewhat on the application and the industry using them. See Table 2-1 for some of the standard ratings.

2-2 Contactors

Some relays, though operating basically the same as control relays, are made heavier, for special applications, such as turning on the field lights at a football

Table 2-1 Most Common Relay Voltages

12-V AC/DC	24-V AC/DC
48-V AC/DC	120-V AC/DC
230-V AC	480 AC

night game. A good example is a *contactor,* which is a relay with much higher contact current ratings than a normal control relay. A **contactor** is a special type of relay designed to handle heavy power loads that are beyond the capability of control relays. The voltage for the football stadium lights will be higher than 120 V, and the amount of current needed will be way out of the range of a control relay. A control relay can pick up a contactor that is built to handle the heavy current and higher voltages. Programmable controllers have I/O capable of operating the contactor, but they do not operate the lights directly.

Programmable controllers cannot replace relays in general, however. The programmable controller is designed to replace the physically small control relays that make logic decisions but are not designed to handle heavy current or high voltage. Often, programmable controller advocates will claim that relays are obsolete, which is not the case. The major relay companies, such as Allen-Bradley, Cutler Hammer, General Electric, Westinghouse, and Square D, all sell programmable controllers. Relay sales are a healthy part of their business, making up more than 50% of their total sales in the early 1990s. The PLC does not compete with the heavy-current high-voltage relays, nor was it designed to do so.

2-3 Motor Starters

The motor starter is another example of a special-application relay. It is a contactor in series with an overload relay. The overload relay will open the supply voltage to the contactor if it detects an overload on a motor. It does this by putting heaters in series with the contactor supplying the voltage to the motor. When these heaters are heated by the current, their heat indirectly heats a eutectic element such as a bimetal strip, which trips a mechanical latch. Tripping this latch opens a set of contacts that are wired in series with the supply to the contactor feeding the motor. The characteristics of the heaters can be matched to the motor so that the motor is protected against overload. Thus, the **motor starter** is a relay specially designed to provide power to motors that has both a contactor relay and an overload relay connected in series and prewired so that if the overload operates, the contactor is deenergized. These starters come in various standard National Electric Manufacturers Association (NEMA) sizes and ratings (see Figure 2-5). They can work with motors from fractional horsepower to over 1000 horsepower. When a control relay or a PLC needs to control a large motor, it must work in conjunction with a starter. The input requirements to the starter coil are within the capabilities of the I/O of the PLC.

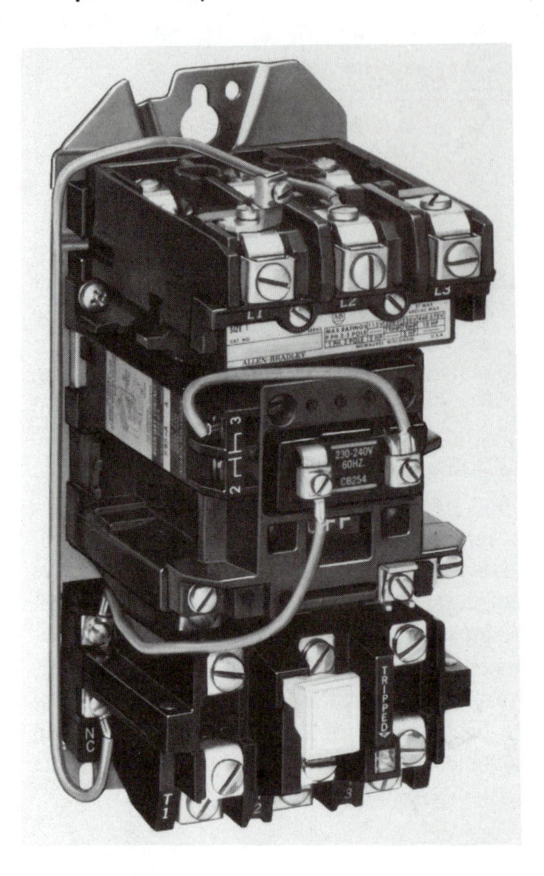

Figure 2-5 NEMA Standard Motor Starter *(Courtesy of Allen-Bradley)*

2-4 Solenoids

A **solenoid** is a very simple electromechanical device used to get mechanical displacement from an electrical signal. Its construction is straightforward. Basically, a solenoid is constructed by winding a coil on an iron core, as shown in Figure 2-6a. The iron core has one of its sides hinged and held open by a spring. Attached to the hinged piece of core is an operating rod that will move back and forth as the hinged piece moves. Applying a voltage to the source magnetizes the iron core, and the hinged piece is then pulled to the closed position, as in Figure 2-6b.

This type of device comes in many other configurations, but all work on the same principle as the one shown. The operating rod can be used to turn valves off and on or to open and close a door in a bin so that a chemical can be released. It can also operate clamps to hold a piece of work in place for a manufacturing process. It could be used to engage and disengage gears or to release a landing gear on an airplane. Solenoids frequently are used to operate valves when fluids are being controlled. It is easy to think of hundreds of applications. The solenoid is a common device that a programmable controller would operate as an output. It

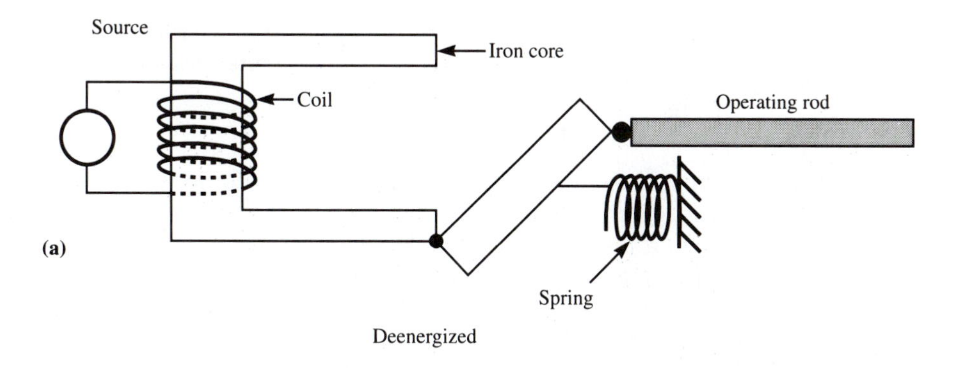

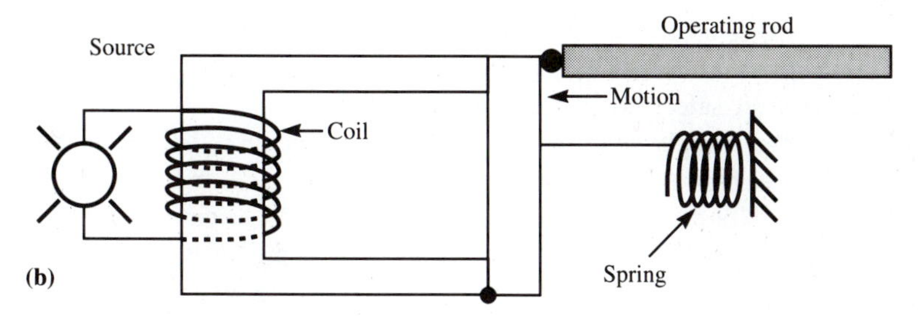

Figure 2-6 Solenoid

is mentioned in this chapter because of its frequent use. And we will employ it in our control example.

2-5 Indicating Lamps

Indicating lamps are common in relay controls, to show the person watching the control which devices are on and which are off, to indicate a dangerous situation, or to act as an alarm when necessary. Many different types of indicating lamps are available for various voltages, both AC and DC. Some have dropping resistors in series with the lamp to reduce the voltage to the lamp; others use small transformers. Different types of lamps are available, such as neon and incandescent, and they come in various sizes and styles. Some have a lamp test feature: pushing on the lens will test the lamp. Indicating lamps are mentioned here because you will hardly find a control scheme without them. They will be shown in control schemes as a circle with a letter inside. The letter indicates the color of the lens, such as R for red, G for green, etc. An example of the schematic symbol is shown in Figure 2-7.

2-6 Ratings

When purchasing any type of relay, it is important to consider ratings. Let us take a brief look at a few of the important ones.

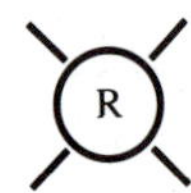

Figure 2-7 Schematic Symbol for an Indicating Light

The **rated voltage** is the voltage at which the device will operate properly when this voltage is applied continuously. Indicating lamps will have their life substantially shortened if this rating is exceeded.

The **rated current** is the amount of current the device can carry continuously without damage. It is particularly important that contact ratings not be exceeded, for this could cause welding of the contacts due to excessive heat and shorten contact life.

Exceeding voltage or current ratings could decrease the mean time to failure, or cause overheating, malfunctioning, or insulation failure. If voltage or current values are too excessive, safety could also be a problem.

The type of power supply also must be taken into account. DC is harder for contacts to interrupt than AC. This is because AC signals pass through current zero every half-cycle. The voltage caused by breaking inductive circuits usually is sufficient to establish an arc by ionizing the air, which produces a very low resistance path and requires little current to maintain the arc. Interrupting this arc is easier for an AC device because the arc will go out at every current zero, meaning the contact will be moving apart with sufficient speed to prevent restrike of the arc.

In contrast to AC contacts, DC contacts must be able to interrupt this arc without the help of current zero, and thus they require special design considerations. Special coils called *blow-out coils* create a strong magnetic field that is perpendicular to the current flow through the contacts and that forces the arc into an *arc shoot,* that is a special chamber that stretches and cools the ionized air. This ionized air is the conducting medium supporting the arc, and the stretching and cooling causes deionization. Deionized air is bad conductor, so the arc is extinguished. Contacts rated for AC but used on DC systems can fail to interrupt the arc, because they lack the special blow-out coils and arc shoots required. Continuous arcing will result in complete failure of the device.

Summary

Because of its simplicity and ruggedness, the relay has been employed for over a hundred years for control. It is still used today, though it will undoubtedly play a lesser role now that inexpensive and reliable programmable controllers are available. However, for certain applications and because of safety considerations, relays will still have a place. The next chapter will introduce relay logic, which will make it easier to understand ladder logic when introduced on the PLC. A technician or engineer will be required to understand and work with both relays and PLC's.

Exercises

2-1. Why are contacts shown in the deenergized state in control schemes?

2-2. Give some examples of relays that are not being replaced by programmable controllers. What type of relay *is* being replaced by programmable controllers?

2-3. Describe the difference between a motor starter and a contactor.

2-4. Which is harder to interrupt, AC or DC? Explain.

2-5. Describe some consequences of exceeding the voltage rating on a device.

2-6. Could you start a 50-horsepower motor with the output of a programmable controller or the contacts of a control relay? Explain.

2-7. How is isolation achieved between the source of power operating the coil of a relay and the source of power connected to the relay contacts?

3 Relay Logic

OUTLINE

OBJECTIVES

Upon completion of this chapter, the student will be able to:

- Show how logic functions can be created with control relays to make AND, OR, and EXclusive OR gates.
- Demonstrate basic control-design skills, such as memory techniques utilizing sealing-in relays.
- Use Joint International Congress (JIC) standard drafting practices and symbols to construct a control scheme.
- Generate a wiring diagram via the point-to-point method of hard-wiring a control scheme.
- Design a simple control, including correct labeling and symbols.

SIMPLE examples of ladder logic will be presented to show the function of a control relay and how it can be connected to make logic decisions. The connection required to make AND, OR, and EXclusive-OR gates will be explained and demonstrated. The basic skills for utilizing relays for memory will be shown. Joint International Congress (JIC) standards for drafting relay logic will be explained and demonstrated. The exercises will involve designing simple control problems, including correct use of labeling and symbols.

3-1 Control Relay

The term *relay* is used very loosely by people in industry and needs to be clarified. In this chapter—and the rest of the book—we will be talking only about the general-purpose control relay or the industrial control relay. (These simple relays with pickup coil and some contacts were shown in Figure 2-1.)

The **general-purpose control relay** is a simple electrical device with a coil that, when energized or deenergized, causes the states of its associated contacts to change from closed to opened or vice versa. The **industrial control relay** is similar to a general-purpose relay except that it is constructed more stoutly and its contacts generally are rated to withstand greater voltage and current. The industrial control relay is designed to operate reliably in industrial environments and thus is physically bigger and tougher. In Figure 3-1, note stout construction of this type of relay.

3-2 Logic Gates: AND, OR, EXclusive OR

Relay logic is discrete control implemented through the use of relay coils and contacts. We can create logic gates from relays by making the coils of the relays the input and arranging their associated contacts in the configurations shown in Figure 3-2. First let us look at an AND gate.

The **AND gate** is a device requiring all its inputs to be satisfied before providing continuity to its output. Figure 3-2 shows that this requires the contacts to be in series and that coils A and B both be energized. We would connect one side

Figure 3-1 Industrial Relays *(Courtesy of Allen-Bradley)*

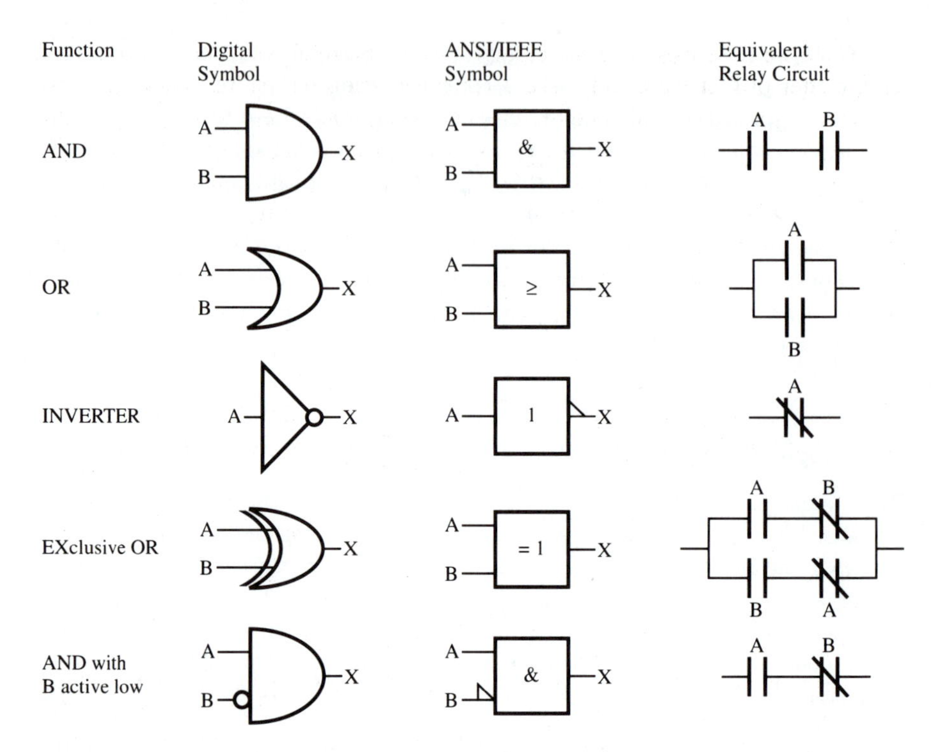

Figure 3-2 Logic Gates

of this series arrangements to 120-V AC and the other side to a device. If we energized both A and B, the device would turn on because power would be applied through the series contacts. Note that we need two relays and that we connect their normally opened contacts in series. Let us see what kind of decision we might make with this control circuit.

Suppose we were working on a control for an automatic car wash. In order for the water to be turned on to wash a car, the car would have to be in the starting position and a person would have to pay for the wash. A limit switch could be used to detect if the car was in the correct starting position, and a dollar-bill machine could detect if the money was paid. These would energize the appropriate relays if the requirements were met, and these contacts would operate a solenoid to turn a valve that would turn on the water to start the wash.

OR Gate

Changing the contacts so they are in parallel will make an OR gate. The **OR gate** device requires one or the other or both of its inputs to be satisfied before providing continuity to its output. If one or the other relay is energized, the power can turn the device on. An example application is an alarm. Say we are at the car wash again and need to let an attendant know when we are out of soap or wax. We could have detectors on either of these supplies, and when one or the other runs out, the detectors could energize a relay. The contacts could ring a bell if we wanted an audible alarm.

Once we have these two logic gates, we can make all kinds of decisions, and as a matter of fact we could make a computer using relays. To match what can be made with today's microchips, this computer would have to be huge in size and weight, and in addition it would use enormous amounts of electricity. Historically, sequential control using relays has been uncomplicated as far as decision making is concerned. The gates shown in Figure 3-2 generally are not combined into complicated circuits. Sequential control tends to make decision making straightforward and easy to follow because one thing follows another in an orderly fashion.

INVERTER

The INVERTER is implemented by using a normally closed contact. An **INVERTER** device causes the opposite condition of its input to occur at the output. When a relay is energized, this contact can be used to turn a device off by opening the circuit and thus removing power to it.

EXclusive OR

The EXclusive OR is a handy decision-making gate. The **EXclusive-OR gate** is a device that requires one or the other but not both of its inputs to be satisfied before providing continuity to its output. In Figure 3-2, the condition for turning a device on would be that either relay A or relay B, but not both A and B, could be energized. This is accomplished by putting a normally opened contact from one relay in series with a normally closed contact from another relay. Being in series ensures that only one relay can be active at a time. If we parallel a complementing set of these series contacts, in essence we are saying one or the other relay can be energized but not both.

An example of this is a standby generator at a remote transmitter for a radio station. Assume that we want to switch to standby power when the commercial power goes off, or back to commercial power when it is available again. We do not want the two power sources connected together, because they will not be in synchronization and one could be badly out of phase with the other, causing excessive current and subsequent damage. We could use an Exclusive OR in this application. Also, we might add a mechanical interlock, to further ensure that the two power circuits do not mix. This can be accomplished easily with relays by having the relays mounted side by side so that when the coil operates on one of the relays, a mechanism using the relay pickup motion moves a bar that blocks the pickup of the other. The opposite condition moves the bar to block one relay and permits the other to pick up. This type of mechanical device interlock gives additional protection against failure of the control circuit.

AND with Active Low Input

Finally, Figure 3-2 contains an AND gate with an active low input, to show that either active low or high signals can be used. Notice that with this gate we have to energize relay A and deenergize relay B to turn a device on.

Figure 3-2 shows that we can make any type of logic desired using relays. The relay control was the backbone of sequential control before the programmable controller came along. It still has a place in certain critical applications where power failure on the control system can create unsafe conditions. The relay control would always go to its deenergized state, which is completely predictable. The programmable controller must rely on a battery backup to ensure that control remains unchanged during a power failure, meaning the battery backup must not fail under these conditions. The relay control does not rely on a battery backup or solid-state memory to maintain the original control circuit. Its circuits are hardwired and will not change.

Figure 3-2 also gives the American National Standards Institute/Institute of Electrical and Electronic Engineers (ANSI/IEEE) standard symbols for the logic gates. These symbols were developed in the mid-1980s, and the industry's changeover to these standards is still going on. It probably will be many years before these standards are universally employed instead of depicting gates in the way they have historically been shown. The advantage of the new symbols is that they are easier to draw and interpret.

3-3 Memory (Seal-in)

In addition to the logic gates, we need some way of remembering when a condition has occurred. Figure 3-3, which shows the control circuit for memory, depicts how this can be accomplished with relays.

1. Push button 1 (PB1) is pushed, energizing control relay 1 (CR1).
2. CR1's normally opened contact in parallel with PB1 closes.
3. PB1 is released and its contacts open. But since a now-closed CR1 contact is in parallel with PB1, the coil of CR1 is still energized. The coil of CR1 will remain energized until the control power is turned off.

Note that once the normally opened contact of CR1 closes, it keeps the coil of CR1 permanently energized. This process is called *sealing in a relay,* and in effect it memorizes if that particular relay was ever picked up. This seal lets us make a decision or change decisions if a particular event has occurred. It also means that one cell of memory will require a relay to be sealed in. If many conditions need to be memorized, it will take many relays, which would be expensive. There is really no comparison between the cost of relay memory versus programmable memory, which is solid state. If memory requirements are to cover more than a few events, the programmable controller is obviously much more economical.

3-4 Joint International Congress (JIC) Standards

The long history and heavy use of relays for sequential control prior to the introduction of the programmable controller resulted in standards being developed for showing control schemes. There needed to be standard ways of showing devices as well as the control itself. The standard symbols and notation are given in Appendix A.

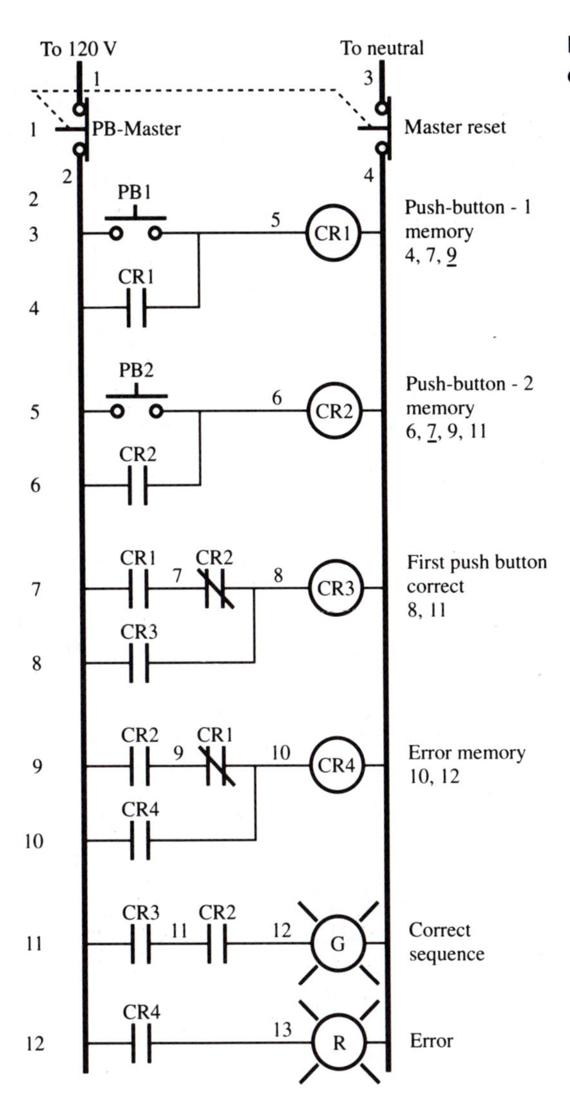

Figure 3-3 Typical Relay Ladder Diagram

The control circuit is normally fed by a control power transformer. The nongrounded supply from this transformer is shown as a vertical line on the left-hand side of the control scheme, and the grounded conductor is shown as a vertical line on the right. (Refer to Figure 3-3.) These will be the two nodes to which the control circuits will be connected. The nodes in the control scheme will be given numbers as node designations. The nongrounded power supply from the transformer is designated as 1 and the grounded supply as 2. The first line of control is drawn horizontally between the vertical supplies. The control-relay coils and indicating lights will be on the right, with one side connected to the nongrounded supply and the other connected to relay contacts. The contact and devices that will operate the control relays are shown between the nongrounded conductor and the coil or indicating lights.

Notice that the nodes are then labeled consecutively in numerical order. The control is then shown, one line after another, in this fashion. The resulting control when drawn looks like a ladder; hence, this type of control is aptly named *relay ladder logic*. Each rung of this ladder logic is labeled in consecutive order on the left side of the control. To the right of each control relay is a brief description of its function and its contact locations, which are designated by giving the line or rung on which they are shown. If the contact in the rung is normally opened, no special designation is required; if it is normally closed, then the number is underscored. This orderly way of showing the circuits helps in understanding and interpreting these schemes. The node designations are used to label wires when hard-wiring this control, and these labels aid in troubleshooting and testing.

EXAMPLE 3-1
RELAY SCHEME

Control Problem: Design a control scheme that requires a person to push two push buttons, one at a time in the proper sequence, to operate a particular machine. Once the two buttons have been pushed correctly, the machine will stay on until a reset button is pushed. The correct sequence is PB1 followed by PB2. If the wrong sequence occurs, a relay will detect this and light an error light. The error light will stay on until the control is reset. Once an error has occurred, no operation of the push buttons can turn on the machine.

Notice in Figure 3-3 that the relay coils and the indicating light are always on the right-hand side of the control and are in parallel. Two coils or a coil and a light are never connected in series. This is because these are voltage devices that require full voltage to operate properly. If they are put in series, they will divide the voltage according to their impedances and will not work properly because they will have less than their rated voltage at their terminals.

3-5 Wiring Diagrams

There are several methods used by industry for implementing the necessary wiring to get a control scheme operational. We will cover the most popular one, called the *point-to-point method*.

A **wiring diagram** shows the way electrical conductors are connected to devices to implement the desired control shown on the schematic. A structured approach is necessary, both to avoid wasting wire and to document the hard-wiring. Having documentation that shows how the wiring was accomplished is essential for troubleshooting. Thus, technicians and engineers must be able to read wiring diagrams. Likewise with student projects: If such projects are going to be passed from one class to another, the students need to document what they have done.

The point-to-point method of wiring is easy to learn and follow. It will be required wherever hard-wiring is used. A programmable control will *not* eliminate the need for a certain amount of hard-wiring, because many control cables must come into and out of the controller. For example, a car has many computers in it for control, but if you look under any dash you will see many bundles of wiring.

You need some initial drawings to start this process; we will go through what is required. The first drawing you will need is a control schematic showing the necessary connections between components. The nodes on the schematic need to be labeled. A node is a junction of two or more components. There are two types of node designations: those that give an indication of the function at that node and those that do not.

Let's look at the example schematic in Figure 3-4. There are two distinct portions. One is the power 3-line, which is composed of incoming power terminals, circuit breaker, starter, overloads, and motor. The other is the control portion, which is the part of the schematic fed by the step-down transformer. Notice in the power 3-line that we have used "live designations" because they tell us something about the node. An example of this type of node designation is node A, which is actually phase A of a three-phase source, or L1, which is line 1 of the incoming power. When we go to the control portion, the node designations are simple numbers that tell us nothing about what that particular node does. The node designation merely gives a label for identifying a particular node.

When the panel is wired, it is common practice to put labels on the wires that identify what nodes they are used for. This is essential for testing and later troubleshooting. Most control panels are wired with the same color of wire throughout, which means that without the labels you would have no way of knowing one wire from another. Color-coded wire can be employed, but this becomes prohibitively expensive in controls that must have large wire sizes to handle currents of 10 amps or more.

The second required drawing needs to show how the components will be mounted as well as the layout in the control panel. Figure 3-5 is an example, although this can be done any number of ways. From this drawing will be generated a wiring diagram layout.

Once we have both the control schematic with the nodes labeled and the panel layout, we can generate a wiring diagram as follows: First we create a third drawing—a one-dimensional layout drawing showing what a person will see when facing the components to be wired. In Figure 3-6, notice that component location is the same as in Figure 3-5. We have, however, added some information: a grid system similar to that on any road map. With these horizontal and vertical designations we can specify the location of any device. For instance, the UV relay is located at E3, which means at the intersection of column E and row 3. E3 is referred to as the *address* of the device. Notice that a symbol for the particular device at that location is shown and that its terminals are shown as circles. Next to the device is the device designation, which gives a clue to the function of that device in the control scheme. The necessary terminal blocks are shown and given address designations of TBX. *TB* stands for terminal block, and *X* will be a number to distinguish one terminal block from another. These terminal blocks are used

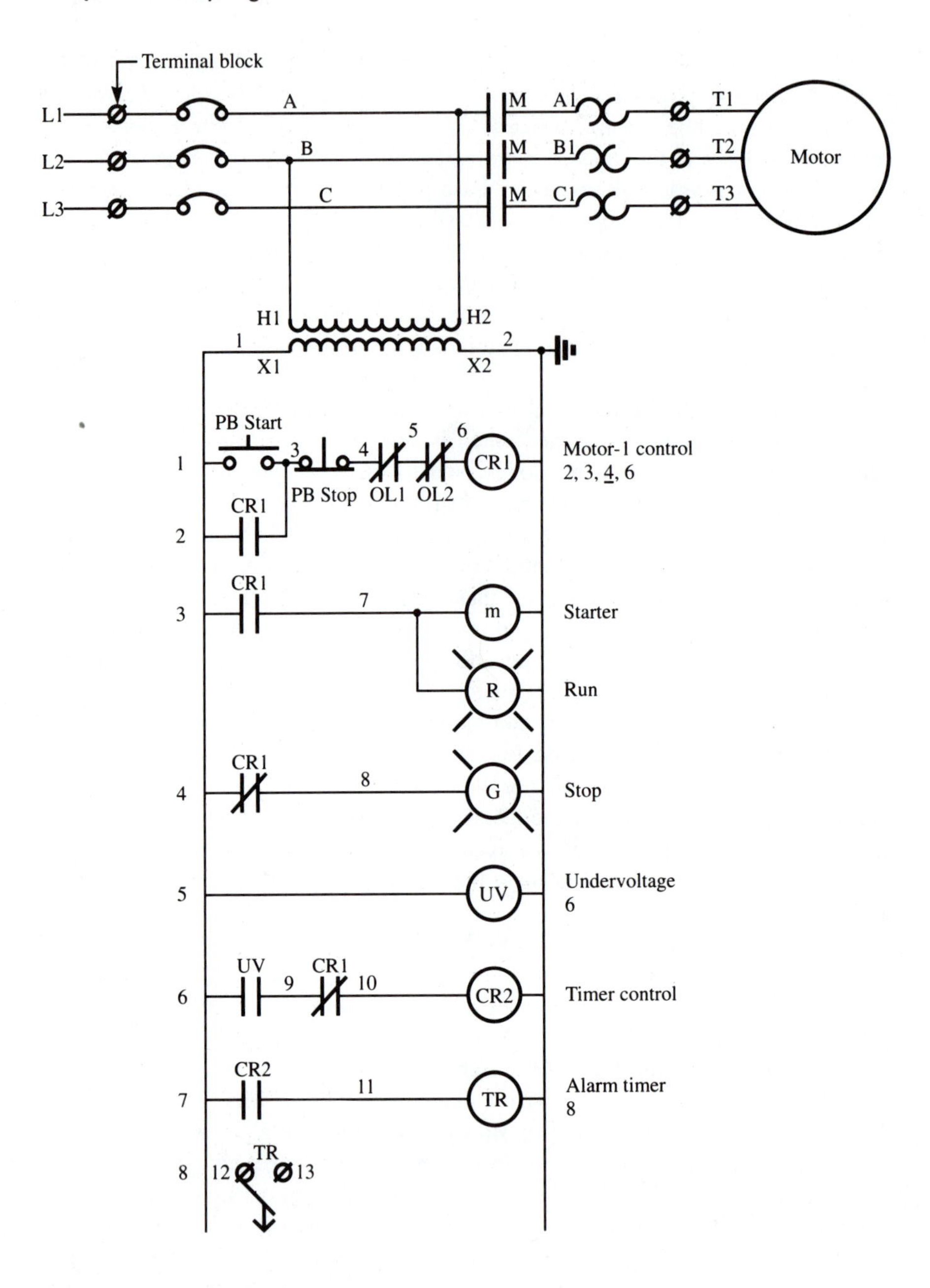

Figure 3-4 Control Schematic

to get wires into and out of the control panel in an organized fashion. Their addresses could be given by the grid system, but usually they are not because they are easy to find and have strips for labeling. The node designation goes on these strips.

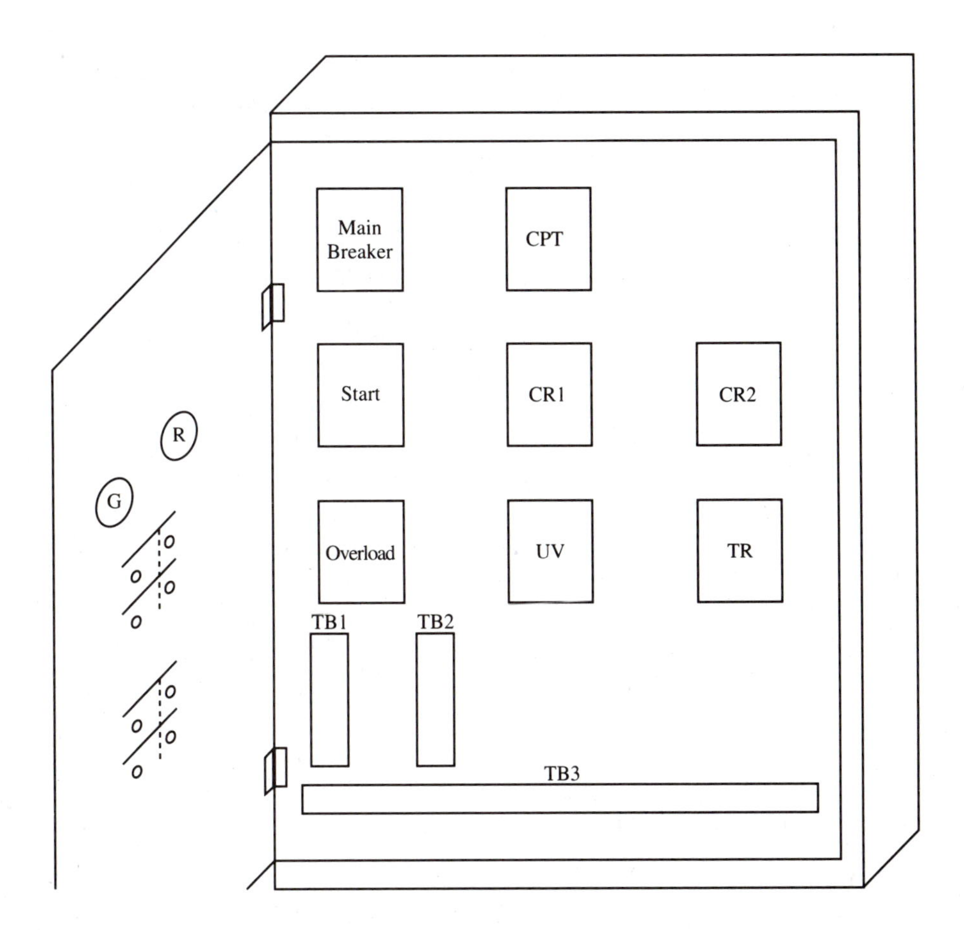

Figure 3-5 Control Panel Layout

Second, with these three drawings we will generate a wiring diagram (see Figure 3-7). The terminals that have wires connected are indicated by one or two lines coming from the terminal, followed by the node designation and finally by the address for the wires' destination. The addresses are shown in parentheses, to distinguish them from node designations. The number of lines coming from a terminal indicates how many wires are to be on that terminal. Most specifications for control wiring will permit only two connections per terminal.

In Figure 3-7, let's see how node 7 was wired. The schematic (Figure 3-4) shows a normally opened contact from CR1 wired to the coil of the starter and the red light. In Figure 3-6 the addresses of these devices are as follows:

Device	Function	Address
CR1	Control relay	E2
M	Starter	D2
R	Red light	B2

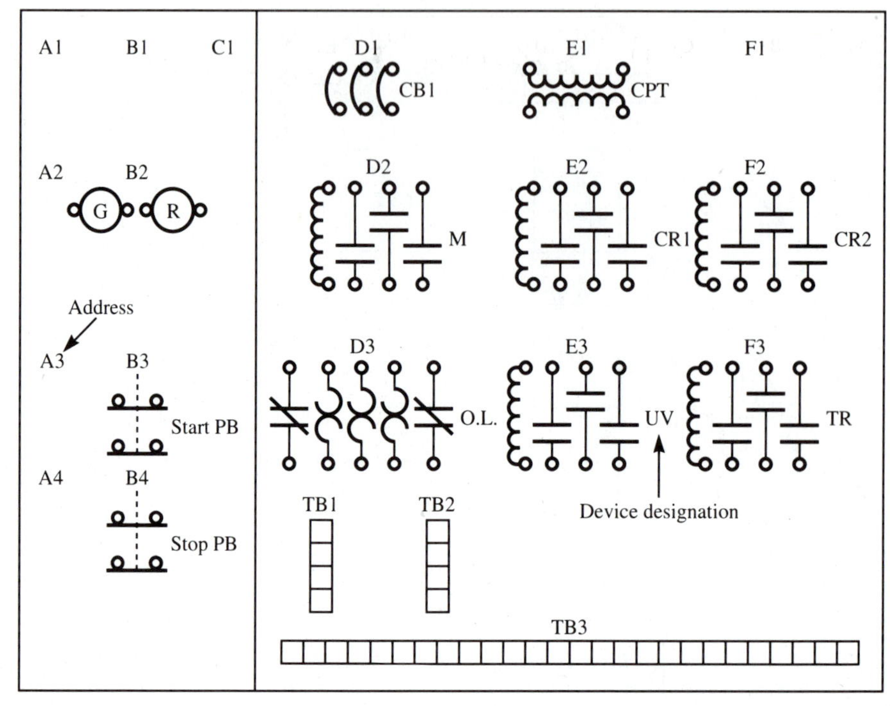

Figure 3-6 Wiring Diagram Layout

Go to each of these locations in Figure 3-7 and find node 7. Note how this would be wired. Let's start at E2. You will see one wire on the bottom of a normally opened contact on CR1; next to that terminal you will see 7(D2). This tells us to put a label 7 on a wire and connect it to this terminal, then to take the other end of the wire to address (D2) and connect it to the terminal designated as node 7. When we go to (D2), we find node 7 at the top of the coil for the starter, with two wires connected to it and the terminal (B2)(E2)7 next to it. The (E2) address, which is where we just came from, tells us that a wire is to go there. The wire we brought should then be connected. We take a second wire, put a label 7 on it, and connect it to this same terminal of the starter. The other end of the wire goes to address (B2). When we go to (B2), we find a node 7 with an address (D2), which is where we just came from. We terminate the wire on this terminal. The wiring is complete for node 7. Node 7 is a simple node.

Let's look at node 1 on the schematic (Figure 3-4) and see where it must go. Table 3-1 presents this information. There are many ways this node could be wired, but the best one will use the least amount of wire. In Figure 3-6, find node 1 at each of these devices. Follow the wiring through in Figure 3-7. Note the jumpers shown on CR1. Where two terminals are side by side, jumpers, rather than point-to-point notation, are sometimes used to indicate wiring. This obvi-

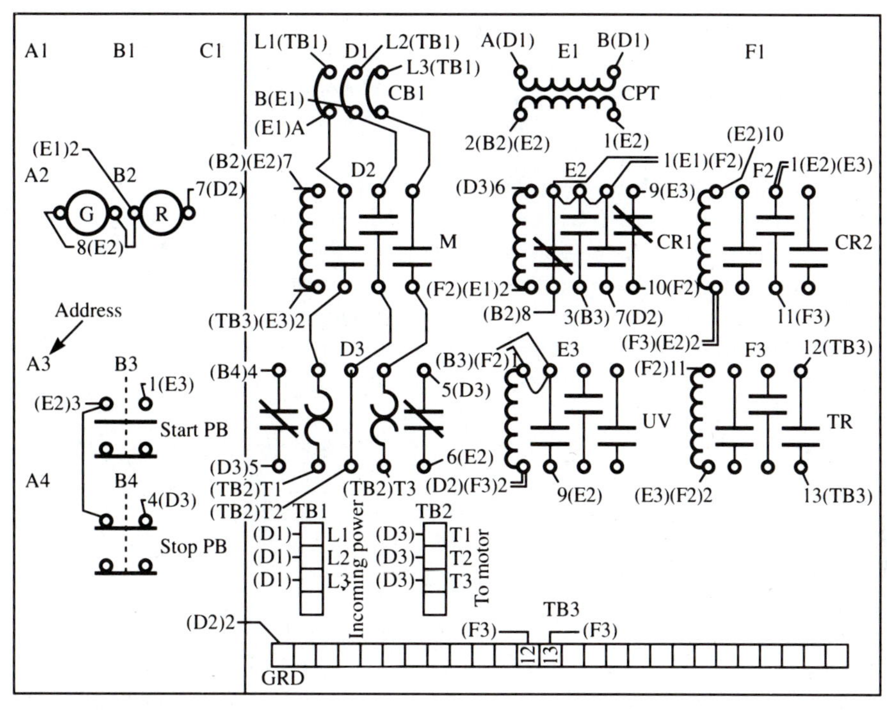

Inside view of door Inside view of back panel

Figure 3-7 Wiring Diagram

ously shows what needs to be done and is convenient and saves time. Notice that the jumpers are not shown as straight lines but are exaggerated. This is so they won't be missed by the person doing the wiring. You will find other jumpers throughout the wiring diagram.

Finally, let's look at the wiring of the time-delay off-timer contact, which is at nodes 12 and 13 on the schematic (Figure 3-4). At address (F3) in Figure 3-7

Table 3-1 Devices Connected to Node 1

Device	Function	Address	Node
CPT	Control transformer	E1	1
PB Start	Push button	B3	1
CR1	Normally open contact	E2	1
CR1	Normally open contact	E2	1
CR1	Normally open contact	E2	1
UV	Undervoltage	E3	1
UV	Normally open contact	E3	1
CR2	Normally open contact	F2	1

these nodes are next to a normally opened timer contact. TB3, the address where these wires will go, stands for "terminal block 3." When you go to TB3 in Figure 3-7, you will see the node designation on the terminal block symbol. This is because the terminal blocks have marking strips on which you can write the node designations.

Study the control schematic (Figure 3-4) and the wiring diagram (Figure 3-6) until this process becomes clear. You may notice that whereas in Figure 3-5, which is the layout used to generate the wiring, the contacts for all the control relays are shown as normally opened, Figure 3-6 shows some of these same contacts as normally closed. This is because industrial control relays have reversible contacts that can easily be changed in the field with a screwdriver to either normally opened or normally closed. The field person can change the contacts as needed while wiring the panel. Thus, someone using the layout of Figure 3-5 to make a wiring diagram can simply cross a contact if it needs to be shown as normally closed.

Summary

We have shown that all the logic gates—AND, OR, EXclusive OR, and INVERTER—can easily be made using relay coils for inputs and contacts arranged in series, parallel, or series parallel for outputs. This type of logic was used for decades before solid-state gates and computers came along. As long as the circuits are simple, it is easy to generate the necessary control with relays. Relay logic is simple to teach and not hard to troubleshoot. Its major disadvantage is bulkiness and the difficulty in making changes, since it must be hard-wired. Now that we have a background in this type of logic, we will introduce the solid-state and PLC technology that is replacing it. Since the symbols of relay ladder logic are employed in programming PLC's, knowledge of relay logic will make it easier to understand PLC's.

Exercises

3-1. What is the purpose of labeling nodes on a schematic?

3-2. What does the term *control relay* mean, and what makes control relay different from other, similar devices, such as starters, contactors, and solenoids?

3-3. What are the characteristics of a ladder diagram versus other types of schematics?

3-4. List the drawings and notations required to make a point-to-point wiring diagram.

3-5. Using Appendix A, show the JIC standards symbols for the following:

indicating light
relay coil
relay contact
overload devices
push button
circuit breaker

3-6. List the basic digital gates, show the relay circuits necessary to create these gates, and explain how the relay gates work.

3-7. Using the wiring diagram in Figure 3-6, list the locations of node 2 by giving the addresses where node 2 is wired.

3-8. Why shouldn't two relay coils or a relay and a light be connected in series?

3-9. Design a control circuit that will turn on a light if one or the other of two push buttons is pushed. If both push buttons are pressed simultaneously, an alarm operates that can only be shut off by removing power to the control circuit. Show the control using JIC standards.

3-10. Make a control schematic that requires you to press PB1 and then PB2 in the correct order to turn on a green light that indicates correct operation. If you push the buttons in the wrong order or both push buttons are pressed simultaneously, then a relay seals in and turns on a red light indicating an error. The power must be turned off and back on to turn off the error light. Show the control, using JIC standards.

4 Discrete Digital Logic

OUTLINE

OBJECTIVES

Upon completion of this chapter, the student will be able to:

- Explain the operation of the basic digital gates AND, OR, and EXclusive OR.
- List some of the available integrated circuits that can be used to implement control using AND, OR, and EXclusive-OR gates.
- Use Boolean algebra to simplify a logic function or a simple control.
- Generate a truth table for a logic gate or a control.
- Use a Karnaugh map to simplify a logic function or a simple control.
- Use standard, inexpensive logic integrated circuits to implement a simple control.

DISCRETE logic devices became available for control with the invention of the transistor. The transistor could be used as an electrically controlled switch, and replaced the relay as the means of implementing logic gates. Logic gates originally were made with individual transistors and components put together on a printed circuit board. A few years after the invention of the transistor, a photo-etch process made it possible to create extremely small transistors. Several individual transistors could be connected together on a microscopic level using specially doped silicon laid down in layers. These layers then were etched through masks made using photography. That is, the masks were negatives reduced photographically to microscopic size. The new device was called an *integrated circuit,* or *IC*. This technology, which made it possible to put many gates in one small, economical package, was very successful. A significant part of control is still implemented via digital IC's.

This chapter gives an overview of digital logic and covers some basic theory so that the reader can appreciate the influence that digital logic has had on PLC's. If in-depth knowledge is desired, there are many good books on digital logic on the market. The chapters that follow will show how to duplicate this type of control on a PLC. You will need a basic understanding of this technology to use the logic commands available on programmable controllers. If you already have a basic understanding of digital logic, you can skip this chapter.

4-1 Digital Gates

The transistor basically acts as an electrically controlled switch to form the various logic gates. Refer to Figure 4-1a. The transistor has three terminals, referred to as *emitter, base,* and *collector.* If the emitter and collector have a DC voltage with proper polarity impressed on them, and the base is isolated or open-circuited, then either no current or an extremely small amount of current will flow

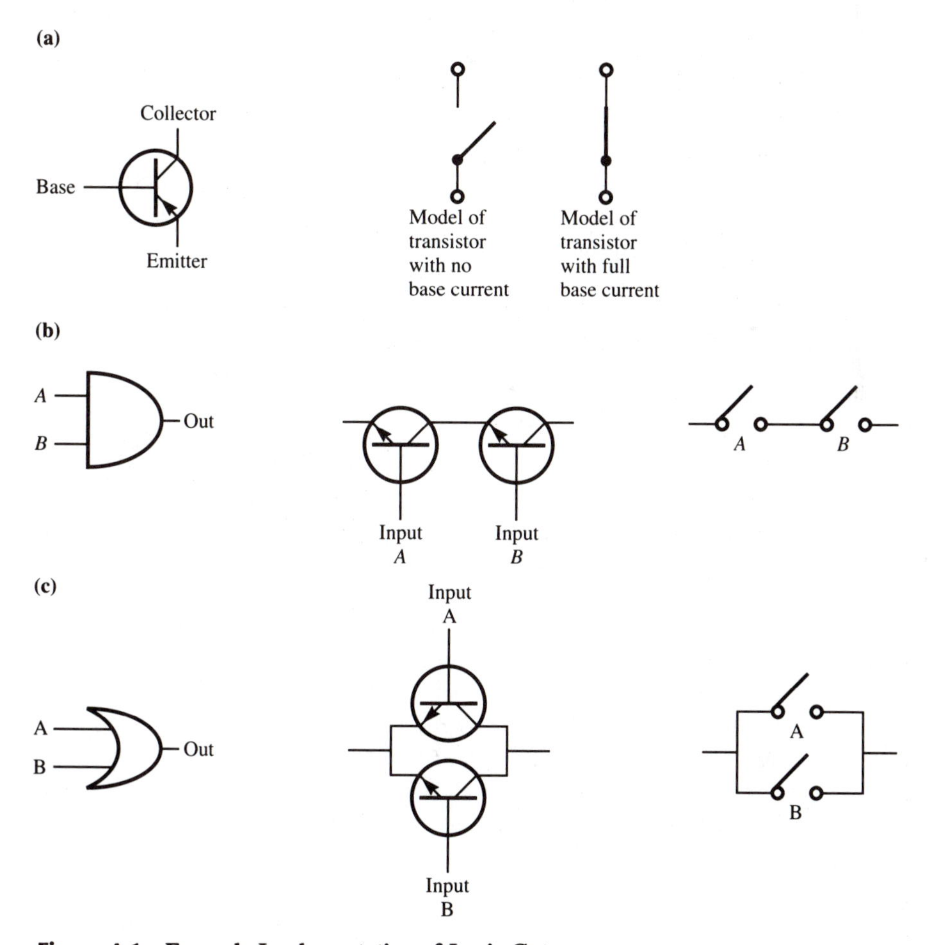

Figure 4-1 Example Implementation of Logic Gates

through either the emitter or the collector leads of the transistor. Conduction, however, will occur through the emitter and collector leads if a sufficient amount of current is sent through the base in the proper direction. The proper amount of base current, called *full base current,* is actually very small (in the microamp range). The transistor then can be modeled as a switch: If there is no base current, the transistor is off; if there is full base current, the transistor is on. Refer to Figure 4-1b. If two transistors are placed in series, we can make an AND gate, with the input to the gate being the base of the transistor. Refer to Figure 4-1c. If two transistors are placed in parallel, we can make an OR gate. (The actual gate circuits are more complicated than this, but the basic idea presented here holds.)

There are two major categories of transistors used to make digital gates. The first is *bipolar transistors,* made by joining two types of semiconductor materials together, material *N* and material *P.* There are two possible arrangements: *PNP* and *NPN*. Either material *N* is placed between two blocks of material *P,* or material *P* is placed between two blocks of material *N*. The second major category of transistors is *unipolar transistors,* made of metal oxide semiconductors referred to as *MOS transistors.* The major advantage of gates made from bipolar transistors is speed of operation. The major advantage of MOS technology is packaging density: A million MOS transistors can be placed on one IC.

4-2 Available Integrated Circuits

Literally hundreds of standard IC packages for gates have been developed, and they are manufactured in quantities large enough to make them inexpensive and readily available. These IC's can be put together on a printed circuit board to accomplish a particular control. If you have a control that isn't going to change implementation, this way is economical. However, if the control has to be changed very often to handle new requirements, a PLC might be the preferable device. We will describe a few of the commercially available standard transistor–transistor logic (TTL) gates and then use them in the next section to implement a control problem. We will only look at a few standard IC's. The lead designation and the contents of four standard IC's are shown in Figure 4-2. If you would like to review all of the available standard logic chips, consult any of the logic data books available from chip manufacturers.

Relay logic was the first type of technology utilized to solve control problems. Then came digital logic, and after that the computer and programmable controllers. Present-day programmable controllers incorporate programming techniques and instruction to duplicate implementation via all of these types of technology. You can program a PLC in relay logic, digital logic, or a high-level computer language. The digital-logic solution requires familiarity with Boolean algebra. The next section explains Boolean algebra and its use in simplifying digital control circuits.

4-3 Boolean Algebra

Discrete logic can be represented by writing Boolean algebraic equations. Such equations differ from regular algebraic equations in several ways. First, the

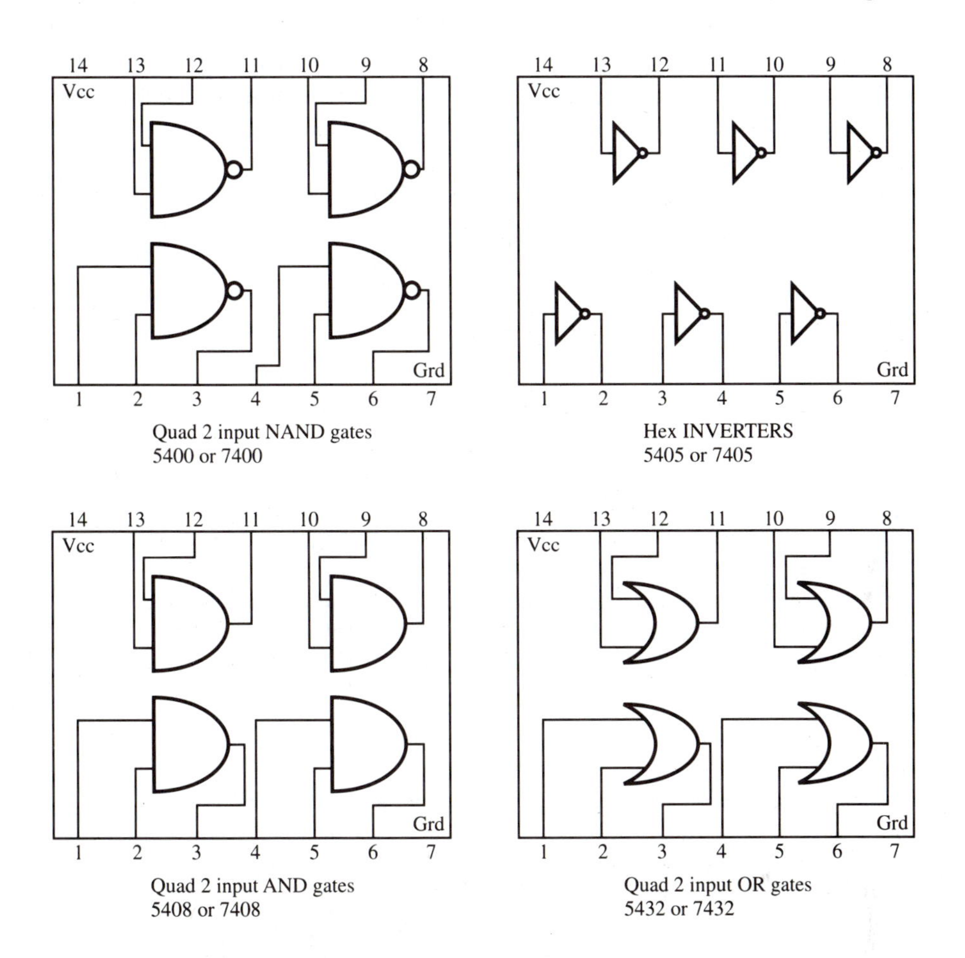

Figure 4-2 Some Standard Logic Integrated Circuits

operations indicated by the standard algebraic symbols differ: When addition is indicated, it represents ORing the variables together rather than adding. When multiplication is indicated, it represents logically ANDing the variables together rather than multiplying. A bar over a variable does *not* mean a vector; instead it means the variable is to be complemented. The first task in studying Boolean algebra is to learn the meaning of the operations indicated and their symbols. What follows are examples of Boolean functions and their meanings.

Function	Meaning
AB	AND
$A \cdot B$	AND
$A \times B$	AND
$A + B$	OR
$\bar{A}$	Complement
$A \oplus B$	EXclusive OR

A second important difference between Boolean and standard algebra is that Boolean algebra allows only two values for the variables and a variable can only have a value of 0 or 1.

Boolean algebra is similar to regular algebra in the way expressions can be manipulated. Once written, a Boolean equation can be rearranged and factored like any ordinary algebraic equation, meaning we can simplify complicated expressions. To get the simplest results, we factor and use the postulates and theorems of Boolean algebra, which follow. *Postulates* are self-evident truths. *Theorems* are statements that can be proven.

Postulates

1a	$A = 1$ if A is not 0	1b	$A = 0$ if A is not 1
2a	$0 \times 0 = 0$	2b	$0 + 0 = 0$
3a	$1 \times 1 = 1$	3b	$1 + 1 = 1$
4a	$1 \times 0 = 0$	4b	$1 + 0 = 1$
5a	$\bar{1} = 0$	5b	$\bar{0} = 1$

Theorems

6a	$A = 1$ if A is not 0	6b	$A = 0$ if A is not 1
7a	$A \times 0 = 0$	7b	$A + 0 = A$
8a	$A \times 1 = A$	8b	$A + 1 = 1$
9a	$A \times A = A$	9b	$A + A = A$
10a	$A \times \bar{A} = 0$	10b	$A + \bar{A} = 1$
11a	$\bar{\bar{A}} = A$	11b	$A = \bar{\bar{A}}$

Study the postulates a little and you will see why they are self-evident, for example, postulate 2a: If we put a low signal into both inputs of an AND gate, obviously the output will be low, since both inputs must be high to get a high output from an AND gate. The theorems take a little more thought, but they can be figured out by considering all possible combinations, for instance, theorem 9b: If we take a variable and connect it to both inputs of a two-input OR gate, then if the variable is 0, both inputs will be 0 and the output will be 0; if the variable is 1, both inputs will be 1, and the output will be 1. We can conclude that the output will follow whatever the input is, and thus $A + A = A$.

The commutative, associative, and distributive algebraic properties apply to Boolean equations. Basically this means you can factor and rearrange Boolean equations the same way you can regular algebraic equations. The simplification steps using the theorems and postulates listed earlier are shown in Examples 4-1 and 4-2.

EXAMPLE 4-1

	$Y = \bar{A}B\bar{C} + \bar{A}BC + A\bar{B}\bar{C} + A\bar{B}C$
Factoring	$Y = \bar{A}B(\bar{C} + C) + A\bar{B}(\bar{C} + C)$
Using Theorem 10b: $A + \bar{A} = 1$	$Y = (\bar{A}B \times 1) + (A\bar{B} \times 1)$
Using Theorem 10a: $A \times 1 = A$	$Y = \bar{A}B + A\bar{B}$

Since the result is an equation for an EXclusive OR, the four AND gates can be replaced by an EXclusive OR. The equation can be implemented using logic gates as shown in Figure 4-3.

EXAMPLE 4-2

		$B(A+B)(A+B+C)C$
Multiplying $(A+B)$ and $(A+B+C)$ by BC		$(ABC+BBC)(ABC+BBC+BCC)$
Using postulate 9a:	$A\times A=A$	$(ABC+BC)(ABC+BC+BC)$
Factoring out BC		$BC(\mathrm{A}+1)(\mathrm{A}+1+1)$
Using postulate 8b:	$A+1=1$	$BC(1)(1)$
Using postulate 8a:	$A\times 1=A$	BC

Boolean expressions are often written to implement truth tables, which are a means of organizing solutions to control problems. The next section explains truth tables.

4-4 Truth Tables

A **truth table** is a chart showing all possible input variables in all possible combinations, plus the desired output for each of those combinations. It is essential to remember that in discrete logic only two variables are allowed, 0 and 1. A 0 is often referred to as a low, and a 1 as a high. The 0 and 1 can be used to indicate conditions for equipment. For instance, 0 can mean something is on, and 1 that it is off; or 0 can mean a push button is being pressed, and 1 that it is not being pressed.

An example truth table is given in Figure 4-4, in which A, B, and C are inputs and Y is the desired output. Notice that all possible combinations of input

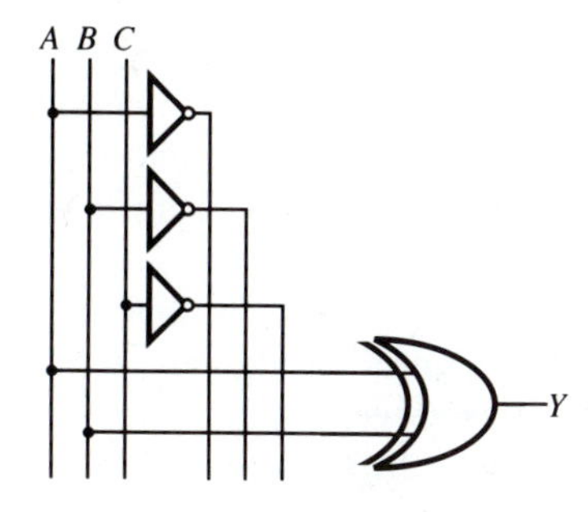

Figure 4-3 Example 4-1 Implemented Using Logic Gates

Truth Table				Fundamental Product
A	B	C	Y	
0	0	1	0	$\bar{A}\bar{B}C$
0	1	0	1	$\bar{A}B\bar{C}$
0	1	1	1	$\bar{A}BC$
1	0	0	1	$A\bar{B}\bar{C}$
1	0	1	1	$A\bar{B}C$
1	1	0	0	$AB\bar{C}$
1	1	1	0	ABC

Figure 4-4 Truth Table and Fundamental Products

states are shown and that the progression is the same as counting in binary. The number of possible combinations is 2^n, where n is the number of inputs. A, B, and C could represent any type of input, for example, three push buttons labeled A, B, and C, respectively. Certain combinations of pushing these buttons will cause the output to be active. Next to these combinations in the truth table, in the Y (output) column, will be the desired state of the output. If a particular combination shouldn't cause the output to be active, there will be a 0 next to that combination; alternatively, if the output is to be active, a 1 will be shown. There are four combinations in Figure 4-4 that will cause the output to be activated.

Once a truth table has been set up, it can be implemented by ORing what is called the *fundamental product*. In Figure 4-4, the fundamental product is listed to the right of the truth table for each input combination. These products are always the variables multiplied together, but the product for each combination is different. When a variable has a truth table value of 1 for a particular row (one of the possible combinations), it is shown uncomplemented in the product for that combination. But if a 0 is shown, then it is complemented in the product. Complementing is indicated by a bar over the variable.

We can easily write the Boolean equation for implementing the truth table in Figure 4-4 by ORing the fundamental products. What does this equation mean in Boolean algebra? Whereas ABC in ordinary algebra means A, B, and C are to be multiplied together, in Boolean it means A, B, and C are to be ANDed together. The Boolean expression for the output of this truth table is implemented by ORing the fundamental products with a Y-value of 1, as follows:

$$Y = \bar{A}B\bar{C} + \bar{A}BC + A\bar{B}\bar{C} + A\bar{B}C$$

Note: If $A = 0$, $B = 1$, and $C = 0$, the AND gate indicated by the first expression, $\bar{A}B\bar{C}$, would produce a high, or 1, at its output. ORing these combinations means we will get a 1 out for each truth table combination with a value of 1, which is exactly what we want. The equation can be implemented using logic gates as shown in Figure 4-5.

The next logical question: Is this the simplest possible solution for this truth table? Though ORing the fundamental products always gives the desired result, generally it is not the simplest logic possible. As we have seen in Example 4-1, this equation simplifies, via Boolean algebra, to $Y = \bar{A}B + A\bar{B}$. The next section

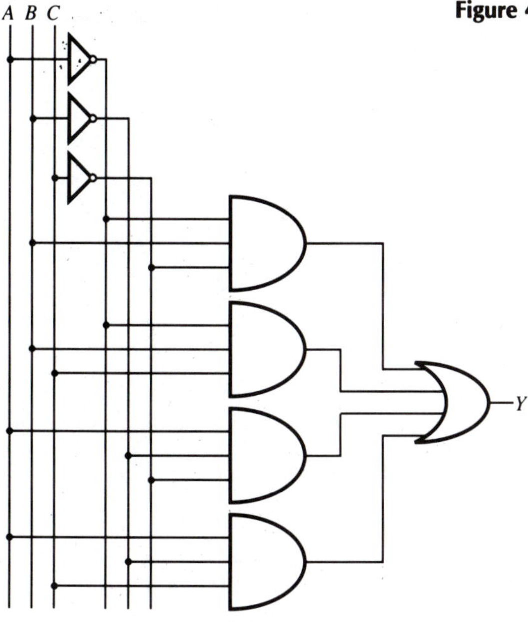

Figure 4-5 ORing Fundamental Products

will introduce a mapping technique for finding groups of variables that can be simplified.

4-5 Karnaugh Maps

A **Karnaugh map** can help us obtain the simplified logic expression for a given truth table by analyzing a special array of the fundamental products. The map's squares forming the array are labeled so that only one variable changes between adjacent squares. This is done by using a special sequence for designating rows and columns. The steps for generating the simplified output equation are:

1. Construct a Karnaugh map with the same number of variables as in the truth table.

2. Place 1's in the squares of the array corresponding to the truth table's fundamental products with a value of 1. The squares for the products with a value of 0 are left blank.

3. The 1's are then placed in groups of one, two, four, eight, or sixteen. The size of the group dictates the number of variables that will be eliminated for that particular group. You keep the variable that doesn't have an associated complement in that group; and if a variable is present in both complemented and uncomplemented form, you throw it out. The grouping is done as follows.

(a) Look for isolated 1's that can't be grouped with any other 1's. These products will be ORed to make the final equation.

(b) Look for pairs of 1's that can't be grouped with any other 1's. These products with one variable thrown out will be ORed to make the final equation.

(c) Look for quads of 1's (groups of four 1's) that can't be grouped with any other 1's. These products with two variables thrown out will be ORed to make the final equation.

(d) Look for octal groups of 1's (groups of eight 1's) that can't be grouped with any other 1's. These products with three variables thrown out will be ORed to make the final equation.

(e) Look for hex groups of 1's (groups of sixteen 1's) that can't be grouped with any other 1's. These products with four variables thrown out will be ORed to make the final equation.

4. Finally, OR the groups found in Step 3.

EXAMPLE 4-3

Look at the Karnaugh map in Figure 4-6. Notice that there are no isolated 1's and that the smallest group, a pair of adjacent 1's, is in row $\bar{A}\bar{B}$ and columns $\bar{C}\bar{D}$ and $\bar{C}D$. The variable to get thrown out is D. The product for this group is $\bar{A}\bar{B}\bar{C}$.

Truth Table

A	B	C	D	Y
0	0	0	0	1
0	0	0	1	1
0	0	1	0	0
0	0	1	1	0
0	1	0	0	0
0	1	0	1	0
0	1	1	0	1
0	1	1	1	1
1	0	0	0	0
1	0	0	1	0
1	0	1	0	0
1	0	1	1	0
1	1	0	0	0
1	1	0	1	0
1	1	1	0	1
1	1	1	1	1

Karnaugh Map

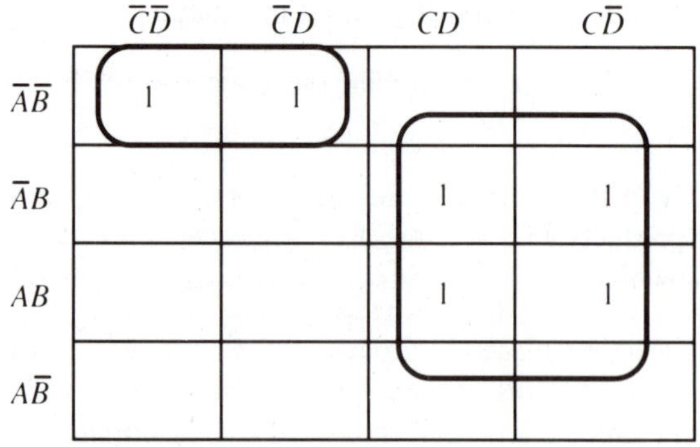

	$\bar{C}\bar{D}$	$\bar{C}D$	CD	$C\bar{D}$
$\bar{A}\bar{B}$	1	1		
$\bar{A}B$			1	1
AB			1	1
$A\bar{B}$				

Figure 4-6 Example 4-3 Truth Table and Karnaugh Map

Notice that the next group, the largest group in the array, is the four 1's in rows $\bar{A}B$ and AB and columns CD and $C\bar{D}$. We will be able to throw out two variables, A and D, since these have both the complemented and uncomplemented versions of each variable. The product for this group will be simply BC. Finally, ORing the products for these groups gives us $Y = BC + \bar{A}\bar{B}\bar{C}$. The groups used are shown in Figure 4-6. If we simply OR the truth table fundamental products with a value of 1, we get:

$$Y = \bar{A}\bar{B}\bar{C}\bar{D} + \bar{A}\bar{B}\bar{C}D + \bar{A}BCD + \bar{A}BC\bar{D} + ABCD + ABC\bar{D}$$

By factoring and simplifying via Boolean algebra, this expression can be reduced to $Y = BC + \bar{A}\bar{B}\bar{C}$. Most people would prefer to use mapping over working through the algebraic solution. The progression for the variable in both the rows and columns must follow a special sequence, known as the *grey code*. (This code is explained in Chapter 6.)

E X A M P L E 4 - 4

Look at the Karnaugh map in Figure 4-7. Notice that there is an isolated 1 and that the product for this group is $\bar{A}\bar{B}\bar{C}\bar{D}$. Notice that the next group, the largest group for this array, is the four 1's in rows $\bar{A}B$ and AB and columns CD and $C\bar{D}$.

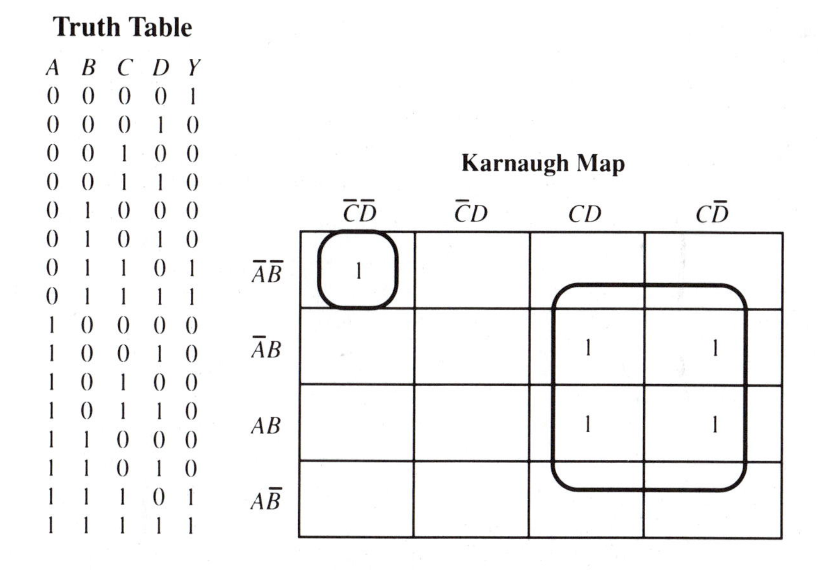

Truth Table

A	B	C	D	Y
0	0	0	0	1
0	0	0	1	0
0	0	1	0	0
0	0	1	1	0
0	1	0	0	0
0	1	0	1	0
0	1	1	0	1
0	1	1	1	1
1	0	0	0	0
1	0	0	1	0
1	0	1	0	0
1	0	1	1	0
1	1	0	0	0
1	1	0	1	0
1	1	1	0	1
1	1	1	1	1

Karnaugh Map

	$\bar{C}\bar{D}$	$\bar{C}D$	CD	$C\bar{D}$
$\bar{A}\bar{B}$	1			
$\bar{A}B$			1	1
AB			1	1
$A\bar{B}$				

Figure 4-7 Example 4-4 Truth Table and Karnaugh Map

We will be able to throw out two variables, *A* and *D*, since these have both the complemented and uncomplemented version of each variable. The product for this group will be simply *BC*. Finally, ORing the products for these groups gives:

$$Y = BC + \bar{A}\bar{B}\bar{C}\bar{D}$$

The groups used are shown in Figure 4-7.

E X A M P L E 4 - 5

Look at the Karnaugh map in Figure 4-8. Initially, it looks like we have three isolated 1's. But we are allowed to fold the map onto itself to form groups, so the four 1's in the corners actually form a quad, and two variables get thrown out. The product from this group is $\bar{B}\bar{D}$. Notice that the next group is the four 1's in rows $\bar{A}B$ and AB and columns CD and $C\bar{D}$. We will be able to throw out two variables, *A* and *D*. The product for this group will be simply *BC*. Finally, ORing the products for these groups gives:

$$Y = BC + \bar{B}\bar{D}$$

The groups used are shown in Figure 4-8.

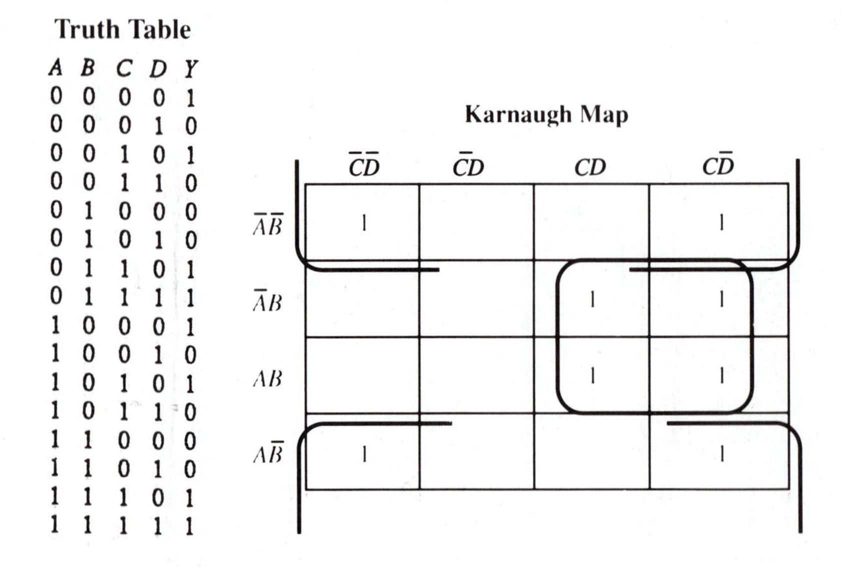

Truth Table

A	B	C	D	Y
0	0	0	0	1
0	0	0	1	0
0	0	1	0	1
0	0	1	1	0
0	1	0	0	0
0	1	0	1	0
0	1	1	0	1
0	1	1	1	1
1	0	0	0	1
1	0	0	1	0
1	0	1	0	1
1	0	1	1	0
1	1	0	0	0
1	1	0	1	0
1	1	1	0	1
1	1	1	1	1

Figure 4-8 Example 4-5 Truth Table and Karnaugh Map

4-6 Example Implementation

In this section we will implement a control problem using standard IC's that are readily available and inexpensive.

E X A M P L E 4 - 6

Design a control that requires a person to press two push buttons simultaneously to operate a particular machine. Once the two buttons have been pushed simultaneously, the machine can only be activated again by releasing both push buttons. Releasing one push button and then pressing it again cannot activate operation. Releasing both push buttons will again allow operation of the machine, but only if the two push buttons are pressed simultaneously. Light a different light to indicate the status of the control as follows.

1. Machine is ready for two push buttons to be pressed simultaneously.
2. Two push buttons have been pressed simultaneously and the machine is operating.
3. Machine is waiting for both push buttons to be released at the same time.

The inputs for this control problem are:

A = pushbutton 1 0 = not pushed 1 = pushed
B = pushbutton 2 0 = not pushed 1 = pushed

The output for this control problem are:

C = light for ready 0 = not ready, light not on
1 = ready, light on

D = light for operate, machine on 0 = not operating, light not on
1 = operating, light on

E = light for waiting 0 = not waiting, light not on
1 = waiting, light on

The solution is shown in Figure 4-9, for which the following integrated circuits are required.

1 Hex INVERTER 7404
1 Quad 2 input AND gate 7408
1 Quad 2 input OR gate 7432
1 Quad 2 input EXclusive OR gate 7408
1 Quad SR latch 74279

The IC's contain more gates than needed for this control, and the extra gates simply are not used. The $\bar{S}\bar{R}$ latch is a bi-stable device, which means the output Q is stable in one of two states: Q is either 1 or 0. The $\bar{S}$ stands for "set"; if a 0 is coming into the $\bar{S}$ input, the Q will go to 1. If the $\bar{S}$ goes to 1, it will not affect

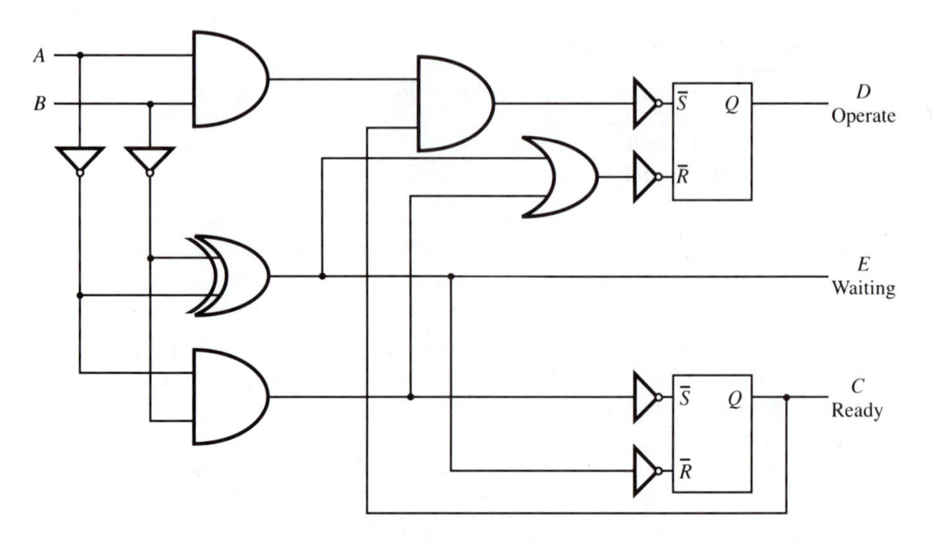

Figure 4-9 Example 4-6 Control Problem Implemented Using Standard Integrated Circuits

the Q output. The $\bar{S}$ input is referred to as an *active low input,* meaning it only responds to a 0 at its input. The $\bar{R}$ input stands for ''reset'' and is also active low. When this input is low, the latch is reset and the Q output goes to zero.

Summary

It is worthwhile to review digital logic because this will help to understand the logic that has become available on programmable controllers. We will also be able to see how this popular control has influenced the features available on PLC's. The PLC itself is actually made using this type of logic, and digital logic is the heart of the central processing unit that is the brains of the PLC. CMOS technology has made it possible to put thousands of transistors on one IC, and to put enormous amounts of digital logic on one chip. We would not have microcomputers without digital logic. We can simplify control circuits by applying theorems and postulates of Boolean algebra or by using Karnaugh maps. This will result in savings in relay control by reducing the number of relays needed, or in the PLC's by making the control easier to troubleshoot. In later chapters we will see that the scan time needed by the PLC can also be reduced by simplifying the control.

This very limited look at digital logic should be sufficient for using a PLC. This type of control merits a complete book, and any well-prepared technician or engineer should spend the time to learn this technology.

Exercises

4-1. Construct a truth table for the following logic problem: The output is to be high if any two push buttons are pressed. There are four push buttons, designated *A*, *B*, *C*, and *D*.

4-2. Given the following truth table, list the fundamental products that would yield a 1 in a Karnaugh map.

A	B	C	Y
0	0	0	1
0	0	1	0
0	1	0	0
0	1	1	1
1	0	0	0
1	0	1	1
1	1	0	0
1	1	1	0

4-3. Based on the following truth table, what is the corresponding Boolean equation for the *Y* output?

A	B	C	Y
0	0	0	1
0	0	0	0
0	1	0	1
0	1	1	0
1	0	0	0
1	0	1	0
1	1	0	0
1	1	1	0

4-4. Write the simplified Boolean expression for the Karnaugh map in Figure 4-10.

	$\overline{C}\overline{D}$	$\overline{C}D$	CD	$C\overline{D}$
$\overline{A}\overline{B}$	1	1	1	1
$\overline{A}B$		1	1	
AB		1	1	
$A\overline{B}$		1	1	

Figure 4-10 Exercise 4-4 Karnaugh Map

4-5. Write the simplified Boolean expression for the Karnaugh map in Figure 4-11.

4-6. Set up the correct Karnaugh map for the following truth table.

A	B	C	Y
0	0	0	0
0	0	1	0
0	1	0	0
0	1	1	1
1	0	0	1
1	0	1	0
1	1	0	1
1	1	1	0

4-7. Find the logic expression for a circuit with three inputs, A, B, and C, and with output Y. Y will be 1 if A or $B = 1$ or if two or more inputs are 1. *Hint:* Make a truth table.

4-8. Simplify the following Boolean equation.

$$Y = BC(A + B + C)(A + B + C)$$

4-9. Simplify the following Boolean equation.

$$Y = \bar{A}\bar{B}(\bar{C} + B\bar{C}) + A\bar{C}$$

4-10. Using the lead designations for the hex INVERTER in Figure 4-2, show how the hex INVERTER would be used by adding the lead designations to Figure 4-9.

4-11. Show the actual logic circuit to implement the following equation:

$$Y = (AB + ACD)C$$

	$\bar{C}\bar{D}$	$\bar{C}D$	CD	$C\bar{D}$
$\bar{A}\bar{B}$	1	1	1	1
$\bar{A}B$		1	1	
AB		1	1	
$A\bar{B}$		1		

Figure 4-11 Exercise 4-5 Karnaugh Map

Safety and Grounding

5

OUTLINE

OBJECTIVES

Upon completion of this chapter, the student will be able to:

- Cite examples of industrial hazards.
- State the three primary reasons for grounding, and give example applications of each.
- Discuss safe-shutdown procedures and precautions when operating energized equipment.
- Explain how to prevent unsafe conditions when the control circuit loses power.
- State the precautions necessary when installing programmable controllers.
- Describe the procedure for installing coaxial cable between a group of transducers and a programmable logic controller.
- Describe the procedure for installing and maintaining a programmable logic controller.

When installing control equipment, it is essential not to create hazardous conditions for normal or abnormal situations. The sections that follow will address proper installation.

Grounding is an area of study that is often neglected and misunderstood. Correct grounding is essential for proper operation of equipment, and we will cover it in some detail. The primary reasons for grounding are safety, establishing a common reference, and preventing interference. We will examine each of these reasons.

Guidelines for installation, maintenance, and troubleshooting will be given. Although this is early in the book to be presenting such information, students already handling PLC's in a laboratory need this background in order to avoid damaging equipment. Problems can occur even in the first labs; reading this chapter will help in troubleshooting.

5-1 Industrial Hazards

Industrial hazards can take many forms, and we need to exercise constant surveillance to prevent personnel from being injured. Here are some examples.

Electric shock

Fire

Explosions

Toxic vapors

Radiation

Heavy machines

You can easily add to this list. When automatic equipment such as a PLC or control equipment is being installed, it is essential that it not operate in a way that might injure personnel due to control equipment response to normal or abnormal conditions. Also, properly installed equipment will ensure that anyone operating the equipment will not be injured. We will not address all possible hazards but instead concentrate on electric shock that can be eliminated via proper installation.

Electric shocks result when a person contacts a device that has created a voltage difference between two parts of the person's body, thereby producing a current flow through the body.

Three factors determine how serious the electric shock will be: Amount of current, the particular body parts involved, and the individual's condition.

Amount of Current An approximate range of current flow and the corresponding effects on a healthy person are as follows.

Current Flow (milliamps)	Effect
0–1	Barely perceptible
1–5	Uncomfortable but not hazardous
10–20	Possibly harmful
20–50	Involuntary muscular contractions
50–100	Pain, fainting, exhaustion, mechanical injury
100–300	Ventricular fibrillation
300+	Myocardial contraction, respiratory paralysis, possible burns

As the table shows, even small amounts of current (in the range of milliamps, or thousandths of an ampere) can result in serious harm. Current flow is a function of the resistance of the body parts across which the voltage is producing the shock. If you put both hands between the terminals of an ohmmeter, the resistance (if you are healthy and dry) will measure over 10,000 ohms. A 120-V circuit through 10,000 ohms of resistance produces 12 milliamps of current, which is just on the threshold of danger. A 120-V source is generally considered safe for a healthy person, but caution must still be exercised. The power feeding control panels that include push-button and operating devices are usually limited to 120 V, and are isolated from the power circuit they control via isolation transformer, relay, and solid-state devices with special isolation built in. Power voltages above 240 V are highly dangerous, since they can easily produce current above 20 milliamps.

The best way to ensure safe electrical operation of equipment is to ground it properly. Equipment receiving power through a standard three-prong plug will be grounded if the receptacle, plug, and power system have been wired according to the National Electric Code. Power equipment often is installed by connecting the hot and grounded conductors separately. Great caution is called for, since power voltages can be lethal. The grounded conductors should always be connected first, for if the equipment is grounded properly *before* power is applied, then the chance of accidental shock will be minimized. The rule is: **Always ground first, before applying power.** The next major section will discuss why this is necessary.

Particular Body Parts Involved The severity of electric shock is also determined by which parts of the person's body a voltage develops across. Vital areas are the heart, lungs, and brain. Obviously, current flowing from the big toe to the little toe of one foot would result in minimum harm. Current flowing through the body as a result of a voltage difference between the left and right hands would maximize risk, since the heart would be in the current's path.

Conditions of the Individual The third factor that affects severity of electric shock is the condition of the person at the time of the shock. A healthy person will have more resistance than someone who is ill. As a result, hospitals go to great lengths to protect patients in operating rooms from electric shock.

5-2 Grounding for Safety

Metal enclosures containing electrical equipment are grounded so that anyone in contact with ground will receive no shock. The ground attached to the metal housing of a device will hold the potential of the housing at zero. Figure 5-1 shows how an electrical device receives power from a standard 120-V AC source. The power company brings in power through a high-voltage distribution transformer and steps it down to 120/240 V. The common connection between the two 120-V windings of the secondary is grounded at the distribution transformer by the power company. Three wires are brought into the distribution panel, which feeds various circuits. Two of these wires are ungrounded and have a potential of 120 V

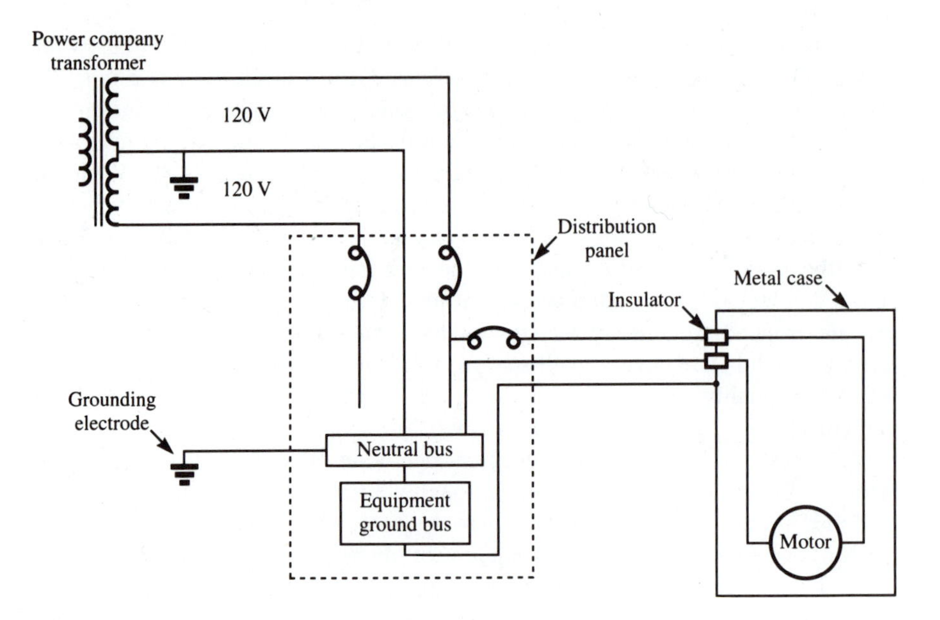

Figure 5-1 Power from a Standard 120-V AC Source

relative to ground. The third wire, called the *neutral*, is grounded, and goes to a neutral bus, which is grounded again at the distribution panel. Loads are fed by connecting them to the ungrounded mains and the neutral bus. The ungrounded conductor goes through a fuse or a circuit breaker in the distribution panel before feeding the load.

The normal current path goes out the ungrounded connection through the load and back to the neutral bus. The ungrounded conductor and the neutral are insulated from the metal enclosure containing the load. An equipment ground connected to the metal enclosure normally doesn't carry current but does act to keep the enclosure at 0 volts relative to the ground.

Figure 5-2 illustrates what happens if the ungrounded conductor comes in contact with the metal enclosure due to insulation failure. Insulation failure creates a short circuit between the equipment ground conductor and the ungrounded wire. The resultant heavy currents, well above normal, will trip a breaker or blow a fuse. The person touching the enclosure receives no shock because the equipment ground holds the enclosure to 0 volts.

Figure 5-3 shows what happens if the equipment ground isn't connected. The current flows from the ungrounded conductor to the enclosure, then from the enclosure through the person to ground, and the person receives a severe electric shock. Metal housings are grounded to prevent such a situation.

Again, the best way to ensure safe electrical operation of equipment is to ground it properly. Doing so *before* power is applied minimizes the chance of receiving a shock. **Always ground first, before applying power.** This is critical when working with voltages above 120 V or with power circuits. If high voltage is

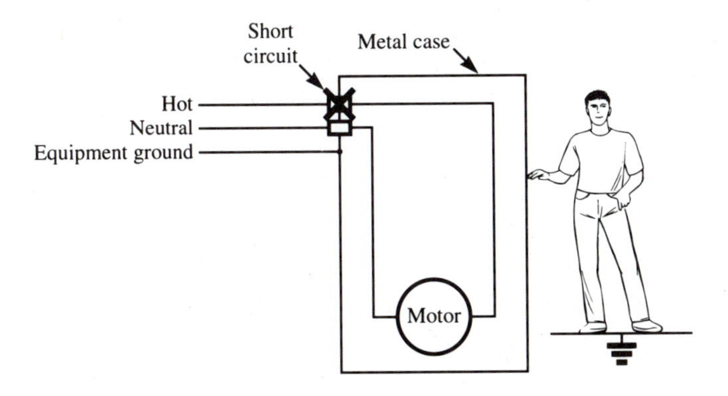

Figure 5-2 Insulation Failure Hazard with Proper Grounding

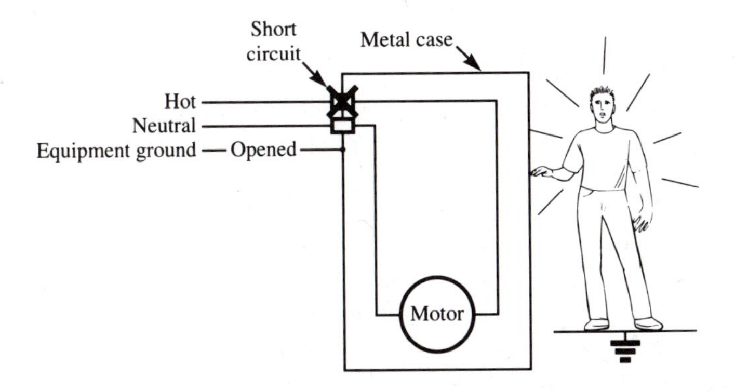

Figure 5-3 Insulation Failure Hazard Without Proper Grounding

going to be applied to equipment through a metal enclosure, it could come in contact with the metal structure. If power is applied *before* the enclosure is grounded and there is an insulation failure, the enclosure could be at the potential of the high voltage. A person contacting the structure under these conditions would probably be killed. However, if the equipment was grounded first, the application of voltage will produce a short and, hopefully, trip a circuit breaker or blow a fuse. Even if the circuit isn't interrupted, proper grounding will hold the potential of the structure at near 0 volts and thus prevent anyone coming in contact with the structure from receiving a lethal shock.

5-3 Grounding for Common Reference

Another reason for grounding is to establish a common reference point for measurements. Measuring voltage difference requires the voltmeter to be connected to two different points. If ground is made one of the reference points, one lead can be permanently connected to ground and we can measure by probing with the other lead, which is more efficient than moving two leads. It also makes it easier for giving reference voltage for troubleshooting. Figure 5-4 shows a typical electronic schematic with many points grounded. The grounds help establish a reference point.

5-4 Grounding to Prevent Interference

Interference is a serious problem if the signals being transmitted are relatively weak or in the millivolt range. Interference that will swamp these values can easily be generated. There are five different ways of generating interference:

Electrostatically

Magnetically

Electromagnetically

Via resistive coupling

Via ground loops

The first three kinds of interference can be cured by surrounding the conductors with a grounded metal shield. An electric field will be shorted to ground by this shield and therefore cannot couple to the signal-carrying wire. Magnetic fields set up eddy currents in the shield that counter their effects. And electromagnetic fields cannot propagate through a conducting shield. The usual way to protect against such interference is to use coaxial cable as shown in Figure 5-5. Coaxial cable is made so that an inner conductor that will be carrying the signal is completely surrounded by an outer, braided conductor. This shield is grounded and prevents electrostatic, magnetic, and electromagnetic interference.

Resistive interference results when several devices are fed from the same source. The source has some input impedance, so when one of the devices is operated, it affects the others. A classic example is the dimming of the house

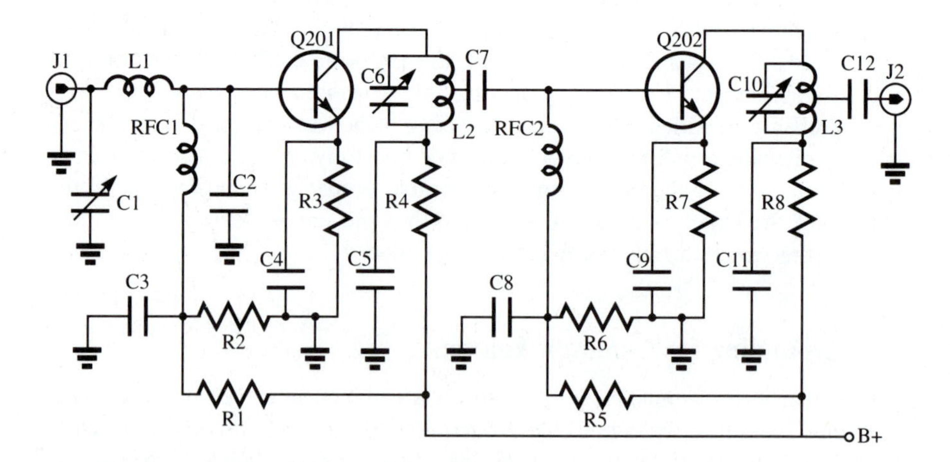

Figure 5-4 Grounding for Common Reference

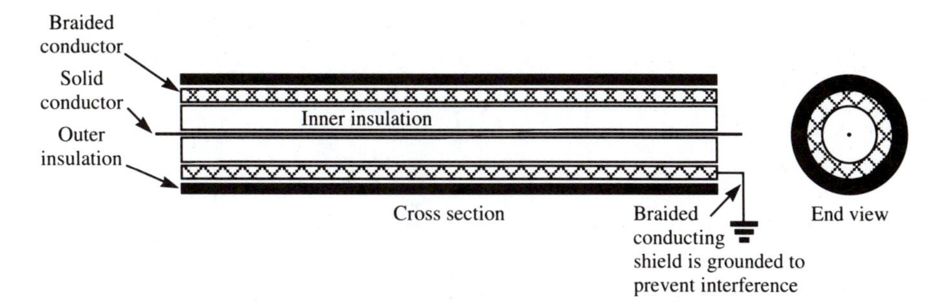

Figure 5-5 Grounding to Prevent Interference

lights when the refrigerator compressor starts up. The start-up current for the compressor causes above-normal currents to flow for a short period, and this current surge produces a voltage drop at the supply due to the internal impedance at the source. The low voltage will be present on all circuits connected to this source.

Ground-loop interference results from multiple grounds that form loops ideal for picking up interference. If a varying magnetic field passes through one of these loops, a voltage is produced and current flows in the loop. There also can be differences of potential between one ground point and another due to transient conditions in the power system. For example, when a short circuit occurs to a chassis on some piece of equipment, heavy currents could flow in the ground connection. Since there is some resistance in these connections, the current will produce differences of potential through the ground circuit. If a ground loop is established between two ground points, this difference will generate a current flow in the loop. When signals are in the millivolt range, such conditions will produce sufficient interference to swamp these signals. The solution to this problem is to ground the coaxial cables only at one end. Figure 5-6a shows how ground loops are formed when low-signal devices are connected to multiple grounds. Figure 5-6b shows the correct way to connect the ground so no ground loops occur.

5-5 Safe Shutdown

PLC's are manufactured to be reliable and safe; however, a malfunctioning PLC could produce unpredictable outputs, a potentially serious safety problem for certain types of machines. For example, a hydraulic ram that crushes automobiles into small cubes could be controlled by a PLC. If the ram needed to be serviced or worked on, it is obvious that the ram shouldn't be under the control of a device that might initiate operation if a failure took place. There needs to be an emergency-stop push button to override the PLC. Figure 5-7 presents an example circuit.

(a)

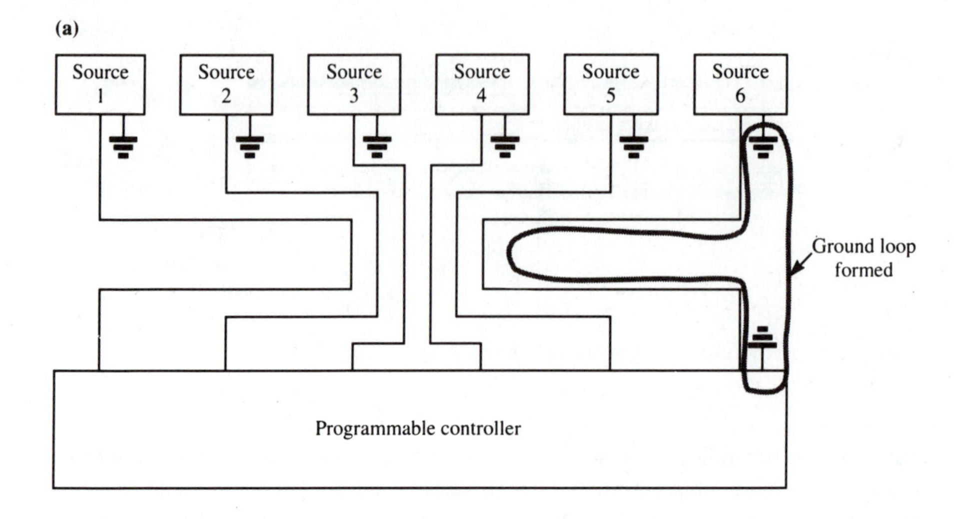

(b)

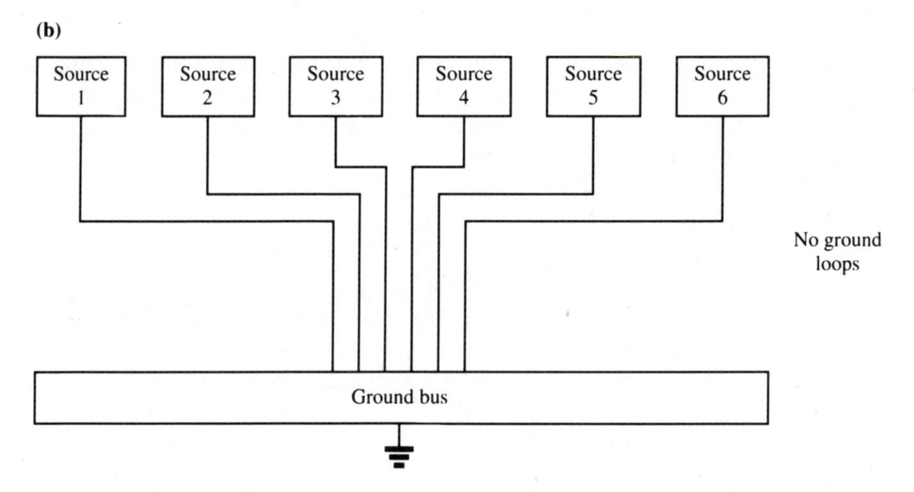

Figure 5-6 Ground Loops

5-6 Precautions When Installing Programmable Controls

Placement

The following precautions should be considered when placing a PLC.

1. Place the PLC near the equipment it will control. This will greatly aid the checkout and troubleshooting of the control programs. If the equipment being controlled can be observed while the programs are running, troubleshooting is made much easier. Remote locations will require either running back and forth between the PLC and the equipment or two persons talking over some type of communications system.

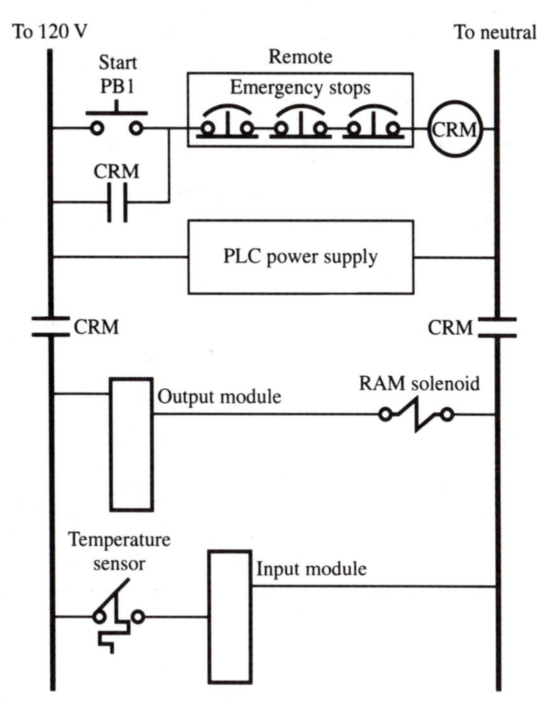

Figure 5-7 Safe-Shutdown Circuit

2. Put the PLC in an enclosure if the area in which it will be mounted is unfriendly. Examples of unfriendly conditions requiring a special enclosure are:
   (a) *Corrosive atmosphere.* This could be excessive dust or dirt, excessive moisture, oil, etc.
   (b) *Excessive vibrations.* This will require special mounting of the enclosure to insulate the PLC from the vibration. Otherwise the vibration will cause the mechanical connectors to fail.
   (c) *Electrical interference sources.* The most common source of interference is heavy-power circuits, which will produce electromagnetic interference. The enclosure when properly grounded will act as an electromagnetic shield. This doesn't shield against interference from incoming cables (this was covered earlier in the chapter).
   (d) *Abnormal ambient temperature.* Temperature control must be provided when the ambient temperature is outside the range recommended by the PLC manufacturer. Excessive heat will require air conditioning, in the form of a fan in the simplest requirement or an actual air conditioning system if the problem is severe. Louvers must be provided in the enclosure for air flow. Lower-than-normal temperatures will require heaters.
   (e) *Outside locations.* Temperature cycling from day to night produces condensation that will cause failure due to power leakage over moist surfaces. A thermally controlled heater located in the enclosure will avoid this problem.

(f) *Adequate work space.* The PLC's enclosure must have adequate working room around it, ideally a minimum of three feet of clear space in front of the enclosure. There also must be adequate clearance to fully open the enclosure door. The enclosure should have a power disconnect so that, when required, the PLC can be worked on with the power off. Also, a viewing window is desirable so that the indicators and the PLC and modules can be viewed without having to open the enclosure during normal operation.

Installation

Obviously, one should follow the steps outlined in the instruction manual for the particular programmable controller selected. Here are some general rules to follow when installing any PLC.

1. Make sure the control is grounded properly by verifying that the ground electrode is a true ground potential. Where cables are connected to the ground stud on the control, make sure the ground stud is clean and free of paint or anything else that could prevent the connecting ground cable from making good contact. Use star washers on either side of the connector to ensure good metal-to-metal contact. Figure 5-8 shows the components of a properly grounded system.

2. Make sure that shielded cables have their shields grounded at one end only. The best place to do this is usually at the PLC and not at the transducer end, because a good ground may not be available at the transducer site.

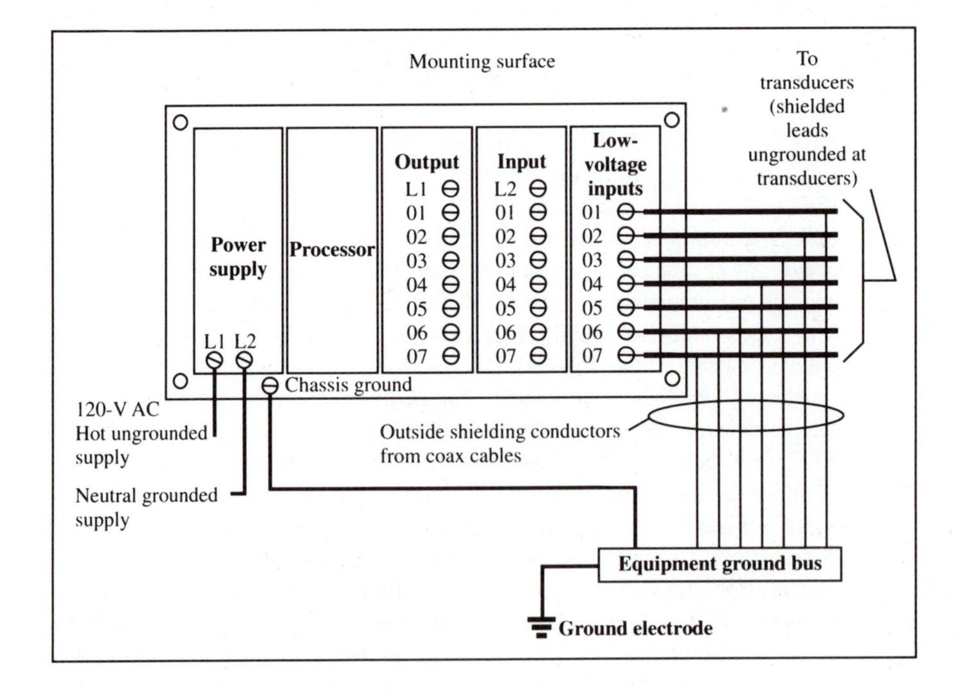

Figure 5-8 Properly Grounded System

3. When handling the main unit, modules, or plug-in circuit boards, make sure you are touching a good ground so that any static charge your body might have accumulated is drained off. Even after you have done this, avoid touching exposed connections. When modules aren't plugged in, they should be stored in antielectrostatic plastic bags. The electronics in the modules and main unit contain components that can be damaged by electrostatic electricity.

4. When inserting a unit or module into the main unit, make sure the power is off. This is necessary because if the pins are misaligned during insertion, they could produce improper connections that will damage the unit.

5. Never force a part into a connector. Check for keys that must be aligned for proper insertion. Obviously, forcing a unit into an inappropriate connector will cause damage.

6. Most PLC's have a switch for determining what the outputs will do in case the PLC detects an error. Normally you will want the outputs to be turned off if an error is detected. There usually is a switch that enables you to leave the outputs in their last state before the error is detected when the PLC goes to automatic shut-down. Make sure you are not creating a hazardous condition by your choice of operation. If you choose the last state before error, an output could be unpredictable, since an error occurred.

7. Power circuits should be free from interference. The most serious type of interference that will not be taken care of by proper grounding is resistive interference, which occurs when equipment sharing the power circuit requires heavy current surges. Such surges increase voltage drops through the self-impedance of the source, and severe voltage swings occur. Examples of equipment requiring large current swings are big motors, compressors, welding machines, and arc furnaces. The PLC should not be on a power source with this type of equipment, and a separate source may be required.

8. Low-voltage signals going to the modules must be kept separate from power circuits. The power circuits will produce magnetic and electric fields that couple to the low-signal cable if not kept separate. If the low-voltage signals are in the millivolt range, the interference produced by coupling will swamp these signals.

9. Document all programming by making hard copies of the control diagrams. This documentation should show all I/O, with addresses to indicate which module terminals are being used. Outside devices connected to these terminals should be noted on the hard copy. Any changes made during start-up and checkout must be documented.

10. The programming and termination of the external device should be checked before the PLC is placed in the run mode. This is done with the PLC in the test position. PLC programming terminals will show a portion of the control and indicate the effects of external inputs and resulting changes in outputs. The outputs will not be physically turned on while the processor is in the test mode. The programs should be thoroughly reviewed and tested before going to the run mode.

11. Finally, the programming must be checked in the run mode. Since failures and unexpected operations are most liable to occur on the first test run, take the following precautions before going to the run mode:
    (a) Determine where the power can be turned off in case of a siltation that will result in damage to equipment or personnel.
    (b) Make sure all personnel not necessary for checkout are clear of equipment under the control of the PLC. Alert all personnel needed for testing that a test run is going to be conducted.

Some companies use lock-out and tag procedures to make sure that equipment does not operate while maintenance and repairs are being conducted. Equipment operation could result in severe injury or damage to both personnel and equipment. Basically, a tag—usually red—is placed on the power source for the equipment and the PLC, and can only be removed by the person who originally placed the tag. In addition to the tag, a lock can also be placed so that equipment cannot be energized. Removing a red tag or operating the equipment under these conditions may be grounds for firing.

5-7 Troubleshooting

During Installation

The following steps are necessary for efficient troubleshooting when first installing a PLC.

1. Prepare a well-documented control schematic. This can be either a hard copy of the control scheme or a computer and monitor with the necessary software to let you connect to the PLC and do on-line monitoring. The latter is more desirable but also more expensive.

2. Check input and output labeling. Input and output devices must be clearly labeled, and the labeling must be coordinated with the control schematic.

3. You need a means to document required changes as soon as they occur. This can be done either by using several hard copies of the control scheme that can be marked up as changes are made or by a computer and monitor with the necessary software to let you connect to the PLC. The computer will, with minimum effort, enable you to make changes readily and to save changes as new files.

While Checking Out an Application

Troubleshooting is largely an art, and may vary from one person to the next. Assuming the program has been loaded into the PLC, the following steps are typical when checking out an application of a PLC.

1. Place the PLC in the test mode before turning on the power, to enable you to see the effects the inputs will have on the control without actually operating the outputs.

2. Check that the PLC and modules have proper power supplied for operation.
3. Check that PLC-status indicating lights are correct for proper operation. Example: A battery indicating light will show if the batteries are okay.
4. Verify that inputs are connected to the modules per the control scheme, for example, operating a push button and observing whether the correct terminal indicating light lights on the appropriate module.
5. Verify that outputs are connected to the modules per the control scheme. If you have a computer and monitor with the necessary software to let you connect to the PLC, you can use the forcing function to check outputs. **Take all necessary precautions to protect personnel and equipment during forcing, since the output can be forced so that particular piece of equipment will be energized.**
6. Verify that inputs affect the contacts in the control as designated, by operating inputs and observing the indicator on a monitor. You must have a programming terminal with monitor display capability to do this. If contacts do not respond to inputs per the control scheme, check for errors in addressing.
7. Verify the operations of the control scheme by operating inputs and monitoring outputs to see if they are responding as desired. If remote I/O is not readily available, the forcing function can be used to change inputs.
8. Once you are satisfied that you have worked out the bugs in the test mode by going through the previous steps, you are ready to check the control in the run mode. You will need to be able to watch how the actual equipment being controlled is responding during the run test, so you can respond to undesired equipment operation that could damage equipment or produce unsafe conditions with potential for injuring personnel. If there are problems at this point, the source could be the design of the control itself. There are three steps for fixing the control:
   - (a) Determine what symptoms are indicating improper operation.
   - (b) Given the symptoms, determine which section of the control scheme is malfunctioning.
   - (c) Given the symptoms and section, narrow the problem down to the components causing the problem, commonly improper addressing at this stage of troubleshooting.
9. Make certain any changes have been documented. This requirement cannot be overemphasized, since it is the key to efficient troubleshooting and equipment maintenance.

5-8 Maintenance

Modern PLC's are relatively maintenance-free. One of the major advantages of PLC's over relay logic is the reliability resulting from having solid-state devices without moving parts. However, they do require some care.

1. Periodically check the environment surrounding the PLC to make sure it is not being subjected to excessive moisture, corrosive atmospheres, dust, soot, and the like.

2. Filters required for PLC's located in enclosures need to be changed so proper air circulation is maintained for cooling. How often depends on how much dust is in the enclosure area.

3. Periodically, dust and soot should be cleaned from the PLC to avoid potential power leakage problems. Dust and soot can also affect the proper operation of heat sinks.

4. Terminals should be checked periodically for tightness. Normal thermocycling or vibrations will sometimes produce loose connections and possible malfunctions.

5. Check to make sure no extraneous items are placed on PLC's that might close vents designed for cooling, for example, rags and tools.

6. New equipment being installed near the PLC should be monitored to see that it will not introduce electrical interference to cause a malfunction. The power circuit should be checked to ensure that no equipment is added that will create excessive resistive-coupled interference.

7. Modules should be checked to see that they remain seated properly and that locking mechanisms are in place. This would be particularly important with equipment that produces excessive vibration.

8. A set of spare parts should be available. The number and type of spare parts needed is largely a function of the number of PLC's and the importance of the operations being controlled. For instance, an automobile assembly line will use many PLC's and modules, necessitating that commonly used processors and modules be available as spare parts. A small operation with one PLC may not warrant any extra modules or processors.

Summary

Proper grounding is important so that personal safety as well as correct operation of equipment is ensured. Needless damage to equipment can be avoided by following the procedures outlined in this chapter. Interference can easily be prevented via proper grounding, thereby preventing unwanted signals to the PLC. The information for proper grounding can also be found in the installation manuals for various PLC's; it is wise to read and follow such instructions before applying power to a PLC or other devices.

One of the keys to efficient troubleshooting is good documentation. This is often neglected, which results in longer downtime, wasted hours trying to determine what was previously done, and, as a consequence, needless additional expense.

Exercises

5-1. What are the three primary reasons for grounding? Give an example of each.

5-2. State three precautions you would take when installing a PLC.

5-3. What is the purpose of the shield wire on a coaxial cable?

5-4. Describe the procedure for installing coaxial cable between a group of transducers and a PLC.

5-5. Ground loops are a problem for what range of signals?

5-6. Will attaching multiple grounds to the chassis of a device create any problems?

5-7. Explain how electrostatic charge is a problem, and give some methods of reducing this hazard.

5-8. What is the purpose of using keying with connectors when installing a plug-in unit? What precautions are required during insertion?

5-9. Give an example in which setting error detection to leave the outputs in the last state could create a problem.

5-10. What precautions must be taken when forcing an output?

5-11. List five factors that would affect the location of a PLC.

5-12. What are the three main steps required to troubleshoot a control program, assuming the hardware is connected properly?

6 Programmable Controller

OUTLINE

OBJECTIVES

Upon completion of this chapter, the student will be able to:

- Explain the similarities and differences between programmable logic controllers and personal computers.
- Name and explain the functions of the essential parts of a programmable logic controller, such as the central processing unit, memory, and registers.
- Describe how a programmable logic controller scans and executes a program.
- Make conversions between base-2, -8, -10, and -16 number systems.
- Encode and decode in binary-coded decimal code and in grey code.

A PROGRAMMABLE controller will be explained by showing the similarities and differences between it and personal computers. The discussion of essential parts such as the central processing unit, memory, and registers will be done on a generic basis. Numbers systems also will be covered, and will be addressed on a generic basis. The emphasis will be on giving simple examples so that a novice to programmable controllers can learn the basics of their use.

6-1 Definition of Programmable Controller

A **programmable controller** is a computer that has been hardened to work in an industrial environment and is equipped with special input/output (I/O) and a control programming language. The special control program continuously checks what is happening with the inputs, processes these inputs via a user program, and then alters the outputs. The common industry abbreviation for these devices is *PLC*, which stands for *programmable logic controller.*

6-2 Central Processing Unit (CPU)

The heart of a modern programmable controller is a micro chip called a **central processing unit (CPU)**. By adding to the CPU the necessary buses to access memory and input/output, we make a computer. As Figure 6-1 shows, the CPU is connected to the memory and I/O by three buses: The *control bus* lets the CPU control whether it is to receiving or to send information from either I/O or memory; the *address bus* lets the CPU address specific areas in I/O or memory; and the *data bus* lets the CPU and I/O or memory exchange information. Figure 6-1

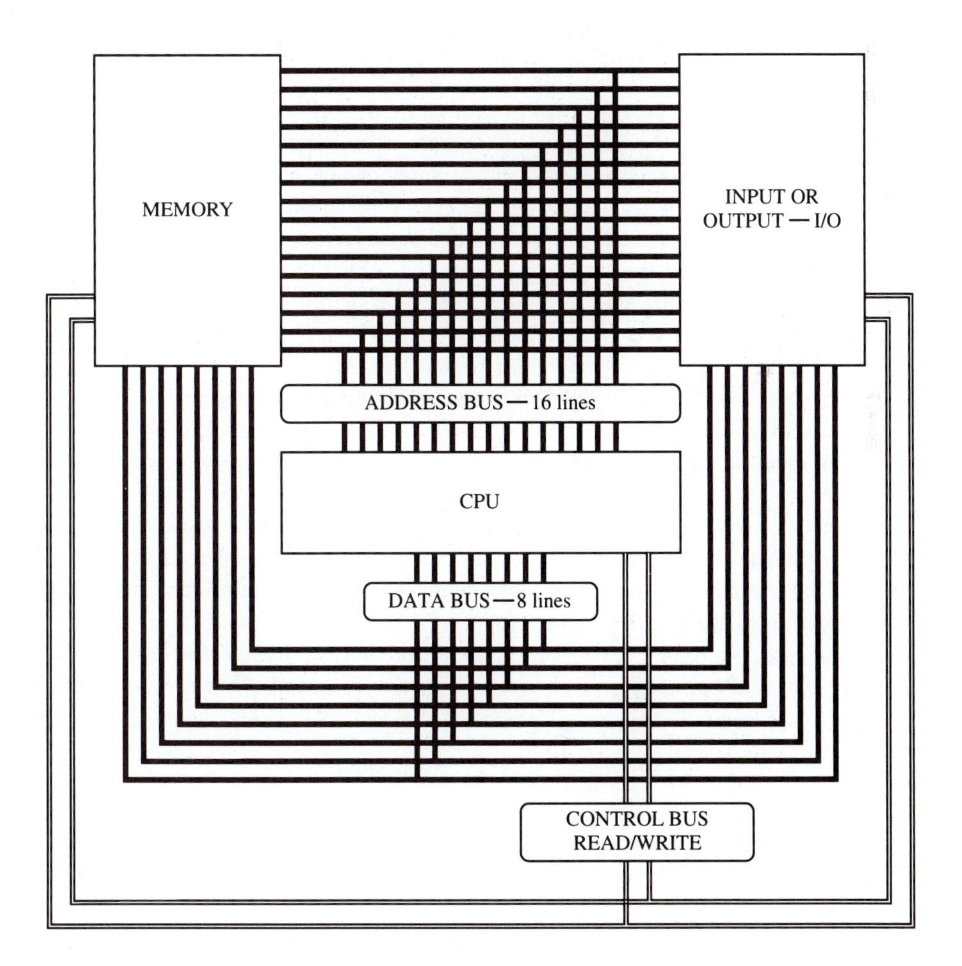

Figure 6-1 Typical Computer

shows a 16-bit address bus and an 8-bit data bus. On large machines, these buses might have more bits. Instead of 16 and 8 bits, we could have 32 and 16. The number of lines in the address bus basically determines how many specific memory locations can be addressed.

The language of the CPU is binary, and is referred to as **machine language.** The CPU will send out a pattern of 0's and 1's to address memory or I/O. With 16 bits, we can send out 2^{16} binary combinations, or 65,536 unique addresses. This is refered to as 64K in the computer industry because 1K is equal to 2^{10}, or 1024.

The size of the data bus determines how many binary bits of information can be passed between the CPU and the memory and I/O. A larger data bus usually means that more binary information can be exchanged in less time. The standard terms used in the computer industry for the number of binary digits are shown in Table 6-1.

6-3 Memory

The memory contains two different types of memory: RAM and ROM. **RAM,** *random-access memory*, can be read or altered by writing new data. The choice of RAM as the acronym for this type of memory is unfortunate, since both RAM and ROM are random-access memory. (A better choice might have been *RWM*, meaning *read–write memory*.) **ROM**, or *read-only memory*, is written permanently and cannot be altered. All computers must have this type of memory so that they can be started by a bootstrap program, which then enables one to communicate with the computer. When you turn on your personal computer, a bootstrap program must be used to get things started. It has to get the computer up and running and ready to receive information from the keyboard. This program is critical, and must always be present; therefore, the manufacturer makes it permanent. Also, if special languages are needed, as in programmable controllers, programs must be permanently present that can understand and work with these languages.

Some memory into which we can write our programs must also be present, and this memory is known as RAM. Basically, RAM can be altered as needed and will be used by the programmable controller to hold control programs and information used by those programs.

There are two types of RAM: volatile and nonvolatile. **Volatile RAM** is memory that will be lost if power to the computer is turned off, and **nonvolatile**

Table 6-1 Terms for Bit Groups

Number of Digits	Term
1	Bit
4	Nibble
8	Byte
Two or more bytes	Word

RAM is memory that will not be lost if the power is turned off. Nonvolatile RAM is usually protected by a battery. It is essential that a PLC not lose its programming if power is interrupted briefly, and this is accomplished with battery backup for when AC power is lost.

There are several types of read-only memory available to manufacturers of PLC's, and we will briefly discuss how they are usually employed.

When a programmable controller is first being designed, the manufacturer puts the bootstrap program in an **erasable programmable read-only memory (EPROM)** or **electrically erasable programmable read-only memory (EEPROM)** because changes in this program may be necessary. The EPROM can be programmed with a special inexpensive device. Sine the EPROM is an integrated circuit, it can be changed by simply taking it out of its socket, reprogramming it, and putting it back. This EPROM is erased by exposing it to ultraviolet light, and it can be changed over and over again.

EEPROM can be used in the same manner as EPROM. The only difference is that it can be erased and programmed electrically either with a special device or by having a computer control voltages issued to the EEPROM. The EEPROM is unique because it can be modified electrically in the field under the control of another computer. If a group of PLC's had EEPROM's and were under the control of a main computer, the main computer could modify the bootstrap program. Currently, this is not normally done, but may be in sophisticated future applications of PLC's.

The second phase of production may be a trial production run wherein a limited number of units are released for trial. A **programmable read-only memory (PROM)** can be used for this stage. The PROM is programmed by a special device that selectively destroys parts of the circuitry of the integrated circuit (IC) so that once a memory cell is changed, it can't be changed back. These IC's are cheaper than the erasable memory chips but are still relatively expensive since they must be individually programmed.

The third production phase is when the manufacturer goes to full production runs, when the normal practice is to go to ROM for the bootstrap program. The advantage of ROM is that it is permanent and much cheaper than erasable memory chips. The ROM must be manufactured in large quantities to be inexpensive, and ROM's are manufactured in such a way that they cannot be altered. These constraints require the manufacturer to be sure of the design at this point. An example of how ROM's are permanently altered is when a base connection to a transistor is left off so that it can never be turned on. This is done when the IC is manufactured.

Table 6-2 summarizes the types of memory and their applications.

Memory Maps

How the memory is organized is often shown on a memory map. The programmable controllers need four areas: one section of ROM and three sections of RAM. A typical memory map is shown in Figure 6-2. These maps are very handy when programming, because it is much easier to show the memory layout on a map than to explain it verbally.

Table 6-2 Types of Memory

Type	Meaning	Application
RAM	Random-access memory	Read/write
ROM	Read-only memory	Cannot be altered
PROM	Programmable read-only memory	First production run
EPROM	Erasable programmable read-only memory	Prototype design
EEPROM	Electrically erasable programmable read-only memory	Prototype design

1. The executive section of memory is the area containing the programs that show how the PLC responds and operates. On the map in Figure 6-2, this area is between addresses 512_{10} and 2047_{10}. It is in control of the CPU and determines how the rest of the memory will be organized. This section of ROM contains the permanent memory required to run the PLC. This programming enables the PLC to start up when the power is initially turned on and also enables it to interpret the instructions it receives from a keyboard or from the user section of memory.

2. One section of RAM is required as a kind of scratch pad or work area that enables the executive to temporarily store the changing information it needs. In figure 6-2, this area is in two sections: between addresses 000_{10} and 007_{10} and addresses 064_{10} and 071_{10}.

3. A RAM-dedicated storage area that is defined by the executive is shown in the two sections between addresses 008_{10} and 063_{10} and addresses 072_{10} and 127_{10} in Figure 6-2. This area will hold information about the status of inputs, out-

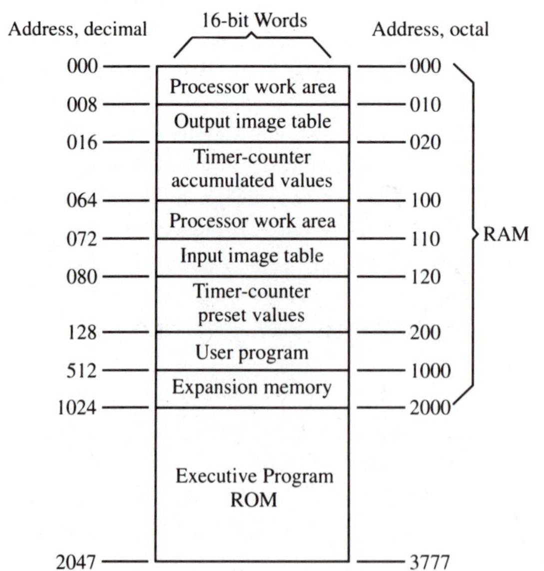

Figure 6-2 Typical Memory Map

puts, timers, and counters as well as constants that are needed for timers, counters, and other functions.

4. There has to be a section of RAM where the user puts the program that is to be executed. This is the area between addresses 128_{10} and 511_{10} in Figure 6-2.

6-4 Registers

A **register** is a collection of memory cells, and is used for temporary storage of binary information. Figure 6-3 shows a 16-bit register. The number of bits stored by a register is usually in multiples of eight. Registers can be in part of the main memory or in the CPU itself. The CPU has special registers it uses to keep track of the information it needs. An example is when the CPU is adding two numbers: First, it must fetch one of the numbers from the data bus. It does this by reading a memory location, and it can only address one location at a time. The CPU must store the first reading somewhere while it is getting the second number. It puts the first number in a special register called the *accumulator* and then gets the second number and adds them together. It then needs some place to store the result, and again it uses the accumulator. Registers are also needed in the programmable controller. The status of input and output devices is required for the control program. Thus, this information must be continuously read, updated, and stored in registers. Counters and timers used need to have their status stored temporarily.

6-5 Scan

The programmable controller is designed to monitor the status of several inputs and initiate changes to its outputs as dictated by the control program entered by the user. This requires that the PLC constantly scan the input in order to update the status of these devices. Once it has the latest status, it must make those changes in the control program and then finally make the necessary changes in the outputs. Since the inputs can change at any time, the PLC must carry on this process continuously. This requires the scan cycle shown in Figure 6-4. The length of time it takes to scan is a function of the length of the control program. Scan can take from about 1 to 20 milliseconds. An equivalent relay logic scheme would generally require more time to respond, since each control relay takes several milliseconds to operate.

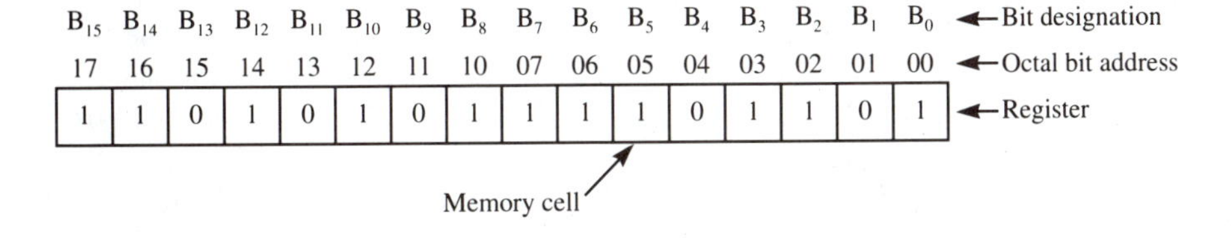

Figure 6-3 16-Bit Register

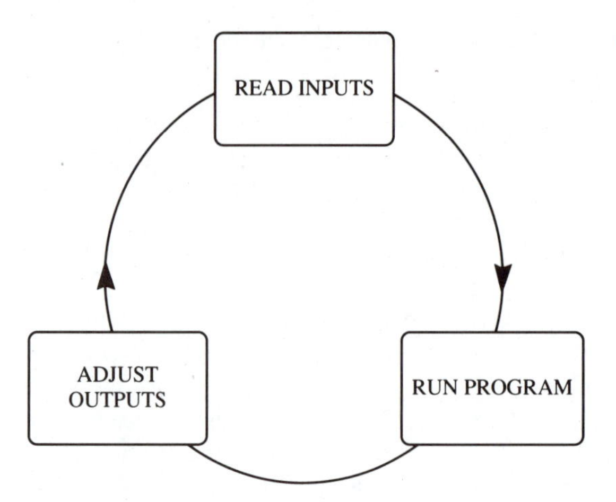

Figure 6-4 Scan Cycle

The time required for a relay-logic control scheme varies, depending on what is happening. The more relays that have to respond sequentially, the longer it takes to get the final output. Special emergency conditions can be handled quickly by having a relay operate and shut down a process depending on only one contact closure. This is a case where the relay logic control would respond faster than a PLC unless the scan is altered. The scan can be changed to handle this situation. The PLC can be programmed to look at that particular input more often; if that input changes, the PLC will immediately change the appropriate output.

6-6 Number Systems

Using programmable controllers requires us to become familiar with other number systems besides decimal. We will need to learn these machine-oriented systems. All the numbering systems we will deal with are weighted number systems, which means that the place a digit holds in a particular number has a specific value. This weighting is a function of the powers of the base we are using.

Looking at the generalized weighted number system (B^n) shown at the top of Figure 6-5, notice the point with numbers to the left and right. The first place (or position) to the left of the point is the base to the zero power (B^0), the second position to the left is the base to the first power (B^1), the third position to the left is the base to the second power (B^2), etc. The first position to the right is the base to the minus 1 (B^{-1}), the second position to the right is the base to the minus 2 (B^{-2}), etc. The count in a particular position is equal to the weight of that position times the digit in that position. We are so familiar with decimal systems that we hardly think about it. The number 987.0 means there are 7 units, 8 tens, and 9 hundreds, which gives a total count of nine hundred eighty-seven.

$$B^n \ldots B^4 \quad B^3 \quad B^2 \quad B^1 \quad B^0 \cdot B^{-1} \quad B^{-2} \quad B^{-3} \quad B^{-4} \ldots B^{-n}$$
$$P_n \ldots P_5 \quad P_4 \quad P_3 \quad P_2 \quad P_1 \cdot P_{-1} \quad P_{-2} \quad P_{-3} \quad P_{-4} \ldots P_{-n}$$

B = Base, P = Position

	Binary	Octal	Decimal	Hexadecimal
	0	0	0	0
*	1	1	1	1
**	10	2	2	2
***	11	3	3	3
****	100	4	4	4
*****	101	5	5	5
******	110	6	6	6
*******	111	7	7	7
********	1000	10	8	8
*********	1001	11	9	9
**********	1010	12	10	A
***********	1011	13	11	B
************	1100	14	12	C
*************	1101	15	13	D
**************	1110	16	14	E
***************	1111	17	15	F
****************	10000	20	16	10

Figure 6-5 Weighted Number Systems

EXAMPLE 6-1

$$789 = ?_{10}$$

$$7 \times 10^0 = 7 \times 1 = 7$$
$$8 \times 10^1 = 8 \times 10 = 80$$
$$9 \times 10^2 = 9 \times 100 = \underline{900}$$

TOTAL COUNT 987

The possible numbers in the decimal system are 0, 1, 2, 3, 4, 5, 6, 7, 8, and 9. When we have counted up to the number 9, we have run out of digits. The next count is 0, and a count of 1 is carried to the next position to the left. This procedure is the same when we reach the maximum digit for a number system with a different base that is similarly weighted. The maximum digit depends on the base for that particular number system; but when a count in a column reaches that last digit, the next count will cause a 0 to be placed in that column and a carry of 1 to the next. The number systems we will be concerned with are *binary* (base 2), *octal* (base 8), *decimal* (base 10), and *hexadecimal* (base 16). The digits for these are as follows:

Binary	0, 1
Octal	0, 1, 2, 3, 4, 5, 6, 7
Decimal	0, 1, 2, 3, 4, 5, 6, 7, 8, 9
Hexadecimal	0, 1, 2, 3, 4, 5, 6, 7, 8, 9, A, B, C, D, E, F

The asterisks in the first column of the bottom portion of Figure 6-5 represent anything we want to count, such as stones, apples, houses. In each system, when a count reaches 1 less than the base, the next count is 0 and a 1 is added to the next column. Also, the total count is the sum of the digits in each column times the weighting for that column. The weighting for each number system is as follows:

Binary	$2^n \ldots 64 \quad 32 \quad 16 \quad 8 \quad 4 \quad 2 \quad 1 \quad . \quad \frac{1}{2} \quad \frac{1}{4} \quad \frac{1}{8} \quad \frac{1}{16} \ldots (\frac{1}{2})^{-n}$
Octal	$8^n \ldots 4096 \quad 512 \quad 64 \quad 8 \quad 1 \quad . \quad \frac{1}{8} \quad \frac{1}{64} \quad \frac{1}{512} \quad \frac{1}{4096} \ldots (\frac{1}{8})^{-n}$
Decimal	$10^n \ldots 1000 \quad 100 \quad 10 \quad 1 \quad . \quad \frac{1}{10} \quad \frac{1}{100} \quad \frac{1}{1000} \ldots (\frac{1}{10})^{-n}$
Hexadecimal	$16^n \ldots 4096 \quad 256 \quad 16 \quad 1 \quad . \quad \frac{1}{16} \quad \frac{1}{256} \quad \frac{1}{4096} \ldots (\frac{1}{16})^{-n}$

We need to develop a way of telling the base of a number when we show numbers, because 10 usually means "ten" to us. But how will we know when 10 represents a number in another system? Numbers shown in this book *without* a subscript are to the base 10, that is, are defaulted to decimal. For all other numbers, we will use a subscript to represent the base. Thus, 10_2 represents a *binary* number (specifically the number 2 in decimal).

Converting Any Base to Decimal

We can convert any number to the decimal system by adding the product of each digit times its weight in that number system. What, for example, is 1234_8 in decimal?

EXAMPLE 6-2

$$1234_8 = ?_{10}$$

$$\begin{aligned}
4 \times 8^0 &= 4 \times 1 &&= 4 \\
3 \times 8^1 &= 3 \times 8 &&= 24 \\
2 \times 8^2 &= 2 \times 64 &&= 128 \\
1 \times 8^3 &= 1 \times 512 &&= \underline{512} \\
\text{TOTAL COUNT} & &&= 668
\end{aligned}$$

We can use this method to go from any number to decimal.

Converting Decimal to Binary, Octal, or Hexadecimal

We can convert from decimal to any of the binary, octal, or hexadecimal number systems by the following process.

1. Divide the number by the base. The result will have two parts: a whole number and a decimal. Take the decimal portion and multiply it by the base. Record this number; it is the remainder stated as a whole number.
2. Using the whole-number portion of the answer from Step 1, repeat Step 1.
3. Continue this process until you get only a decimal portion. Take the decimal portion and multiply it by the base; record this number.
4. The answer is the numbers recorded, but written in order with the least significant digit being the first number recorded and the most significant digit being the last number recorded.

Let's look at some examples.

EXAMPLE 6-3
DECIMAL TO BINARY

$$75_{10} = ?_2$$

$$2\overline{)75}\ \ 37.5 \qquad 0.5 \times 2 = 1$$

$$2\overline{)37}\ \ 18.5 \qquad 0.5 \times 2 = 1$$

$$2\overline{)18}\ \ 9.0 \qquad 0.0 \times 2 = 0$$

$$2\overline{)9}\ \ 4.5 \qquad 0.5 \times 2 = 1$$

$$2\overline{)4}\ \ 2.0 \qquad 0.0 \times 2 = 0$$

$$2\overline{)2}\ \ 1.0 \qquad 0.0 \times 2 = 0$$

$$2\overline{)1}\ \ 0 \qquad 0.5 \times 2 = 1$$

Ans. 1001011

Check

$$1 \times 2^0 = 1$$
$$1 \times 2^1 = 2$$
$$0 \times 2^2 = 0$$

$$1 \times 2^3 = 8$$
$$0 \times 2^4 = 0$$
$$0 \times 2^5 = 0$$
$$1 \times 2^6 = \underline{64}$$
$$75$$

EXAMPLE 6-4
DECIMAL TO HEXADECIMAL

$$751_{10} = ?_{16}$$

$16 \overline{)751} = 46.9375$ $\quad 0.9375 \times 16 = 15_{10} = F_{16}$

$16 \overline{)46} = 2.875$ $\quad 0.875 \times 16 = 14_{10} = E_{16}$

$16 \overline{)2} = 0$ $\quad 0.125 \times 16 = 2_{10} = 2_{16}$

Ans. $2EF_{16}$

Check

F	$15 \times 16^0 =$	15
E	$14 \times 16^1 =$	224
2	$2 \times 16^2 =$	512
		751

EXAMPLE 6-5
DECIMAL TO OCTAL

$$1867_{10} = ?_8$$

$8 \overline{)1867} = 233.375$ $\quad 0.375 \times 8 = 3_{10} = 3_8$

$8 \overline{)233} = 29.125$ $\quad 0.125 \times 8 = 1_{10} = 1_8$

$8 \overline{)29} = 3.625$ $\quad 0.625 \times 8 = 5_{10} = 5_8$

$8 \overline{)3} = 0.375$ $\quad 0.375 \times 8 = 3_{10} = 3_8$

Ans. $1867_{10} = 3513_8$

Check

3	$3 \times 8^0 =$	3
1	$1 \times 8^1 =$	8

$$\begin{array}{lll} 5 & 5 \times 8^2 = & 320 \\ 3 & 3 \times 8^3 = & \underline{1536} \\ & & 1867 \end{array}$$

Converting Fractions of Any Base to Decimal

Fractions are handled in a similar manner to whole numbers, except instead of dividing by the base we multiply. The steps are as follows:

1. Multiply the decimal number by the base. Record the whole portion of the result.
2. Using the decimal portion of the answer from Step 1, repeat Step 1.
3. Continue this process until you get 0 in the decimal portion or you have the resolution you are looking for.
4. The answer is the numbers recorded written in order, with the most significant digit being the first number recorded and the least significant digit being the last number recorded.

Let's look at some examples.

E X A M P L E 6 - 6
DECIMAL TO BINARY

$$0.3521_{10} = ?_2$$

$$\begin{array}{ll} 0.3521 \times 2 = 0.7042 & 0 \\ 0.7042 \times 2 = 1.4084 & 1 \\ 0.4084 \times 2 = 0.8168 & 0 \\ 0.8168 \times 2 = 1.6336 & 1 \\ 0.6336 \times 2 = 1.2672 & 1 \end{array}$$

Ans. $0.3521_{10} = .01011$ (to five places)

Check

$$\begin{array}{l} 2^{-1} \times 0 = 0.0 \\ 2^{-2} \times 1 = 0.25 \\ 2^{-3} \times 0 = 0.0 \\ 2^{-4} \times 1 = 0.0625 \\ 2^{-5} \times 1 = \underline{0.03125} \\ \qquad\qquad\;\; 0.34375 \end{array}$$

Note that there is a difference between the checked value and the original: $0.3521 - 0.34375 = 0.00835$. The size of the difference will depend on how many places we obtain for our binary result. This error will diminish if we continue the conversion process to get more places.

$$0.3521_{10} = ?_2$$

$$0.3521 \times 2 = 0.7042 \quad 0$$

$$
\begin{array}{ll}
0.7042 \times 2 = 1.4084 & 1 \\
0.4084 \times 2 = 0.8168 & 0 \\
0.8168 \times 2 = 1.6336 & 1 \\
0.6336 \times 2 = 1.2672 & 1 \\
0.2672 \times 2 = 0.5344 & 0 \\
0.5344 \times 2 = 1.0688 & 1 \\
0.0688 \times 2 = 0.1376 & 0
\end{array}
$$

Ans. $0.3521_{10} = .01011010_2$ (to eight places)

Check

$$
\begin{aligned}
2^{-1} \times 0 &= 0.0 \\
2^{-2} \times 1 &= 0.25 \\
2^{-3} \times 0 &= 0.0 \\
2^{-4} \times 1 &= 0.0625 \\
2^{-5} \times 1 &= 0.03125 \\
2^{-6} \times 0 &= 0.0 \\
2^{-7} \times 1 &= 0.0078125 \\
2^{-8} \times 0 &= \underline{0.0} \\
& 0.3515625
\end{aligned}
$$

Now the difference between the checked value and the original has become $0.3521 - 0.3515625 = 0.0005375$. The error has diminished; and if we continue the conversion process to get more places, the error will eventually be zero. Clearly, the accuracy depends on the number of places, and how far we want to go depends on how much error we are willing to tolerate.

The percentage of error can be calculated with the following formula:

$$\% \text{ Error} = \frac{\text{Converted value} - \text{Original value}}{\text{Original value}} \times 100$$

The numbers that have powers of 2 as their bases are very easy to convert to and from binary. The procedure is described next.

Octal to Binary Starting on the right of the octal, write the three-bit code for each octal digit, putting the three-bit groups in the same order as in the octal number. The following example demonstrates this procedure.

EXAMPLE 6-7

$$3567_8 = ?_2$$

Convert each octal number to its three-digit binary equivalent.

$3 = ?_2$	$5 = ?_2$	$6 = ?_2$	$7 = ?_2$
011	101	110	111

$$3567_8 = 011101110111_2$$

Binary to Octal Starting on the right of the binary number, break the number into three-bit groups. If the last group doesn't have three bits, add the necessary 0 to the left to make it complete. Change the three-bit groups to octal, with the order maintained. The following example demonstrates this procedure.

EXAMPLE 6-8

$$1011110101_2 = ?_8$$

Starting from the right side, break the binary number up into three-bit patterns.

1011110101_2

1 011 110 101

One at a time, convert these segments to octal. Add an extra 0 to complete the last segment.

001 011 110 101

1 3 6 5

Ans. $1011110101_2 = 1365_8$

Hexadecimal to Binary Starting on the right of the hexadecimal, write the four-bit code for each hexadecimal, putting the four-bit groups in the same order as in the hexadecimal number. The following example demonstrates this procedure.

EXAMPLE 6-9

$$F5_{16} = ?_2$$

Convert each hexadecimal number to its 4-digit binary equivalent.

$F = ?_2$ $5 = ?_2$

$F = 1111$ $5 = 0101$

Ans. $F5_{16} = 11110101_2$

Binary to Hexadecimal Starting on the right of the binary number, break the number into four-bit groups. If the last group doesn't have four bits, add the necessary 0 to the left to make it complete. Change the three-bit groups to octal, with the order maintained. The following example demonstrates this procedure.

EXAMPLE 6-10

$$1011110101 = ?_{16}$$

Starting from the right side, break the binary number into four-bit segments.

10 1111 0101

Convert these segments to hexadecimal, one at a time. Add an extra 0 to complete the last segment.

0010 1111 0101

2 F 5

Ans. $1011110101_2 = 2F5_{16}$

This book will deal with discrete logic, which means we will be working in two states, that is, in a *binary* system. The binary system has two digits, 0 and 1, and is the logic-control language of computers. It is easy to generate two states electrically by turning a switch off and on; and in solid-state devices, a transistor turned off and on can simulate a switch. The internal language of a programmable controller is the binary system. Humans are not good at handling binary numbers because such numbers are long and it is hard to readily know their value after you get past five bits. Since octal and hexadecimal convert so easily to binary, these systems are used as a compromise between humans and computers. Octal has evolved as one of the most-used number systems for PLC's. It may well progress to hex as new machines are developed because hex is more efficient (you need fewer digits to represent binary numbers).

Note on Calculators

Having presented this material on converting from one number system to another it should be said that modern scientific calculators are available to convert numbers back and forth between decimal, binary, octal, and hexadecimal. They are inexpensive and easy to use, for example, in converting a number displayed in decimal to one in binary. This simply involves one key stroke to change the display mode from decimal to binary. More expensive calculators have programming capability, and can store programs the user generates. The user would make up a conversion program, store it, and call it up for use when necessary. If you are going to be involved with PLC or single-board computers, it will be much easier if you purchase a calculator with such a conversion capability.

Binary-Coded Decimal Numbers

As you can see from going through the number systems, there is no easy way to get from binary to decimal and back. It is much easier to go from hex to octal to binary and back. Since people have used the decimal system for so long, they are not about to give it up. Thus, an easy way to convert from binary to decimal needed to be devised.

Binary-coded decimal (BCD) was a solution that evolved. In this system, the decimal digits are each given a code that is binary. The most popular code uses the natural binary counting to represent the nine digits. This is referred to as the *8421 code*, since 8421 is the natural binary progression. You can represent 0 through 15 with four binary digits. But the decimal system only needs ten of the sixteen combinations; six of the combinations are not used. The natural BCD code is shown in Table 6-3.

Converting from decimal to BCD is simple. For each decimal digit, you write its four-bit code, putting the four-bit codes in the same order as in the decimal number. The next example will help you see this procedure.

EXAMPLE 6-11

$$98765_{10} = ?_{BCD}$$

9	8	7	6	5
1001	1000	0111	0110	0101

Ans. $98765_{10} = 10011000011101100101_{BCD}$

This procedure is reversed in the next example.

EXAMPLE 6-12

$$1011000100010_{BCD} = ?_{10}$$

Expand the BCD number into four-digit patterns, starting with the least significant digit and working to the right.

1 0110 0010 0010

The last group may contain fewer than four digits. To change this to a four-bit pattern, add the necessary leading zeros.

0001 0110 0010 0010

Finally, change these four-bit codes to their decimal equivalents.

0001	0110	0010	0010
1	6	2	2

Ans. $1011000100010_{BCD} = 1622_{10}$

Table 6-3 Natural BCD Code

Code	Digit	Code	Digit
0000	0	0101	5
0001	1	0110	6
0010	2	0111	7
0011	3	1000	8
0100	4	1001	9

BCD is employed by industry in both computers and programmable controllers. There are many other types of BCD codes besides the 8421 code we have been describing, but this code is the one most common in programmable controllers.

Grey Code

Sometimes a computer or PLC needs a device to input angular position, for example, controlling a motor that turns a steering wheel. To do this properly we need feedback on its angular position so we know when to turn the steering motor off and on. Since the computer understands only binary, or machine, language, we need a device to change angular position to a binary code. This is done using a special decoding wheel that rotates on a shaft. It is made so that it will pass light through some areas but block light in others. One way of doing this is to make a disk out of clear plastic or glass and to paint a pattern on it using black paint, which will not let light through. A rod is then put through the center of this disk, with the rod at a right angle to the plane of the disk. If the rod is then turned, the disk will rotate. This becomes our decoding wheel. Light sources are placed on one side of this decoding wheel, and a photo sensor or light detectors on the other (see Figure 6.6). Light not getting through will result in a 0 and light getting through will result in a 1.

One light source and detector will give us only two possible combinations. Two light sources and detectors will give us four possible combinations. The number of combinations we will be able to detect is a function of the number of light-detector pairs. The number of combinations is equal to 2^x, where x equals the number of light-detector pairs.

Each pair of light sources and photo sensors will sweep a concentric path. The pattern on the decoding wheel is made in concentric rings, which will line up with one of the detectors. With the pattern shown on the left in Figure 6.6, we have three light detectors. We will get eight different binary codes as we rotate the wheel. This enables us to break one revolution, or 360°, in eight different sectors. One sector is $360/8 = 45°$ degrees wide, meaning that we know where the wheel is positioned to within 45°. Hence, the best resolution we can have with three detectors is $\pm 45°$. If we want better resolution, we have to increase the number of detectors. Commercial decoding wheels will have at least eight detectors, which would yield a resolution of (360/256)°.

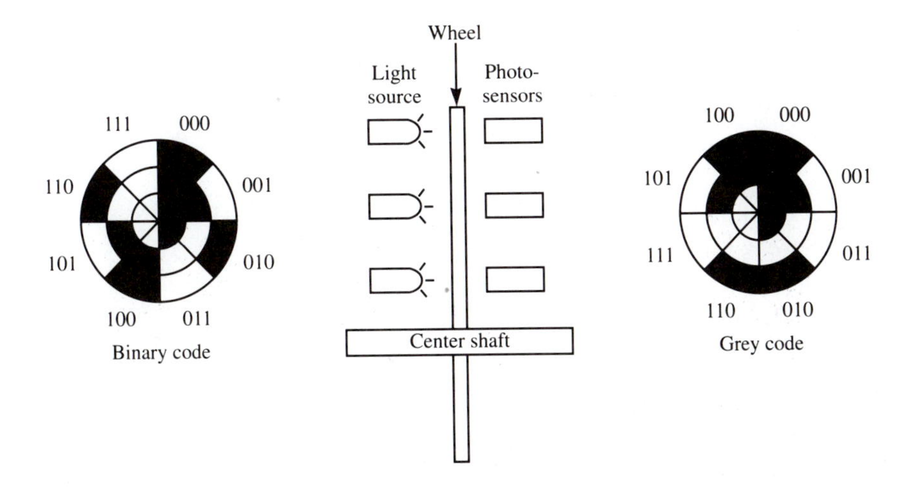

Figure 6-6 Grey Code

Any code desired can be put on the decoding disk. However, here we will look at only two possibilities. In Figure 6.6, the least significant bit (LSB) is generated by the outermost concentric pattern, and the most significant bit (MSB) is the innermost concentric pattern. The decoding wheel on the left has a natural binary progression, and the wheel on the right has what is called the *grey code*. A special problem area called the *grey area* results when the wheel is turned so that a light-detector pair is lined up directly over a spot on a pattern where it is making a transition from black to clear. The result coming from the detector is not predictable, because half the light is getting through and half is being blocked. A 1 or a 0 will result, but this means we will not be able to tell which of these two adjacent sectors we are in. A serious problem results with the natural binary wheel, since all bits will be unpredictable when we are going from 111 to 000. At this point whether we get a 0 or 1 is uncertain for all the bits, and we can get all the possible combinations in this area, making the decoder totally unreliable.

The **grey code** on the wheel on the right in Figure 6.6 makes certain only one bit is ever changing when we go from one sector to another. Thus the error we will have when we are in such grey areas will be limited to only one sector. The resolution of these sectors is a function of the number of concentric rings on the decoding wheel. Our example has three concentric rings, or 2^3 combinations, which divides the 360° into eight sectors, yielding a resolution of 45° per sector. If we increase the number of concentric rings to ten, then we have 2^{10} combinations, which divides the 360° into 1024 sectors, giving a resolution of 0.35° per sector. The grey wheel would be able to determine our angular position within ±0.35°.

The grey-code wheel solves the problems we had with a natural binary wheel but creates another problem. We now must come up with a way of changing the grey code generated to a natural binary progression to make it easy to interpret in computer programs. The procedure for doing this is not hard. We can readily

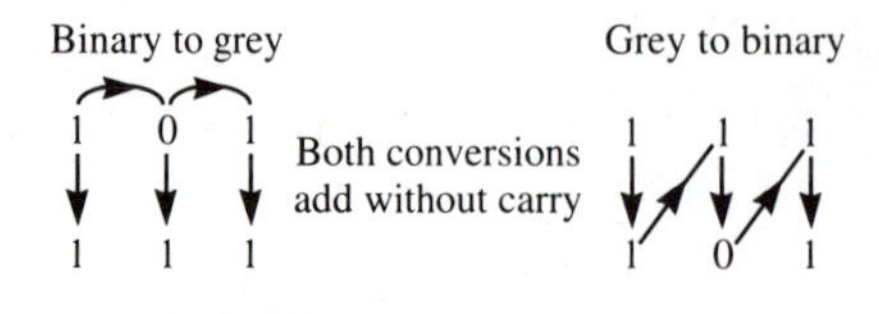

Figure 6-7 Converting To and From Grey Code

convert from binary to grey or vice versa. See Figure 6-7. The process involves the following steps.

Binary to grey

1. Write the binary number and bring down the MSB.
2. Add, without carrying, the adjacent bits, in pairs starting with the MSB, until the pair contains the LSB. The result of each add is brought down below the LSB of the pair that is added. The resulting code is the grey code.

Grey to binary

1. Write the grey number and bring down the MSB.
2. Add, without carrying, the bit just brought down with the next adjacent bit, in pairs starting with the second MSB, until the pair contains the LSB. The result of each add is brought down below the LSB of the pair that is added.

Summary

The heart of a modern programmable controller is a microcomputer, and this chapter covered some of the basics needed to understand these machines. Although you can operate a PLC without understanding the basic concepts, you become a more intelligent user if you have a better understanding of the operation of microcomputers. An essential part of this understanding involves being able to work with various number systems. Whereas people prefer the decimal number system, microcomputers work only in binary. So compromise number systems have evolved that make it easier for people and machines to communicate.

Exercises

Perform the indicated conversions in exercises 6-1 through 6-7.

6-1. (a) $18.75 = ?_2$ (b) $497 = ?_2$

6-2. (a) $FD_{16} = ?_{10}$ (b) $8D.375_{16} = ?_{10}$

6-3. (a) $532_{10} = ?_{16}$ (b) $356.25_{10} = ?_{16}$

6-4. (a) $C9_{16} = ?_2$ (b) $E1F.8_{16} = ?_2$

6-5. (a) $555_{10} = ?_8$ (b) $786_{10} = ?_8$

6-6. (a) $5679_{10} = ?_{BCD}$ (b) $2319_{10} = ?_{BCD}$

6-7. (a) $000110010101_{BCD} = ?_{10}$ (b) $010110110101_{BCD} = ?_{10}$
(c) $110010101_2 = ?_{grey}$ (d) $10010101_{grey} = ?_2$

6-8. Explain the difference between a programmable controller and a home computer.

6-9. What is the most common language used to program programmable controllers?

6-10. Why must a programmable controller continually scan?

6-11. What does *BCD* mean, and why is it used?

6-12. What is a *register*? Give an application.

6-13. What is usually stored in ROM in a programmable controller?

6-14. Explain each of the following terms: *RAM, ROM, nonvolatile memory.*

6-15. Explain the uses for ROM, PROM, EPROM, and EEPROM when manufacturers are designing a new PLC for full production.

7 Devices Commonly Used in Relay Logic and Programmable Controllers

OUTLINE

OBJECTIVES

Upon completion of this chapter, the student will be able to:

- List common devices available for use in relay ladder control, and show how these devices are represented in programmable controllers.
- Explain some of the capabilities and limitations of the common available devices, and analyze simple control schemes with these devices.
- Explain the difference between retentive and nonretentive timers.
- List four ways timing can be accomplished in control.
- Design a relay ladder control scheme using time-delay relays, and analyze an already-designed scheme.
- Explain the function of a latch, and give an example where a latch might be used.
- Explain why inputs/outputs cannot be directly connected to the internal data bus on a PLC.
- Explain the function of an input module and an output module for a PLC.

THIS chapter will discuss the devices most commonly available for use in ladder control, and will show how these devices are represented in programmable controllers. Background in this area will be helpful once you actually start entering instructions into a programmable controller. We will first cover what devices are used in relay logic, and then explain briefly how the same features will be applied in PLC's. The capabilities and limitations of each of these devices will be discussed. This chapter will enable novices to gain the basic knowledge for simple applications, and will act as a review of this knowledge for the experienced.

7-1 Input Devices

Input devices are required during a sequential process and are needed to implement controls that involve relays or PLC's. There is, however, a difference between the wiring of inputs for a programmable controller and the wiring for relay controls. Whereas relay logic (Figure 7-1) incorporates the inputs by wiring them directly to the relay coils, the programmable controller needs to have these devices go through special I/O modules so that the power source (120-V AC or whatever) can be adjusted to TTL-level voltages for the computer inside the PLC. Also, since the PLC multiplexes (time-shares) its internal data bus, it has to determine when the input information can get through. In other words, you cannot connect the inputs directly to the data bus, because the data bus is being used by the PLC to read inputs and to write other necessary information. The voltage applied by the input devices and the voltages on the data bus that are due to the PLC may be incompatible.

As Figure 7-2a shows, the input module is basically a signal-conditioning and -modifying device. The signal enters the module to be processed so that it is compatible with the PLC's data bus. Figure 7-2b presents a block diagram of this

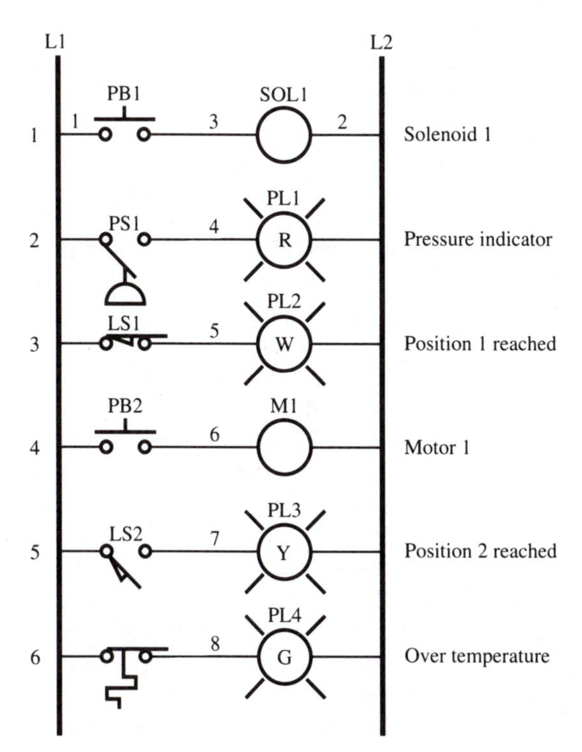

Figure 7-1 Control Schematic for a Relay Ladder

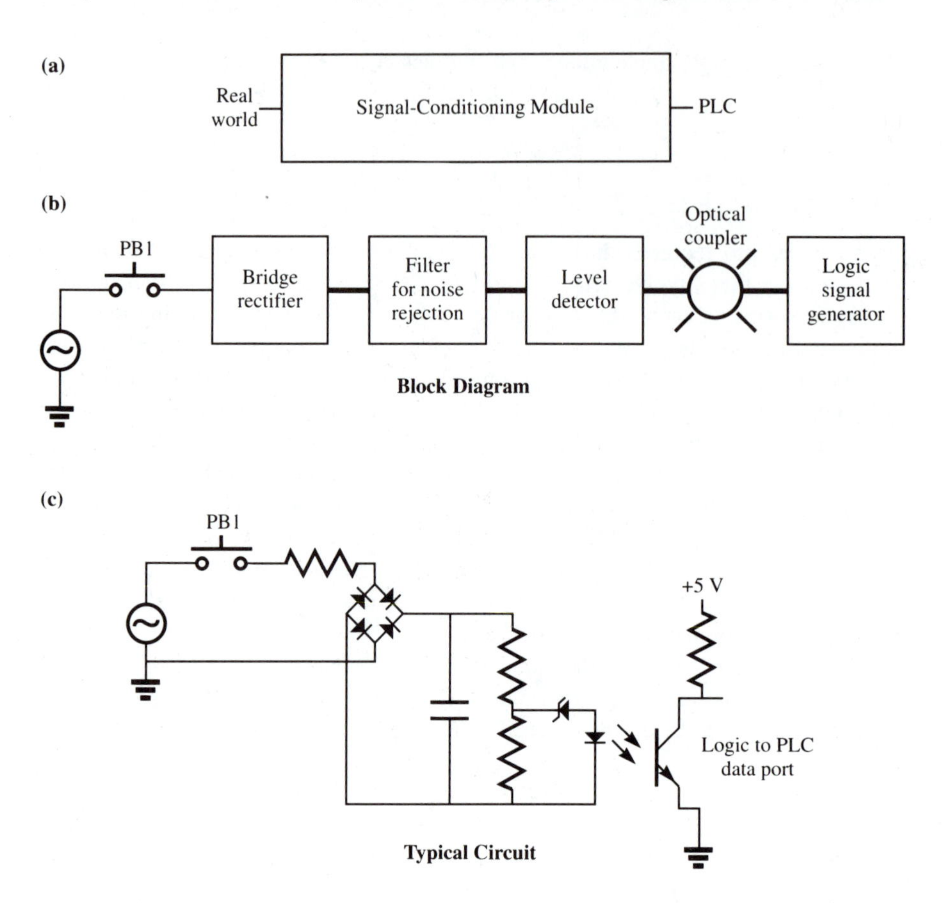

Figure 7-2 Input Module

process for an AC input module: First the signal is rectified by going through a bridge, then it is filtered for noise, and finally it is converted to a level appropriate for operating an optical coupler. The optical coupler provides electrical isolation, which prevents a failure of these first input stages and resultant application of circuit-damaging AC voltage to the PLC. Figure 7-2c presents example circuitry of how the block diagram might be implemented.

Figure 7-3 shows the same control scheme as in Figure 7-1, except whereas the one in Figure 7-1 is implemented via standard relay logic, the one in Figure 7-3 is implemented with a programmable controller. In Figure 7-1, notice that push button PB1 operates solenoid SOL1 directly. In Figure 7-3, we see that the inputs and outputs must go through special I/O modules before their status can be sent for use in the control program. Push button PB1 will apply a voltage to input module 0, terminal 01. Due to internal circuitry in the module, the 001 contact will go high when PB1 is high, thereby energizing output coil 101. Output module 1 will receive this information from the control and apply the necessary voltage to SOL1 through terminal 01. Figure 7-3 is not the standard way of drawing such control schemes. In the chapters that follow, however, these controls will be drawn

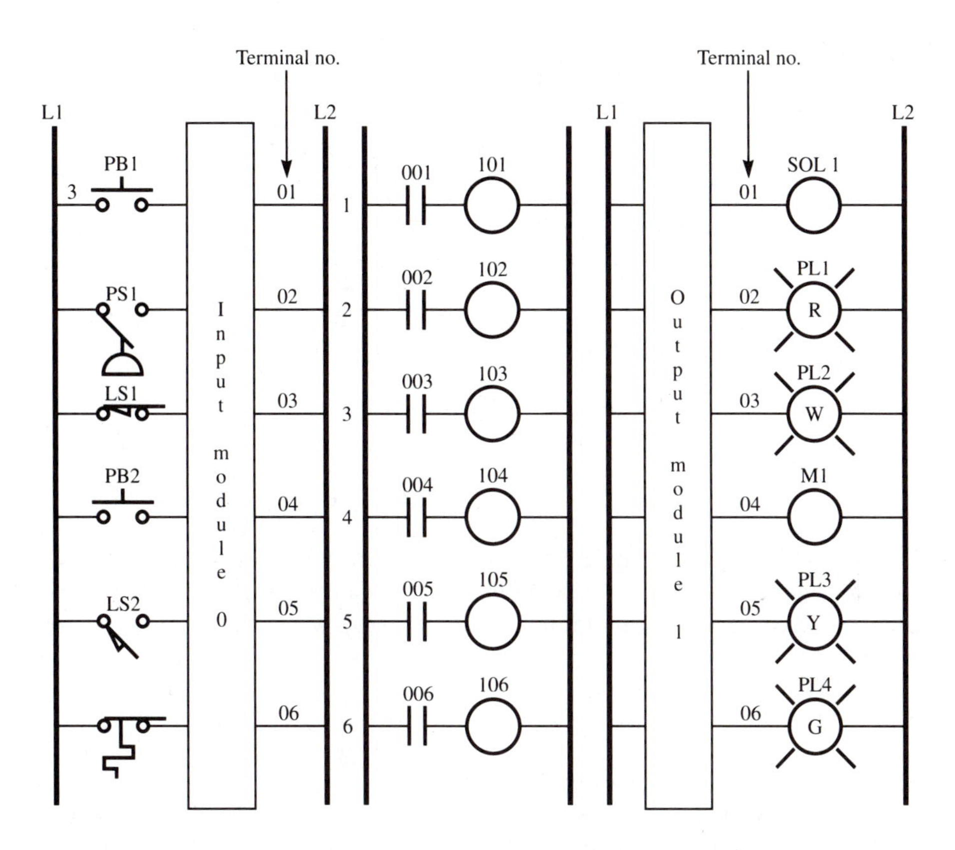

Figure 7-3 Control Schematic for a Programmable Controller

in standard form. Note that only the I/O terminals are shown and that the module designation is part of the terminal designation.

Relay Logic for Input Devices

The most common input device in a control is a push button, of which many types are available. Basically, a push button merely enables a person to inject an electrical signal into a control. Although several voltages are used for control, by far the most common is 120-V AC. This popularity stems from its availability, and it is low enough in voltage to be relatively safe. One side of the push button's contact is fed by a voltage source; the other side is connected to whatever it's feeding in the control.

Another common device is a limit switch, which is activated when something comes into contact with it. It can inject an electrical signal into the control in the same way as a push button.

Solid-state logic can also be used for sensing, and usually generates signals at TTL-level voltages (0–5-V DC). Solid-state relays are available to change the low-voltage information to a higher voltage, such as 120-V AC.

Table 7-1 Common Input Devices

Discrete	Analog
Push button	Radiation sensor
Thermostat	RF receiver sensor
Limit switch	Flow switch
Auxiliary contact	Speed sensor
Proximity switch	Photosensor
Pressure sensor	Level sensor
Relay contact	Weight sensor
TTL device	Moisture sensor
Thumb wheel	Thermocouple
Encoder	Motion sensor

Different types of transducers may generate contact closures or electrical signals. Examples are a pressure transducer that closes a contact when a given pressure is reached, and a thermostat that closes a contact when a given temperature is reached.

We have just touched on possible input devices, and have discussed only *discrete* input devices. It is also possible to inject *analog* signals, which are signals that constantly vary. Table 7-1 shows some of the possible discrete and analog input devices. Basically, any device that can close a contact for some purpose could be used as a discrete input device. Most such devices will supply a contact that is isolated from any voltage; this is referred to as a *dry contact*. Such contacts will then be connected as needed in the control.

Programmable Controller for Input Devices

The input devices for inputs to relay ladder logic will also be used as input for the programmable controller, with one modification: One side of the contact will be connected to a common voltage bus, the other side will be connected to an input module fitted with terminals for this purpose, and the programmable controller will be set up to detect a voltage at these inputs. The voltages used in control vary; Table 7-2 lists the common ones, with the most common being 120-V AC. Although some modules for programmable controllers will accept dry contacts, most frequently, a voltage will be applied to the PLC inputs through the input-device contacts.

Table 7-2 Most Common Voltages Used In Control

12-V AC/DC	24-V AC/DC
48-V AC/DC	120-V AC/DC
230-V AC/DC	TTL-level 0–5-V DC

What follows are some of the modules available that enable input devices to communicate with a PLC. Buying a PLC usually involves buying a chassis to which modular units can be added to customize a PLC to your application.

Discrete-input Module The discrete-input module lets the user make two-state signals available to the PLC's for use in the control program. It isolates the input signal and changes the form and the level so that the altered signals are compatible with the PLC's data bus. Control signals from the PLC's processor determine when the transformed information can be placed on the PLC's data bus to be read. The most popular input is 120-V AC; however, modules can be purchased for various DC and AC levels.

Thumb-wheel Module This module enables the use of thumb-wheel switches for feeding information in parallel to the PLC to be used in the control program. The thumb-wheel information is usually in BCD form, and provides data for registers used in the control program, and enables a person to externally change set points or preset points without modifying the control program.

TTL-input Module This enables devices producing TTL-level signals to communicate with the PLC's processor. It normally requires a separate 5-V source so that the PLC's power supply isn't overloaded. TTL-level signals are in a form the processor can accept, and only buffering is required; so the PLC's processor can control when the data accesses the data bus.

Encoder-counter Module This module enables continual monitoring of an incremental or absolute encoder. Encoders keep track of the angular position of shafts or axes. Grey code is common for absolute encoders, with position determined by decoding the grey code. An incremental encoder issues pulses that can be counted to determine position. The count is proportional to the arc of rotation. Although the encoder-counter module works independent of the PLC, it can be read by the PLC when the control program needs this information.

Grey-encoder Module This converts the grey-code signal from an input device to straight binary. Control signals from the PLC's processor determine when the transformed information can be placed on the PLC's data bus to be read.

ASCII-input Module Such a module converts ASCII code input information from an external peripheral to alphanumeric information that the PLC can understand. It uses one of the standard communication interfaces, such as an RS-232 or an RS422.

Analog-input Module This kind of module converts analog signals to digital signals. It isolates the input analog signal, then changes it to digital form and to the correct level so that the altered signals are compatible with the PLC's data bus.

BCD-input Module This allows the processor to accept four-bit BCD digital codes.

Isolated-input Module The isolated-input module enables the PLC's processor to receive dry contacts as inputs. The module monitors these contacts and produces two-state digital signals that the processor can read.

7-2 Output Devices

Both relay logic and logic implemented on a PLC must be able to manipulate outputs, but again there is a difference in how this is done. A relay ladder logic scheme will show the relay contacts connected directly to the output devices. The PLC can't do this the same way because it works with voltages that, in general, are incompatible with the output devices. Also, the data bus on the PLC is multiplexed, so the voltage would be constantly changing. An output module is required for the PLC so it can output a voltage compatible with the device. This voltage needs to be latched so that it is on continuously when needed. Another problem is that the power requirements to operate the outputs would generally exceed the data bus's capability.

As Figure 7-4a shows, an output module is basically a signal-conditioning and -modifying device. The signal enters the module from the data bus of the PLC and is processed so that it is compatible with the output device. Figure 7-4b presents a block diagram of this process for an AC output module. First the signal is applied to an optical coupler, which provides electrical isolation to prevent any outside, circuit-damaging AC voltage from being applied to the PLC. Figure 7-4c shows example circuitry of how the block diagram might be implemented.

Relay Logic for Output Devices

The output device in relay ladder logic is the coil of a relay. When energized, a coil in a control will operate some other device with its contacts. The most com-

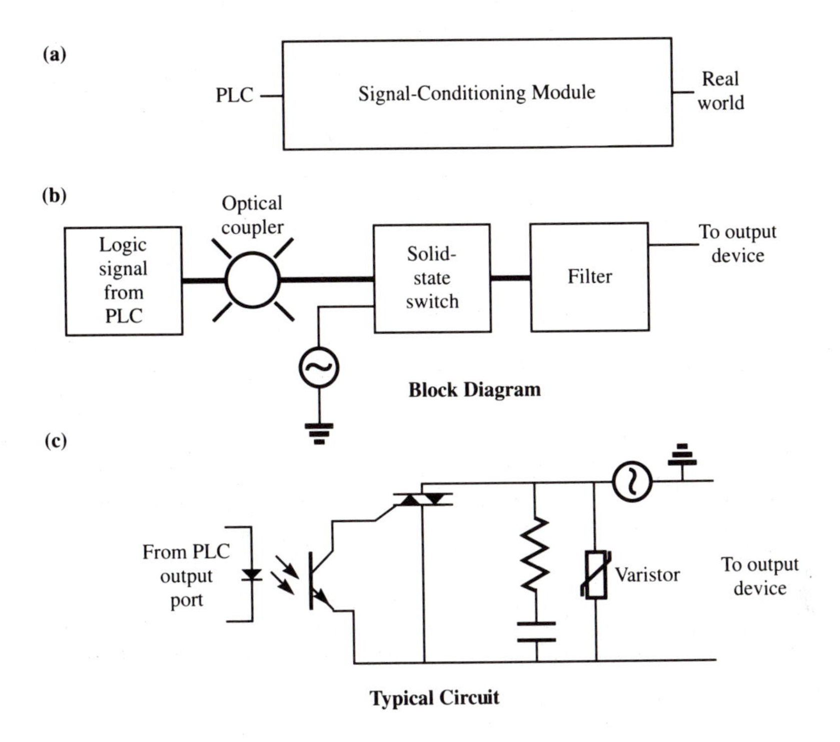

Figure 7-4 Output Module

mon such devices are motors and solenoids. Contacts on industrial relays have current ratings for 10 amps at various voltage ratings, such as 120-V AC and 240-V AC. If the motors or solenoids have current and voltage ratings above these levels, the relay contacts will pick up starters and contactors that will do the heavy work. Table 7-3 lists some common output devices.

Programmable Controller for Output Devices

The programmable controller must be able to operate output devices similar to those given in Table 7-3. This it does by sending voltage out of an output module equipped with terminals for connecting it to the output devices. The amount of current the module can supply to the output device is determined by the electronics

Table 7-3 Discrete Output Devices

Alarm	Light
Motor	Solenoid
Latch	Valves
Contactor	Motor starter
Heater	Fan

in the module. If the output device draws too much current, an auxiliary relay may be needed. The program in the programmable control determines when these voltages will be active, and latches the output voltage as needed.

Discrete-output Module The discrete-output module enables the PLC's processor to control output devices by changing a digital-level signal to the level required by the devices being controlled. The signal going to the output device and the digital signal coming from the processor are optically isolated.

TTL-output Module This module enables a PLC to operate devices that require TTL-level signals. Normally it needs a separate 5-V source so that the PLC's power supply is not overloaded. It provides the necessary buffering between the output device and the PLC.

ASCII-output Module This converts alphanumeric information from the PLC to ASCII code to be sent to an external peripheral via one of the standard communication interfaces, such as an RS-232 or an RS422.

Analog-output Module This kind of module converts digital signals from the processor to isolated analog signals that can drive output devices. Several voltage levels and ranges are available, for example, +10 V through −10 V.

Stepper-motor Module The stepper-motor module provides pulse trains to a stepper-motor translator, which enables control of a stepper motor. The PLC is free to do other tasks once it communicates with the stepper-motor module. This module will send the command to the translator and will not accept a different command until that command has been executed. The commands for the module are determined by the control program in the PLC.

BCD-output Module This type of module enables a PLC to operate devices that require BCD-coded signals.

Dry-contact-output Module This enables the PLC's processor to control output devices by providing a contact that is isolated electrically from any power source. The processor's digital signal operates a relay, and the contacts from the relay are available for use at the output terminals.

Some modules function simultaneously as both input and output devices or have special applications. These are referred to as *intelligent I/O,* since they have their own microprocessors on board that can function in parallel with the PLC. Here are some of the special modules available.

Language Module This module enables the user to write programs in a high-level language. Via a high-level-language interpreter, it converts the high-level commands into machine language understandable to a PLC's processor. The languages available differ from one manufacturer to another. BASIC is the most popular language module. Other language modules available include C, Forth, and PASCAL.

PID Module Proportional integral-derivative closed-loop control lets the user hold a process variable at a desired set point. For example, to maintain the speed of a motor at a set-point rate of 3600 rpm, a transducer would detect the speed at any particular moment and then feed an analog signal to the PID module. The module would compare this signal to the set point and send an output signal to the motor's speed-control device. The greater the difference between the set point and the process variable, the greater the output signal; the smaller the difference, the smaller the output signal.

Servo Module Closed-loop control is accomplished via feedback from the device. The programming of this module is done through the PLC; but once programmed, it can independently control a device without interfering with the PLC's normal operation.

Communication Module This module allows the user to connect the PLC to high-speed local networks that may be different from the network communication provided with the PLC. This unit can interface with the desired network by changing the format of transmitted and received data to the protocol needed for the particular network.

7-3 Timers

The need for timers in sequential control is self-evident. However, the timing function on programmable controllers evolved because of what was available in relay control. We will look first at timers in relay logic, and then at how these same types of timers have been implemented in programmable controllers.

Relay Logic for Timers

Timing relays come in various styles: pneumatic, hydraulic, and solid state. Pneumatic and hydraulic timers can be made rugged, accurate, and impervious to electrical noise. Solid-state timers also can be made rugged and inexpensive, but, because they rely on solid-state devices, they can be damaged by electric power surges. Consequently, care must be taken when using them in high-voltage, high-power environments. The timing can be done on energizing and deenergizing, and the relays are available with both normally opened and normally closed contacts. These relays are sometimes confusing because of all the possible combinations.

Figure 7-5 shows a timing diagram that can help explain the various contacts and what happens as the coil is energized and deenergized. The contact symbols are shown in the no-power-applied state. The *normally opened instantaneous contact* will follow the energizing of the timer coil: When the coil is energized, it closes; when the coil is deenergized, it opens. There is no time delay in the operation of this type of contact. The *normally closed instantaneous contact* is closed when the coil is deenergized and opens when the coil is energized, without any time delay in its operation.

Notice that for *normally opened time-delay-on-energizing contact,* the arrow is pointing up. This indicates time delay on energizing. That is, the contact closes when the timer is energized, but only after some time delay, which can be set for different values, with timing starting as soon as the timer is energized. If the timer is energized and the timer times out, then the contact closes and stays closed until the timer is deenergized, at which time the contact opens instantaneously. The delay occurs only on the timer being energized. The *normally closed time-delay-on-energizing contact* operates the same as the normally opened time-delay-on-energizing contact, except its opening, instead of its closing, is delayed.

The *normally opened time-delay-on-deenergizing contact* operates the same as a normally opened instantaneous contact, except when the timer is deenergized, in which case the contact goes through a delay before opening. Assuming the timer is energized, the delay starts after the timer is deenergized. The *normally closed time-delay-on-deenergizing contact* operates the same as the normally opened time-delay-on-deenergizing contact, except the delay is on closing after the timer has been deenergized. Note that "time delay off" is indicated by the down-pointing arrow. Study Figure 7-5 carefully because this is a case where a picture is worth a thousand words.

Timers also come as retentive or nonretentive. A *retentive timer* will time cumulatively as it is energized and deenergized. For example, if set to time out at 50 seconds (the length of time a timer is initially set for is called the *preset value*), then if the timer was energized for 30 seconds the first time and deenergized, then reenergized for another 10 seconds and deenergized again for some period of time, it would have accumulated 40 seconds towards its time out. The third time it is energized it will take only 10 seconds to time out. Thus, a reten-

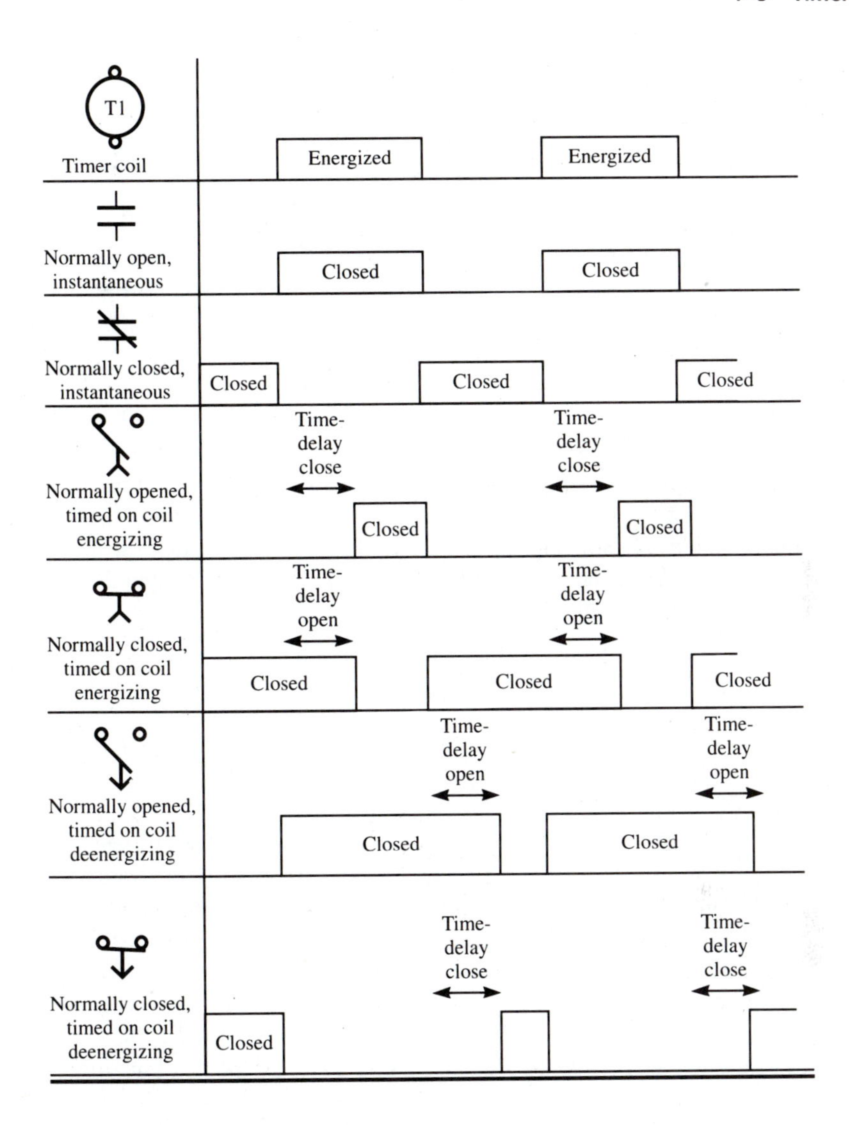

Figure 7-5 Timing Diagram for Various Timer Contacts

tive timer accumulates and retains the time it has been energized. A *nonretentive timer* will time out only if it is energized continuously for its preset value. If deenergized at any time, it will automatically reset to 0. Thus, a nonretentive timer set for 50 seconds must be energized for 50 seconds or it will not time out.

Programmable Controllers for Timers

Since the computer at the heart of a programmable controller must have a clock to operate, timing by programmable controller can be accomplished easily. What is needed is a place for the computer to hold information about the type of timer needed, the time-out value, and the required timing interval, information that usually can be stored in one or two bytes (eight bits) of memory location. Also required is a location to hold the number of time intervals counted as the timer is timing, which usually takes another one or two bytes. These memory locations usually are referred to as *registers.* How many timers a particular controller can have is a function of memory size and how much of that memory is set aside for timers. If a timer takes two bytes of memory, we could have 512 timers in 1K of memory. The programming for the timers would be stored permanently in ROM, while the registers for the timers would be in RAM. Since memory is relatively cheap, we can have timers for pennies. In contrast, 512 mechanical timers would cost several thousand dollars.

Both retentive and nonretentive timers are available on PLC's. The status of the timers can be accessed at any time by reading the address where the information is stored. Typically, a single bit can be addressed to determine when the timer is active and when the timer has reached its limit. The time delay on deenergizing normally is not available, but can be duplicated with relay logic programming available on the PLC's. How to program timers on an Allen-Bradley PLC-5 will be covered in detail in Chapter 10.

7-4 Counters

Counters are very similar to timers, because all timers must have counters inside them. Timers need an internal clock that can generate pulses at a specific rate, and they must count these pulses. We can convert a timer into a counter by removing the internal clock and bringing the input terminal of the internal counter out to external terminals, thereby allowing an external event to generate the pulse that will be counted, for example, a limit switch that gets closed every time a box goes down a conveyor belt. If the limit-switch contacts are in series with a voltage source, a pulse will be generated that can be fed into the counter. It makes no sense for a counter to be nonretentive, because a nonretentive counter wouldn't count. Since counters are naturally retentive, they must be reset if they are to be used to count again. Therefore, counters have reset coils for this function.

7-5 Latches

It is common in a sequential control to have to remember when a particular event takes places and not to permit certain functions once this event occurs. The operator may have to manually reset a device before further operations can occur. Running out of a part on an assembly line may require the shutdown of the process until the needed parts become available. Latches are devices that can be set and that will not reset until a reset signal is sent.

Relay Logic for Latches

In control it is sometimes necessary to hold a device on once it has been energized. Relays are made with mechanical latches that will hold their contacts closed after they have been energized even though the coil itself has been deenergized. For example, a relay energized when the power fails to an oil pump that lubricates bearings can trip a turbine and initiate action allowing the turbine to shut down rapidly so that damage is avoided. We do not want the turbine to attempt an automatic restart, and a latching relay may prevent this. The latching relay could require manual resetting, or it could have a coil that releases the latching mechanism.

Programmable Controller for Latches

Latching in programmable controller logic is accomplished via special instructions entered at the keyboard to cause an output to be latched. This output will stay latched until a reset is made active in the control. The reset also is entered by special instructions. There is a big difference between what happens if the power to the programmer fails and what happens if the power to a relay logic control fails: The 120-V AC from the programmer's output would be lost even though, because of battery backup, the programming itself in the programmable controller would not be lost. The relay latch will hold with or without power.

Safety may be a problem in this case, and each particular control would have to be studied. If it is critical and possibly unsafe, a simple solution would be to have the programmable controller pick up a latching relay instead of using the latch instruction in the PLC. This is because on loss of power the PLC is unpredictable while the latching relay is not.

Summary

Programmable controllers have been designed to work with all the common devices used in relay logic control. These devices remain popular because they are so functional and are essential as input and output devices. Push buttons, limit switches, lights, timers, counters, latches, and power-output modules are so basic that it is unlikely these devices will be replaced in the near future. It will be important to learn how to program the PLC to accept and communicate with these devices. In the chapters that follow we will do just that.

Exercises

7-1. In the control scheme in Figure 7-6, the time-delay relays are set for 5 sec for TD1 and 2 sec for TD2. How will the red and yellow lights respond if PB1 is pushed at $t = 0$ sec?

7-2. In the control scheme in Figure 7-6, the time-delay relays are set for 5 sec for TD1 and 2 sec for TD2. How will the white and green lights respond if PB2 is pushed and held closed at $t = 0$ sec?

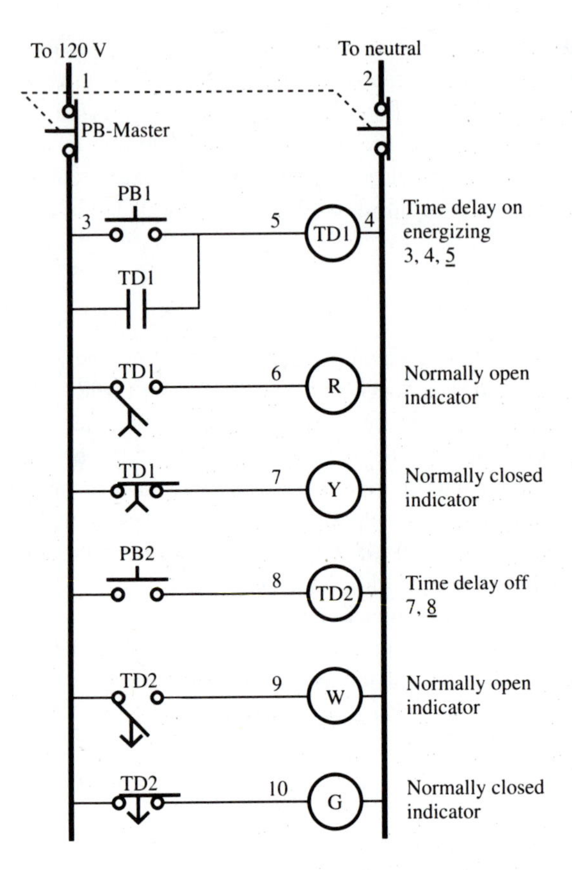

Figure 7-6 Control Scheme for Exercises 7-1, 7-2, and 7-3

7-3. In the control scheme in Figure 7-6, the time-delay relays are set for 5 sec for TD1 and 2 sec for TD2. How will the white and green lights respond if PB2 is pushed and released at $t = 0$ sec?

7-4. Design a relay ladder control scheme using one or two time-delay relays that will cause a light to be on for 1 second then off for 1 second. The light should blink continuously at this rate. Label the control per JIC standards.

7-5. Explain why inputs/outputs cannot be directly connected to the internal data bus on a PLC.

7-6. Could you operate a 25-horsepower motor directly from the terminals of an output module of a PLC? Explain.

7-7. What is a latch? Give an example where it might be used.

7-8. Explain the difference between retentive and nonretentive timers.

7-9. Briefly explain the function of an input module and an output module for a PLC.

7-10. List four ways timing can be accomplished in control.

Processor Memory and Addressing

8

OUTLINE

OBJECTIVES

Upon completion of this chapter, the student will be able to:

- Describe the processor's memory layout, including the breakdown of the program files and data files.
- Give the addressing format of the various data file locations.
- Explain the relationship between the input image table, the output image table, and the hardware addressing.

IN THIS chapter we will determine how the processor's memory is allocated. The memory layout will be based on Allen-Bradley's PLC 5 processor. It will be broken down into its three areas: program files, data files, and unused memory. The processor files and data files will also be subdivided into their individual components. Logical addressing of the data files and the relationship between the data table addresses and the external hardware will also be shown.

8-1 Processor Memory

The processor memory is broken down into three main areas, as shown in Figure 8-1: the data files, the program files, and unused memory. The processor memory is expressed in 1024-word increments (or 1K of memory). A processor with 6K of memory has $6 \times 1024 = 6144$ total words of memory. This would then be divided into the data files, the program files, and unused memory. As you enter program, the program files and data files fill up space from unused memory.

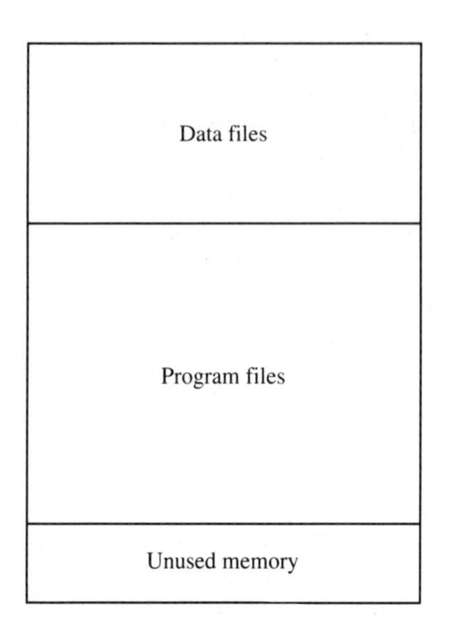

Figure 8-1 Processor Memory

8-2 Program Files

Program files are the area of processor memory where the ladder logic programming is stored. The processor has a potential for 1000 program files. Before you start to enter program, the processor may be set up, through the software, to do either standard ladder logic programming or sequential-function-chart programming. Exactly how the program files are set up depends on whether the processor is to do sequential-function-chart programming or standard ladder logic programming. In either case, the processor can store and execute only one program at a time.

With the processor set up for standard ladder logic, the main program will always be in program file 2, and program files 3 through 999 will be subroutines. Two special subroutine files may be assigned: the selectable timed-interrupt file and the fault-routine file. The **selectable timed-interrupt file** is executed on a time basis, and the **fault routine** is executed when the processor detects a major fault. Subroutines, along with the selectable timed-interrupt file and the fault routine, will be covered in detail in Chapter 14.

Program file 0 stores the processor password and any program identification. Using a password as protection against unauthorized access to the processor is an option. If a password is being used, you must know the password before you can access the program files or data files through the software.

Program file 1 controls the setup of sequential function charts and cannot be used to store ladder logic or for any other purpose.

If the processor is set up to do sequential-function-chart programming, files 2 through 999 store the ladder logic for the steps and transitions of the sequential function chart. Sequential function charts will be covered in Chapter 15.

8-3 Data Files

The data file portion of the processor's memory stores input and output status, processor status, various bit status, and numerical data, and all this information is accessed via the ladder logic program. There is a potential for 1000 data files, numbered 0 through 999. When the processor's memory is cleared, the data files default to nine data file types, numbered 0 through 8. Files 9 through 999 may be assigned as needed. The default data files are shown in Figure 8-2.

Each data file is made up of a number of **elements.** Each element may be one, two, or three words in length. Timer, counter, and control elements are three words in length, floating-point elements are two words in length, and all other elements are a single word in length. A **word** consists of sixteen bits, or binary digits. The processor operates on two different data types, integer and floating point. All data types, except the floating-point files, are treated as integer. Integer values are stored in 2's-complement binary (see next subsection for an explanation of 2's complement), and floating-point vales are stored in a 32-bit binary element.

2's Complement

The PLC needs a way to distinguish positive and negative numbers. To do this it uses the MSB for sign designation. A 0 indicates a positive number and 1 indicates negative.

Address Range		Size, in elements
O:000 – O:037	Output image file	32
I:000 – I:037	Input image file	32
S:000 – S:031	Processor status	32
B3:000 – B3:999	Bit file	1–1000
T4:000 – T4:999	Timer file	1–1000
C5:000 – C5:999	Counter file	1–1000
R6:000 – R6:999	Control file	1–1000
N7:000 – N7:999	Integer file	1–1000
F8:000 – F8:999	Floating-point file	1–1000
	Files to be assigned file nos. 9–999	1–1000 per file

Figure 8-2 Data-File Memory Organization

For Positive Numbers With eight bits, the positive numbers can go from 0 to 127:

True Form	Decimal Equivalent
00000000	+0
00000001	+1
00000010	+2
00000011	+3
.	.
.	.
.	.
0111111	+127

For Negative Numbers Negative numbers usually are not stored in true binary form but in the 2's complement. This makes it easier for the computer to perform mathematical operations. The correct sign bit is generated by forming the 2's complement. To get the 1's complement of a binary number, we simply change 1's to 0's or 0's to 1's on a bit-by-bit basis. The 2's complement is formed by adding 1 to the 1's complement. Forming the 2's complement will automatically place a 1 in the MSB. The PLC knows that a number retrieved from memory is a negative number if the MSB is 1. Whenever a negative number is entered from a keyboard, the PLC stores it as a 2's complement. What follows is the original number in true binary followed by its 1's complement, its 2's complement, and finally its decimal equivalent. The maximum negative number with eight bits is −128.

True form	1's complement	2's complement	Decimal Equivalent
00000001	11111110	11111111	−1
00000010	11111101	11111110	−2
00000011	11111100	11111101	−3
.	.	.	.
.	.	.	.
.	.	.	.
01111111	10000000	10000001	−127
10000000	01111111	10000000	−128

Output Data File

Data file 0 is the output data file. Its length, in elements, depends on the processor used. A **rack,** which is an addressable unit, uses eight words in the output data file and eight words in the input data file. A word in the output data file and its corresponding word in the input data file are called an **I/O group.** A rack is made up of eight I/O groups. Therefore, a processor that can communicate with four racks would have an output data file 32 words long and an input data file 32 words long. The output and input data files are always the same length. The addressing for the elements in the output data file is formatted as follows:

File letter designator: 2-digit rack number, I/O group/bit (octal)

For instance, the address O:012/15 would be in the output image table, rack 1, I/O group 2, bit 15. The O indicates the output file, the colon is a delimiter between the file letter designator and the word address, the 01 is the rack number, the 2 is the I/O group number, and the 15 is the bit number. This also corresponds to a hardware terminal address, located in rack 1, I/O group 2, terminal 15. See Figure 8-3 for the relationship between the bit in the processor memory and its corresponding hardware location. All element and bit addresses in the output and input data files are numbered octally. Element and bit addresses in all other data files are numbered decimally.

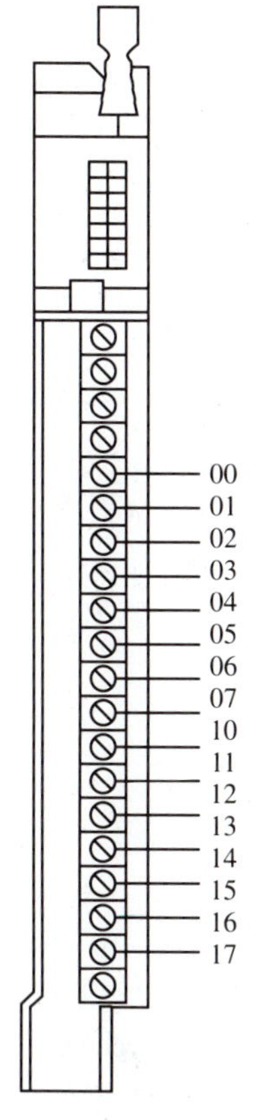

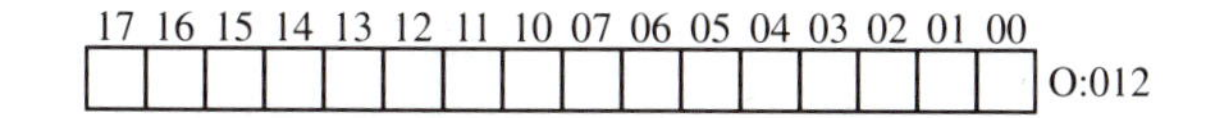

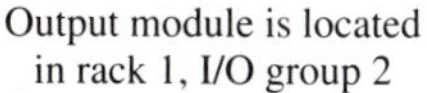
Output module is located in rack 1, I/O group 2

Corresponding data-table word is O:012

Figure 8-3 Hardware—Output Data File

Input Data File

Data file 1 is the input data file. Its length, as in the output data file, depends on how many racks the processor can communicate with: eight elements for every rack with which it has the potential to communicate. The addressing for the input data files is formatted as follows:

File letter designator: 2-digit rack number, I/O group/bit (octal)

For example, the address for a bit located in rack 1, I/O group 3, bit location 17 is I:013/17. The I indicates input, the colon is a delimiter between the file type and the word address, the 01 is the rack number, the 3 is the I/O group number, and the 17 is the bit number (in octal). This address also relates to hardware terminal 17 on an input module in rack 1, I/O group 3. Figure 8-4 illustrates the relationship between the hardware and the input data file.

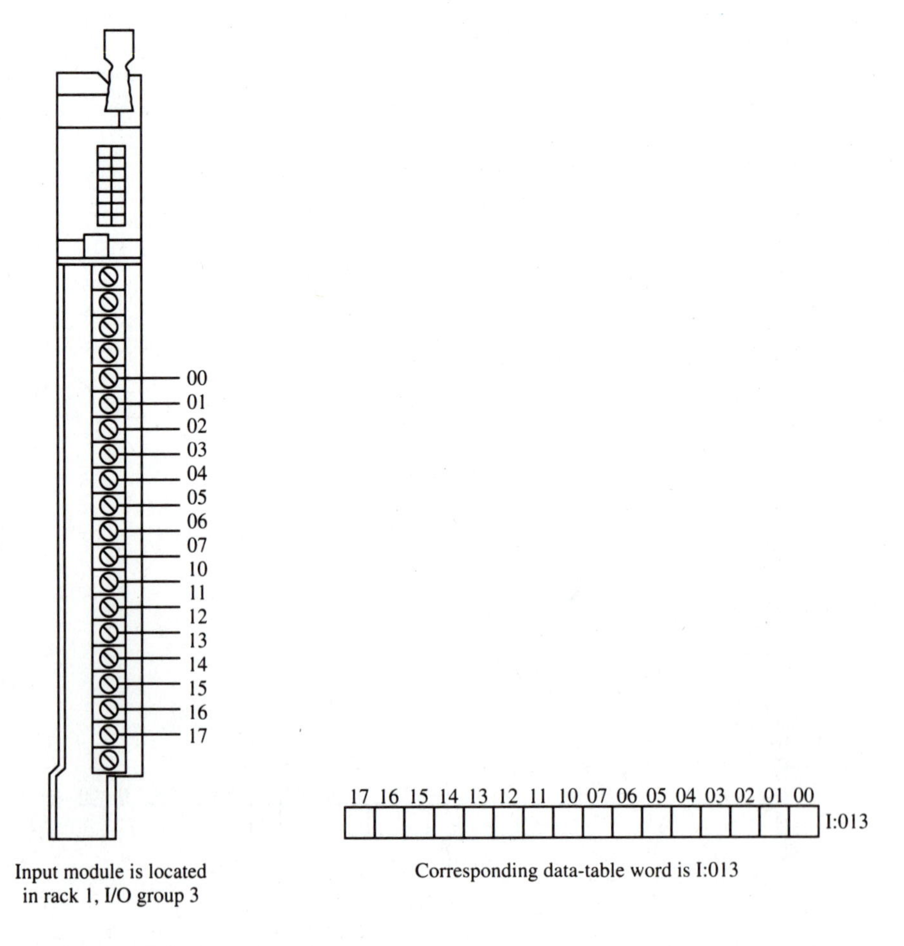

Figure 8-4 Hardware—Input Data File

The addresses in the output data file and the input data file are potential locations for either input modules or output modules mounted in the I/O chassis. These addresses exist whether or not the actual modules are in the I/O chassis, since these locations are for the installation of either input or output modules.

Status Data File

Data file 2 is the status data file, which contains information regarding the processor status. It is 32 words long and is fixed in length. Figure 8-5 shows the layout of the status data file and the assignment of the 32 words in the file.

Words 1 through 14, except for word 11, which is the major-fault word, are intended as read-only words. The major-fault word is often cleared by programming in the fault routine. Words 15 through 31 require that data be entered into them if their functions are to be used. Entering a zero into the word will disable that function. Here are some examples of addressing in the status area:

Addressing word 23: S:023

Addressing bit 9 in word 27: S:027/09

Bit Data File

Data file 3 defaults to the bit file. It is possible to delete this file and reassign it to another type, but this rarely done. The bit data file normally stores bit status and is also used when it is desirable to display the data in 1's and 0's. It frequently serves for storage when using internal outputs, sequencers, bit-shift instructions,

Word	Contents
0	Arithmetic flags
1	Processor status and flags
2	Configuration-switch image
3-6	Data hwy. plus active-station table
7	I/O rack and block-transfer queue status
8	Current program scan, in milliseconds
9	Maximum program scan, in milliseconds
10	Minor-fault flags
11	Major-fault flags
12	Fault codes
13	Program file number of last fault
14	Rung number of last fault
15	Module-specific control file
16-17	Reserved for future use
18-23	Time-of-day clock and calendar
24	Offset value for indexed addressing
25	I/O adapter image
26	Function-chart restart continue
27	I/O rack inhibit and reset
28	Program watchdog timer set point
29	Fault routine file number
30	Selectable timed-interrupt set point
31	Selectable timed-interrupt file number

Figure 8-5 Status File

and logical instructions. There is a potential for 1000 elements in the bit file, numbered 0 to 999. Here are some examples of addressing in the bit file:

Word 200 in the bit file would be addressed as B3:200. Note that the file number is now designated. (The input, output, and status data files are the only files that do not require the file number designator.)

Word 2, bit 15, would be addressed as B3/47, since bit numbers are always measured from the beginning of the file. Remember that bits are here numbered decimally (not octally as they are in the input and output files).

Timer Data File

Data file 4 defaults to the timer file. It is possible to delete this file and reassign it to another type, but this is rarely done. The timer file stores timer status and timer data. A timer element consists of three words: the control word, the preset word, and the accumulated word. See Figure 8-6. The control bits are the enable bit, EN; the timer-timing bit, TT; and the timer-done bit, DN. The function of these bits will be explained in Chapter 10.

The addressing of the timer control word is the assigned timer number. Timers in file 4 are numbered starting with T4:0 and running through T4:999, which allows for a potential of 1000 timers. The addresses for the three timer words in timer T4:0 are:

Control word: T4:0

Preset word: T4:0.PRE

Accumulated word: T4:0.ACC

The enable-bit address in the control word is T4:0/EN, the timer-timing-bit address is T4:0/TT, and the done-bit address is T4:0/DN.

Counter Data File

Data file 5 defaults to the counter file. It is possible to delete this file and reassign it to another type, but this is rarely done. The counter file stores counter status and counter data. A counter element consists of three words: the control word, the preset word, and the accumulated word. See Figure 8-7. The control bits are the count-up enable bit, CU; the count-down enable bit, CD; the done bit, DN; the overflow bit, OV; and the underflow bit, UN. The explanation as to the function of these bits will be covered in Chapter 10.

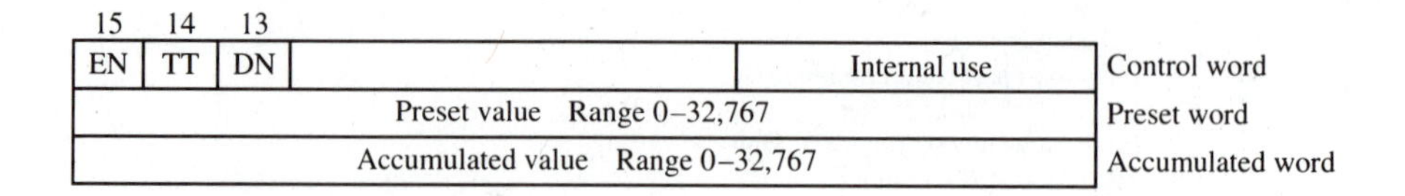

Figure 8-6 Timer Element

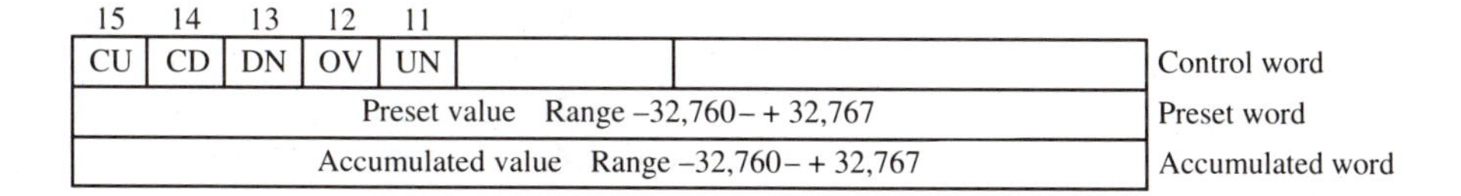

Figure 8-7 Counter Element

The addressing of the counter control word is the assigned counter number. Counters in file 4 are numbered starting with C5:0 through C5:999, which allows for a potential of 1000 counters in file 5.

The addresses for the three counter words in counter C5:0 are:

Control word: C5:0

Preset word: C5:0.PRE

Accumulated word: C5:0.ACC

The count-up-enable-bit address in the control word is C5:0/CU, the count-down-enable-bit address is C5:0/CD, the done-bit address is C5:0/DN, the overflow address is C5:0/OV, and the underflow address is C5:0/UN.

Control Data File

Data file 6 defaults to the control file. It is possible to delete this file and assign it to another type, but this is rarely done. The control file stores the control element's status and data, and is used to control various file instructions, which will be covered in later chapters. The control element consists of three words: the control word, which contains the various control bits; the length word, which stores the file length in either elements, words, or bits; and the position word, which acts as a pointer to indicate the current element, word, or bit being operated on by the instruction. See Figure 8-8.

The addressing of the control's control word is the assigned control number. Control elements in control file 6 are numbered starting with R6:0 and running through R6:999, which allows for a potential of 1000 control elements. The addresses for the three words in control element R6:0 are:

Control word: R6:0

Length: R6:0.LEN

Position: R6:0.POS

15	14	13	12	11	10	9	8		
EN	EU	DN	EM	ER	UL	IN	FD		Control word
Length value									Length word
Position value									Position word

Figure 8-8 Control Element

Here's an example of addressing a control bit: The address of the enable bit is R6:0/EN.

There are numerous control bits in the control word, and their function depends on the instruction in which the control element is used. These functions will be explained in Chapters 8 through 14 in sections covering the particular instructions.

Integer Data File

File number 7 defaults to the integer file. It is possible to delete this file and assign it to another type, but this is rarely done. The integer file stores integer data values, with a range from −32,768 through 32,767. Stored values are displayed in decimal form. The integer element is a single-word element, and there is a potential for 1000 integer elements, addressed from N7:000 through N7:999. A typical element address in the integer file for element 200 would be N7:200. Bit addressing is decimal, from 0 through 15. For example, bit 10 in word 15 would be addressed N7:015/10.

Floating-Point Data File

File number 8 defaults to a floating-point data file. It may be deleted and reassigned to another data type, but this is rarely done. The floating-point element is a two-word element, and the floating-point file has a potential for 1000 elements, addressed from F8:000 through F8:999. Individual words or bits cannot be addressed in the floating-point file. A floating-point element can store values in the range from $\pm 1.1754944e^{-38}$ to $\pm 3.4028237e^{+38}$.

Data Files 9 Through 999

Data files 9 through 999 may be assigned to different data types, as required. When assigned to a certain type, a file is then reserved for that type and cannot be used for any other type. There are two different ways in which a data file may be assigned to a certain type. One is to create an address in which the processor will allocate all addresses in that file up to and including that address. For example, if you were to create address N11:050, then the processor would assign file 11 to integers and create the integer addresses in file 11, from N11:000 through N11:050. This would remove 51 words from unused memory. The second way to assign a type to a data file is by entering an address in an instruction as you are programming. The processor then allocates memory in that data file, up through that address.

There are two additional file types that are used for display purposes only: the decimal file and the ASCII file. The decimal file will display the BCD value for the bit pattern stored at that address, and the ASCII file will display the ASCII representation for the bit pattern stored at that address. The processor does not recognize either BCD or ASCII; if you tried to use these values, you would get erroneous results. Any files from 9 through 999 can be assigned to either ASCII or decimal data types. An example of an ASCII address in file 12, element 10, is: A12:010. An example of a decimal address in file 20, element 11, is: D20:011.

Summary

The processor's memory is divided into three areas: program files, data files, and unused memory. Program files store the ladder logic programming; data files store status and data-value information that is accessed by the ladder logic program. There is a potential for 1000 program files, which may be set up in two ways: either (1) standard ladder logic programming, with the main program in program file 2 and program files 3 through 999 assigned, as needed, to subroutines, or (2) in sequential function charts in which files 2 through 999 are assigned to steps or transitions, as required. There is also a potential for 1000 data files, with a default table of nine assigned files, from 0 through 8. This default table is created when processor memory is cleared. Data files 9 through 999 may be assigned different types, as required.

Three of the data file areas that store status or data are the bit file, the integer file, and the floating-point file. In which of these you store the data depends on your intended use of the data. If it is desirable to have the data displayed in 1's and 0's, and you are using status rather than data, the bit file is preferable. If you are using very large or very small numbers, and require a decimal point, floating point is preferable. The floating-point data type may have a restriction though, as it may not interface very well with external devices or with internal instructions such as counters and timers, which only use sixteen-bit words. In such a situation it may be necessary to use the integer file type.

Exercises

8-1. Name the three areas of the processor memory.

8-2. When the processor is set up to do standard ladder logic programming, to what program file number is the main program file assigned?

8-3. When the processor is set up to do standard ladder logic programming, to what program file numbers may subroutines be assigned?

8-4. List the default data files established when the processor memory is cleared.

8-5. What are the two data file types not established in the default table but used for display purposes only?

8-6. In what number base does the processor store its values in the data files?

8-7. A motor starter coil is wired to an output module located in rack 1, I/O group 4, and is connected to terminal 14. What is its corresponding address in the output data file?

8-8. In which data file would you store the value 40,000?

8-9. How many words make up a timer element? What are they?

8-10. What three data-file assignments cannot be changed?

9 Basic Ladder Logic on a PLC

OUTLINE

OBJECTIVES

Upon completion of this chapter, the student will be able to:

- Describe how to enter program into the programmable controller.
- Define the basic relay-type instructions, including their functions, their mnemonics, and their ladder symbols.
- Explain branching.
- Build an elementary program using the relay-type instructions, including the addressing of the instructions, as related to ''real-world'' input and output devices.

THIS CHAPTER presents the method for entering the ladder logic program into the PLC, including the basic set of instructions (which perform functions similar to relay functions). The following instructions, instruction symbols, and instruction mnemonics will be explained along with the instructions' functions in the ladder logic.

Instruction	Symbol	Mnemonic
Examine on	-] [-	XIC
Examine off	-] / [-	XIO
Output energize	-( )-	OTE
Output latch	-(L)-	OTL
Output unlatch	-(U)-	OTU
Branching		

9-1 Entering Instructions into the PLC

The device that enters instructions into the PLC is commonly called the *industrial terminal.* For programming an Allen-Bradley PLC 5 programmable controller, the industrial terminal is actually a personal computer with software designed to communicate with the PLC. The personal computer, with appropriate software, can also be used to program and monitor the program in the PLC. Additionally, it may allow *off-line programming,* which involves writing and storing the program in the personal computer without its being connected to the PLC, and later downloading it to the PLC. In contrast, *on-line programming* involves programming or entering ladder logic while the terminal is connected to the PLC, and storing the programs either on the computer's hard disk or on floppy disks. The floppy disks allow the transfer of program between terminals.

The PLC can have only one program in memory at a time. To change the program in the PLC, it is necessary either to enter a new program directly from the keyboard or to download one from the hard disk while on-line. Some PLC's have EPROMS or EEPROMS within them that can store a backup to the program entered in the PLC; if the PLC were to lose its program, the program in the EPROM or EEPROM could be transferred to the PLC's memory.

9-2 *Examine On* Instruction, XIC, -] [-

To determine whether the intruction is true, the *examine on* instruction looks for an ON (a 1) at a bit location in the data table. However, if it finds an OFF (a 0) there, then the instruction considered is false. The *examine on* instruction is an input instruction and has a bit-level address. It can examine any bit address in the data files except one from the floating-point data table type. Figure 9-1 shows the

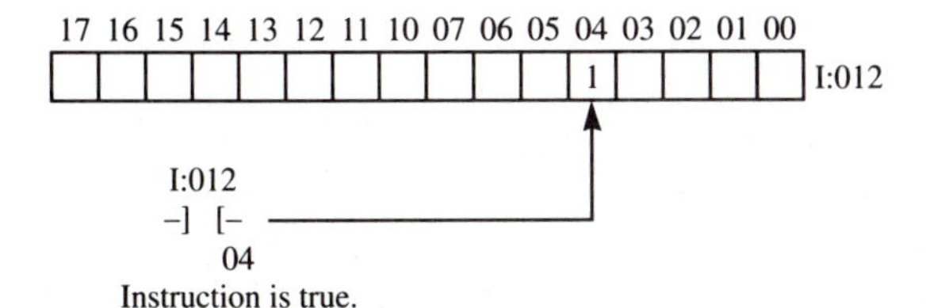

Figure 9-1 ***Examine On*** **Instruction—True**

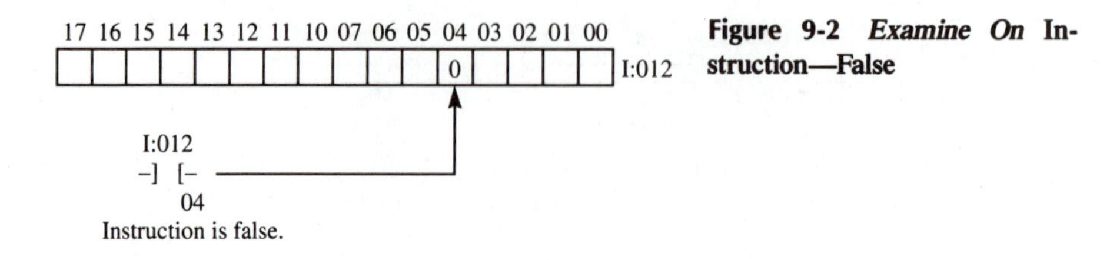

Figure 9-2 ***Examine On*** **Instruction—False**

examine on instruction with its address. The instruction tells the processor to look at its corresponding address in the data table; if it finds a 1 at that location, the instruction is logically true. Figure 9-2 shows the instruction finding a 0 at its address location in the data table, making the instruction logically false.

9-3 *Examine Off* Instruction, XIO, -] / [-

To determine whether the instruction is true, the *examine off* instruction looks for an OFF (a 0) at a bit location in the data table. But if it finds an ON (a 1) there, then the instruction is considered false. *Examine off* is an input instruction, has a bit-level address, and can examine any bit address in the data files except one from the floating-point data-table type. Figure 9-3 shows this instruction with its address. The instruction tells the processor to look at its corresponding address in the data table; if there is a 0 at that location, the instruction is logically true. Figure 9-4 shows the instruction finding a 1 at its address location in the data table, making the instruction logically false.

9-4 *Output Energize* Instruction, OTE, -( )-

Output energize is an output instruction used to control the status of a bit in the data table. When this instruction is true, which means the rung has logic conti-

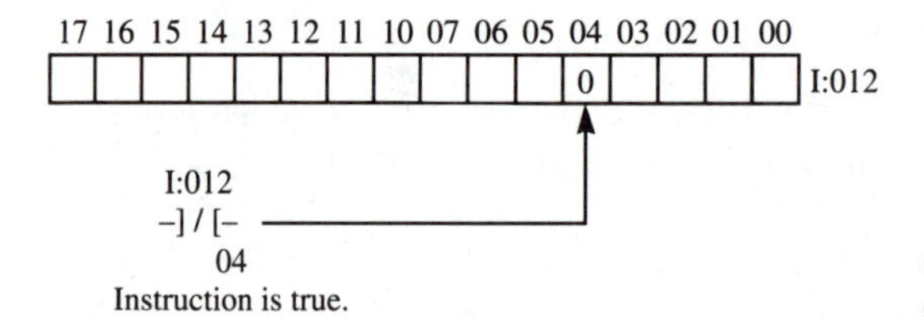

Figure 9-3 ***Examine Off*** **Instruction—True**

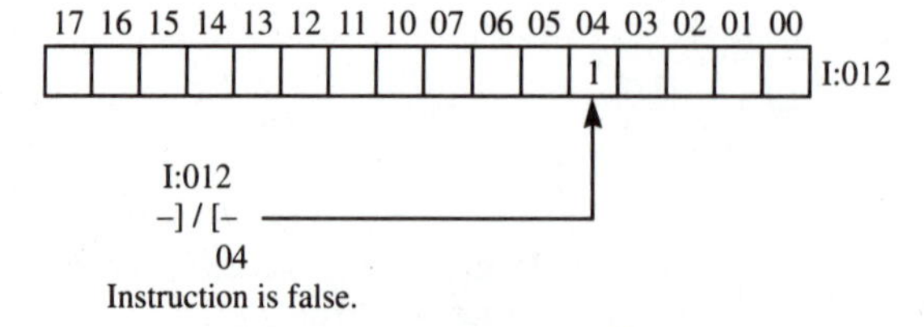

Figure 9-4 ***Examine Off*** **Instruction—False**

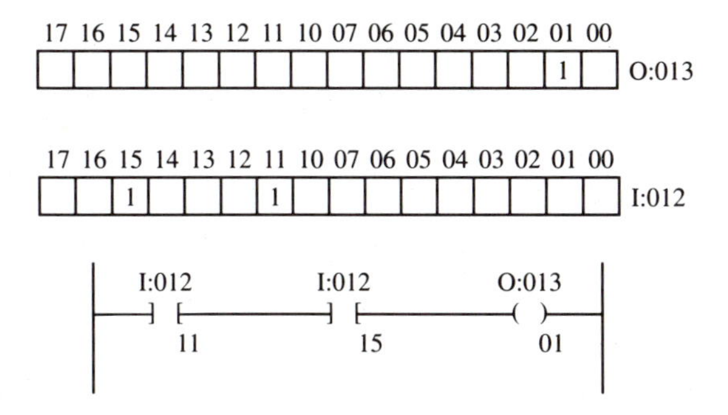

Figure 9-5 *Output Energize* **Instruction—True**

nuity to the *output energize* instruction, a 1 is placed in the data table at the address on the instruction. When there is no logic continuity to the *output energize* instruction, which means the rung is logically false, a 0 is placed at the data table address on the instruction. Figure 9-5 shows an *output energize* instruction in a true state. Inputs I:012/11 and I:012/15 are both true, creating a logically true path to the output, O:013/01. Figure 9-6 shows an *output energize* instruction in a false state. Input I:012/11 is now false, breaking the logical path to the instruction. A 0 is now placed in the data table at the address on the *output energize* instruction.

9-5 XIC and XIO Compared to Normally Open and Normally Closed Contacts

The symbols used for XIC and XIO are -] [- and -] / [-, respectively, which in relay ladder logic indicate the normally open and normally closed symbols. However, even though they have the same symbols, they do *not* mean the same thing. As explained earlier, the instructions XIC and XIO examine bits in the data table.

A normally open limit switch connected to an input module will set a bit in the input image table when it is closed and reset the bit when it is open. That bit location in the data table may be examined if closed or examined if open. Figure

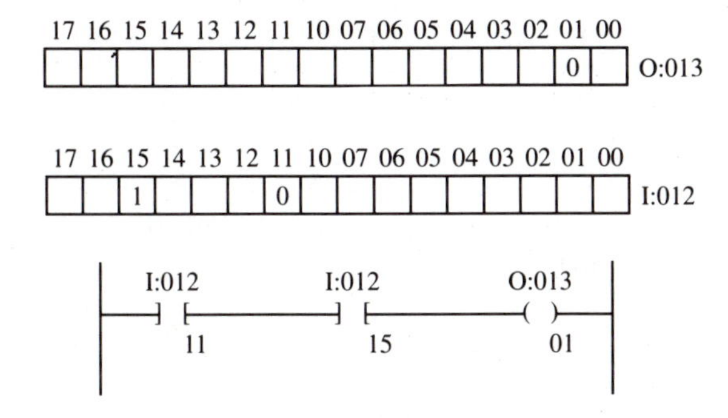

Figure 9-6 *Output Energize* **Instruction—False**

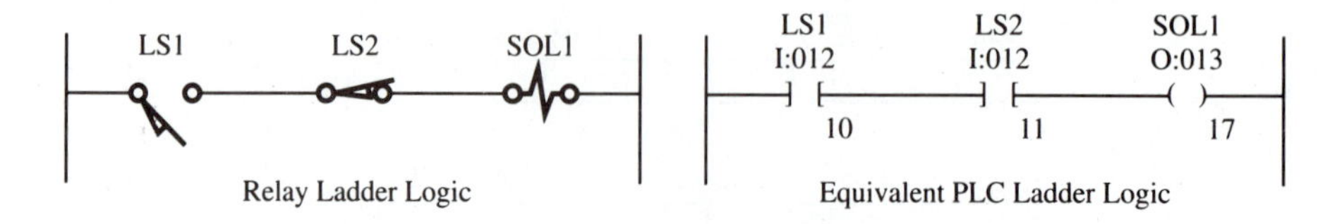

Figure 9-7 Relay Ladder Logic and Equivalent PLC Ladder Logic

9-7 shows a relay logic diagram with two limit switches, one normally open and one normally closed. In order for the valve to be energized, both switches must be closed. Therefore, the equivalent circuit in PLC ladder logic shows two XIC instructions in series connected to an *output energize*. Both logic circuits are logically the same and yield the same results. As can be seen, you can XIC a normally open switch and a normally closed switch.

9-6 Relationship Between Ladder Logic and Hardware

Figures 9-8 and 9-9 show the relationship between the ladder logic and the hardware. In Figure 9-8, push button 1 (PB1) is wired to terminal 14 of an input module located in rack 1, I/O group 7. This makes the terminal address, as well as the input data table address, I:017/14. PB1 is open, which results in 0 status at bit location 17 in word I:017 in the input data file. The XIC instruction in the single rung of the ladder, which examines this bit in the input data file and finds a 0, is thus false. Since the XIC instruction is false, the rung is false and the OTE instruction is false. As a result of the false status of the OTE instruction, the processor places a 0 in the output data file at word location O:014, bit 05. This information is then transferred to the output module, turning off pilot light 1 (PL1).

In Figure 9-9, PB1 is closed, resulting in a 1's being placed in the input data file at word I:017, bit 14. The XIC instruction now finds a 1 and is true. This means the OTE instruction is true, so the processor puts a 1 in the output data file at word O:014, bit 05. This information is then transferred to the output module, turning on PL1.

9-7 *Output Latch* Instruction, OTL, -(L)-, and *Output Unlatch* Instruction, OTU, -(U)-

Output latch is an output instruction with a bit-level address. When the instruction is true, it sets a bit in the data table. It is a retentive instruction, since the bit remains set when the *latch* instruction goes false. In most applications it is used with an *unlatch* instruction, though it may be used by itself to set a bit in the data table.

The *output unlatch* instruction also is an output instruction with a bit-level address. When the instruction is true, it resets a bit in the data table. It, too, is a

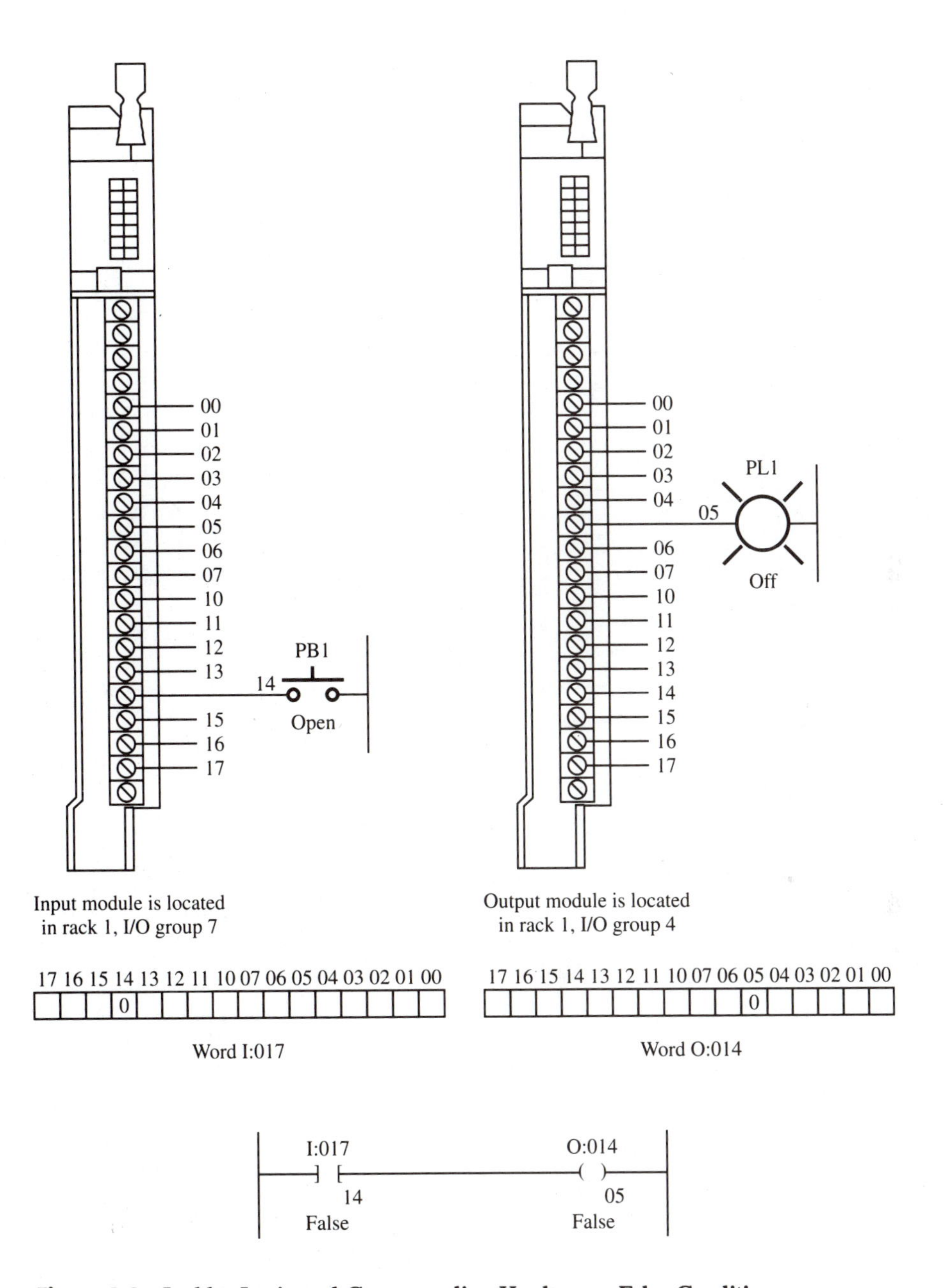

Figure 9-8 Ladder Logic and Corresponding Hardware—False Condition

retentive instruction, since the bit remains reset when the instruction goes false. The unlatch instruction also may be used by itself.

Figure 9-10 shows two rungs, one containing the *latch* instruction and one containing the *unlatch* instruction. Note that both instructions have the same data

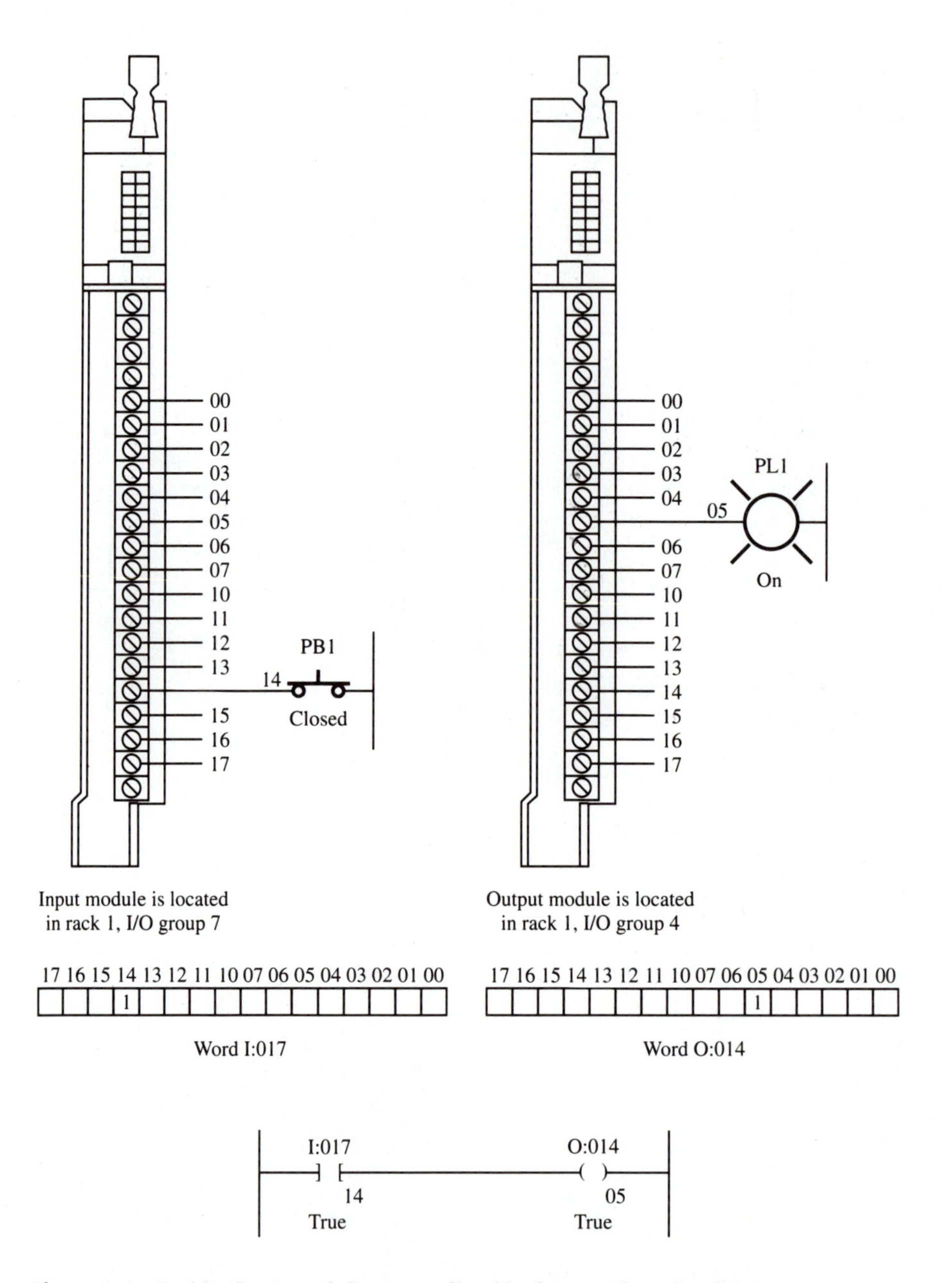

Figure 9-9 Ladder Logic and Corresponding Hardware—True Condition

table address. When input A is true, a 1 is placed in the data table at the bit address on the OTL instruction. When input A goes false, the 1 remains in the data table. When input B goes true, the bit at the address on the OTU instruction (the same bit address as the OTL instruction) goes false. When input B goes false, the bit remains reset.

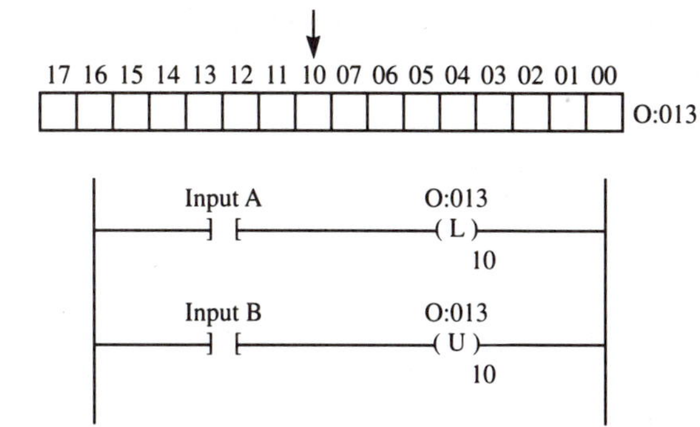

Figure 9-10 ***Output Latch* and *Output Unlatch* Instructions**

9-8 Branching

Branching allows parallel logic paths within a rung. Programmable controllers allow for the branching of inputs and in some processors, including the PLC 5, the branching of outputs.

Figure 9-11 shows an example of input branching. Output D will be true when either input A or inputs B and C or inputs A, B, and C are true. As you can see, parallel paths relate to OR logic.

Figure 9-12 shows branched inputs with a *nested branch,* which is a branch that begins or terminates within another branch. Input D is located in a nested branch since the branch terminates in the branch above, between inputs B and C. Some programmable controllers, including the PLC 5, allow for nested branches, and some do not. If the programmable controller does not allow nested branches, the logic must be arranged to eliminate the nested branch while maintaining the same logic.

Figure 9-13 has a rung that is logically equivalent to the one in Figure 9-12, but the nested branch has been eliminated by repeating input C in series with input D. Logically the two rungs are the same.

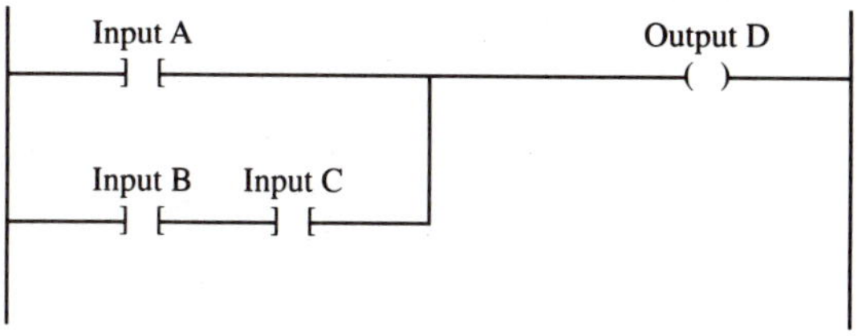

Figure 9-11 Input Branching

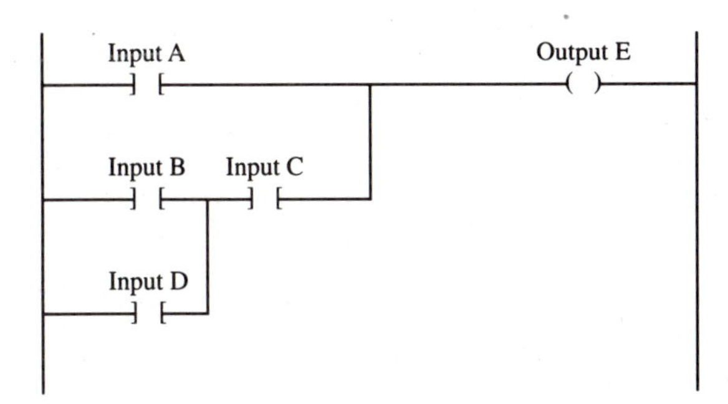

Figure 9-12 Nested Branch

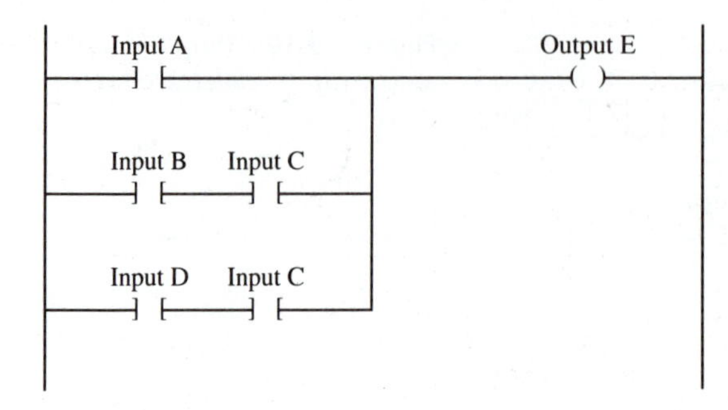

Figure 9-13 Eliminated Nested Branch

In some programmable controllers, outputs, too, may be branched. Figure 9-14 shows two branched outputs, with a series input in the branch containing output D.

Figure 9-15 shows a complex rung with branched inputs, branched outputs, and a nested branch.

Summary

With the six basic instructions covered in this chapter, you should be able to duplicate most relay logic. Mastering the use of these instructions will give you a good start in employing programmable controllers to replace mechanical relays for implementing discrete control. The benefits provided by the PLC make it well

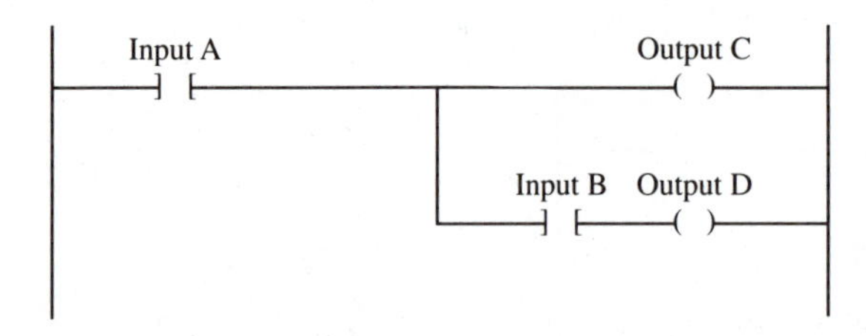

Figure 9-14 Output Branching

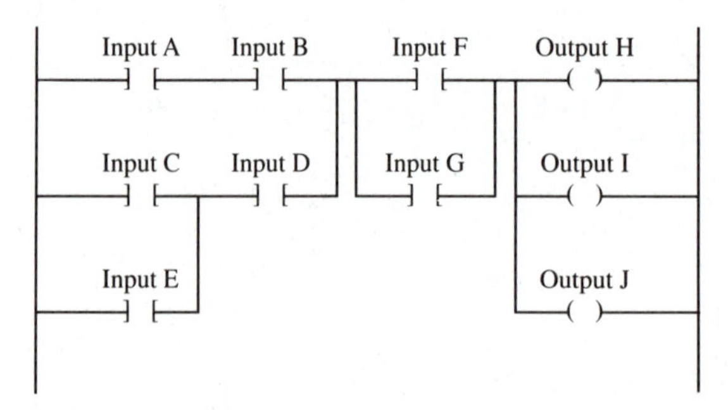

Figure 9-15 Complex Branching

worth the effort. Add to these instructions the timer and counter instructions covered in the next chapter and you will be able to duplicate most mechanical relay controls on a PLC.

Exercises

9-1. Programming or entering ladder logic while the terminal is connected to the PLC is known as what kind of programming?

9-2. What does the *examine on* (XIC) instruction look for at its data file address for it to be logically true?

9-3. What does the *examine off* (XIO) instruction look for at its data file address for it to be logically true?

9-4. Draw a ladder diagram that will cause output, pilot light PL2, to be on when selector switch SS2 is closed, push button PB 4 is open, and limit switch LS3 is open. The devices are wired to the following locations:

PL2: Output module, rack 3, I/O group 7, terminal 17
SS2: Input module, rack 3, I/O group 3, terminal 03
PB4: Input module, rack 4, I/O group 5, terminal 05
LS3: Input module, rack 4, I/O group 1, terminal 04

Show the addressing in the ladder logic.

9-5. Draw a ladder diagram that will cause a solenoid, SOL1, to be energized when limit switch LS 1 is closed and pressure switch PS2 is open. The devices are wired to the following locations:

SOL1: Output module, rack 2, I/O group 3, terminal 04
LS1: Input module, rack 3, I/O group 2, terminal 05
PS1: Input module, rack 4, I/O group 3, terminal 10

Show the addressing in the ladder logic.

9-6. Draw a ladder diagram that will cause output A to be latched when push button PB1 is closed and cause it to be unlatched when either push button PB2 or push button PB3 is closed. Also, do not allow the unlatch to go true when the latch rung is true, nor allow the latch rung to go true when the unlatch rung is true. The devices are wired to the following locations:

PB1: Input module, rack 1, I/O group 3, terminal 07
PB2: Input module, rack 1, I/O group 4, terminal 03
PB3: Input module, rack 2, I/O group 2, terminal 05
Output A: Output module, rack 1, I/O group 2, terminal 10

Show the addressing on the ladder logic.

9-7. Draw a ladder diagram that will cause output D to go true when push button A and push button B are closed or when push button C is closed. The devices are wired to the following locations:

Output D: Output module, rack 2, I/O group 3, terminal 15
Input A: Input module, rack 3, I/O group 2, terminal 12
Input B: Input module, rack 3, I/O group 3, terminal 03
Input C: Input module, rack 4, I/O group 5, terminal 00

Show the addressing in the ladder logic.

9-8. Draw a ladder diagram that will cause SOL2 to go true when PB1 is true and either LS2 or LS3 is true. The devices are wired to the following locations:

SOL2: Output module, rack 3, I/O group 4, terminal 17.
PB1: Input module, rack 4, I/O group 4, terminal 05.
LS2: Input module, rack 5, I/O group 3, terminal 16.
LS3: Input module, rack 5, I/O group 3, terminal 17.

Show the addressing in the ladder logic.

9-9. Draw a ladder diagram that is logically the same as the one in Figure 9-16 but that eliminates the nested branch.

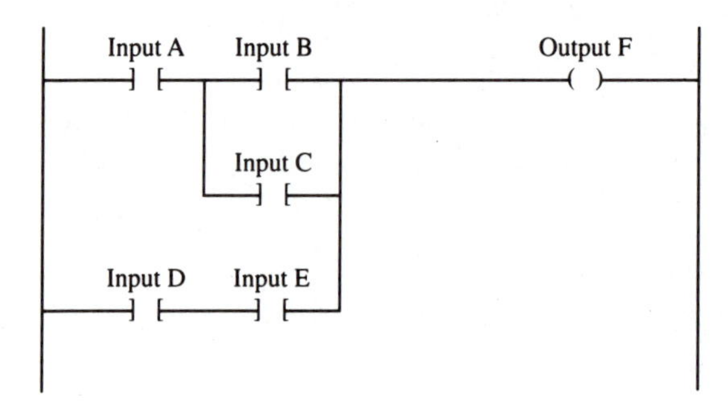

Figure 9-16 Nested Branch

Timer and Counter Instructions

10

OUTLINE

OBJECTIVES

Upon completion of this chapter, the student will be able to:

- Describe the function of the three different timer instructions: the on-delay timer, the off-delay timer, and the retentive timer.
- Program the control of outputs, using the timer instruction control bits, for each of the three different timer instructions.
- Describe the function of the two counter instructions: the count-up counter and the count-down counter.
- Program the control of outputs, using the counter instruction control bits, for each of the counters.
- Develop a program using the up-counter and down-counter together to form an up–down counter function.
- Use the *reset* instruction to reset the appropriate timer and counter instructions.

TIMER and counter instructions, and their function in ladder logic, will be covered in this chapter. How timers time intervals and the way in which they can control outputs will be explained. The method by which counters count events as well as how counters can be employed to control outputs will also be covered.

Timers and counters are output instructions. There are three different timers: the on-delay timer (TON), the off-delay timer (TOF), and the retentive timer on (RTO). There are two different counter instructions: the count-up counter (CTU) and the count-down counter (CTD). The *reset* instruction (RES) acts to reset both the timer instructions and the counter instructions.

Timer and counter instructions are made up of three word elements; as a result, each instruction takes up three words in the data table. The first word in the element is the control word, which stores the control bits. The control bits' function vary with the types of timers and counters; this will be explained with the individual instructions. The second word of the element, the preset word, contains the set-point value. The third word of the element, the accumulated word, stores the accumulated value, that is, either the elapsed time with a timer or the number of events counted by a counter. When the accumulated value reaches the preset value, a control bit changes state.

10-1 Timers

Timer elements consist of three data table words: the control word, the preset word, and the accumulated word.

Bits	15 14 13 12 11 10 9 8 7 6 5 4 3 2 1 0
Control word	EN TT DN
Preset word	Range: 0–32,767
Accumulated word	Range: 0–32,767

The *control word* uses three bits, the enable bit (EN), the timer-timing bit (TT), and the done bit (DN). The *enable bit* is true (has status of 1) whenever the timer instruction is true. When the timer instruction is false, the enable bit is false (has status of 0). The *timer-timing bit* is true whenever the accumulated value of the timer is changing, which means the timer is timing. When the timer is not timing, the accumulated value is not changing, so the timer-timing bit is false. The *done bit* changes state whenever the accumulated value reaches the preset value. Its state depends on the type of timer being used.

The preset value is the set point of the timer, that is, the value up to which the timer will time. The preset word has a range of 0 through 32,767 and is stored in binary form. The preset will not store a negative number.

The *accumulated value* is the value that increments as the timer is timing. The accumulated value will stop incrementing when its value reaches the preset value.

The timer instruction also requires you to enter a *time base,* which is either 1.0 or 0.01 seconds. The actual preset time interval is the time base multiplied by the value stored in the timer's preset word. The actual accumulated time interval is the time base multiplied by the value stored in the timer's accumulated word.

On-Delay Timers (TON)

An on-delay timer can be used to start or stop an event after a specified time delay following another event. Figure 10-1 shows an example of the *on-delay timer* instruction. The information to be entered into the instruction is the *timer number* (which must come from a timer file). The example shown in Figure 10-1

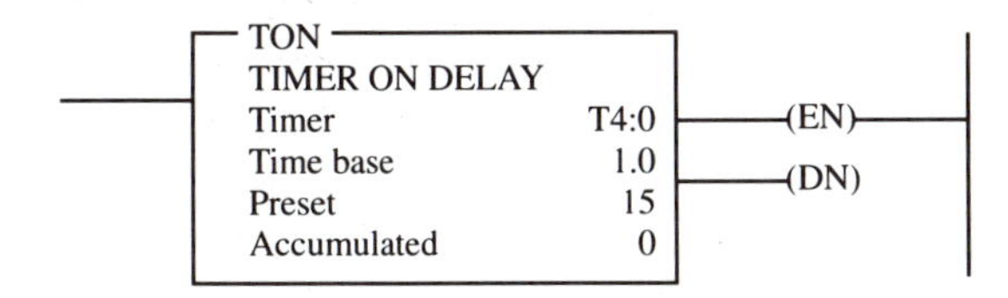

Figure 10-1 Example of the *On-delay Timer* Instruction

is T4:0, which represents timer file 4, timer 0 in that file. There may be up to 1000 timers in each timer file, numbered from 0 through 999. The timer address must be unique for this timer and may not be used for any other timer.

The time base (which is always expressed in seconds) may be either 1.0 or 0.01 seconds.

The timer preset value can range from 0 through 32,767. Negative numbers cause the processor to fault.

The timer's accumulated value normally is entered here as 0, although it is possible to enter a value from 0 through 32,767. Regardless of the value preloaded, the timer will reset to zero whenever it is reset.

The EN and DN bits shown to the right of the instruction in Figure 10-1 are for display purposes and automatically appear when the instruction is inserted in the program.

Figure 10-2 shows the TON timer controlled by input A. The enable bit, the timer-timing bit, and the done bit are each shown controlling an output. The timing diagram for the on-delay timer is shown below the ladder logic. When input A goes true, the timer starts timing and the accumulated value increments. The status of the enable bit follows the status of the timer rung. The timer-timing bit is true while the timer is timing and goes false when the accumulated value reaches the preset value. Let's follow the timing diagram from left to right.

The diagram first shows the timer timing to 4 seconds and then going false. The timer resets, and both the timer-timing bit and the enable bit go false. The accumulated value also resets to 0. (The fact that the accumulated value is not retained means that the TON timer is a nonretentive timer.) Input A then goes true again and remains true in excess of 10 seconds. When the accumulated value reaches 10 seconds, the done bit (DN) goes from false to true and the timer-timing bit (TT) goes from true to false. When input A goes false, the timer instruction goes false and also resets, at which time the control bits are all reset and the accumulated value resets to 0.

Off-Delay Timer (TOF)

The off-delay timer can be used to start or stop an event after a specified time delay following another event. Figure 10-3 shows an example of the TOF instruction. The information to be entered into the instruction is the same as for the TON timer. Again, the address for the timer must be unique and may not be used again for another timer.

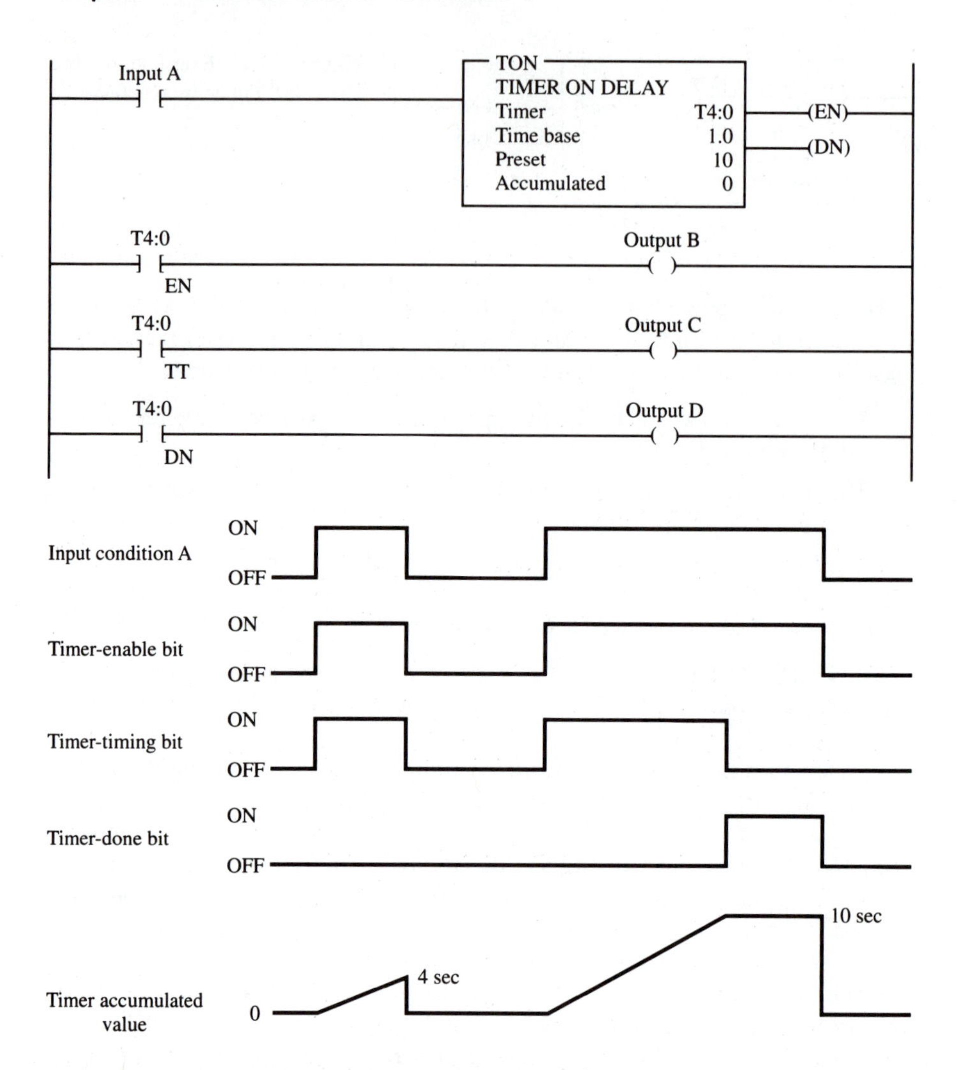

Figure 10-2 Example of an *On-delay Timer* Timing Diagram

Figure 10-4 shows the TOF timer controlled by input A, along with the enable bit, the timer-timing bit, and the done bit controlling outputs. Below the timer rungs is a timing diagram showing the status of the timer and its control bits as the timer times to completion. Notice that the timer is reset when the timer rung goes true, at which time the enable and done bits are true and the timer-

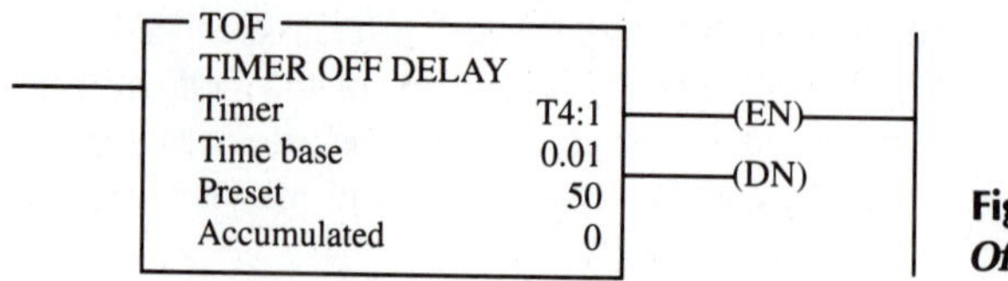

Figure 10-3 Example of the *Off-delay Timer* Instruction

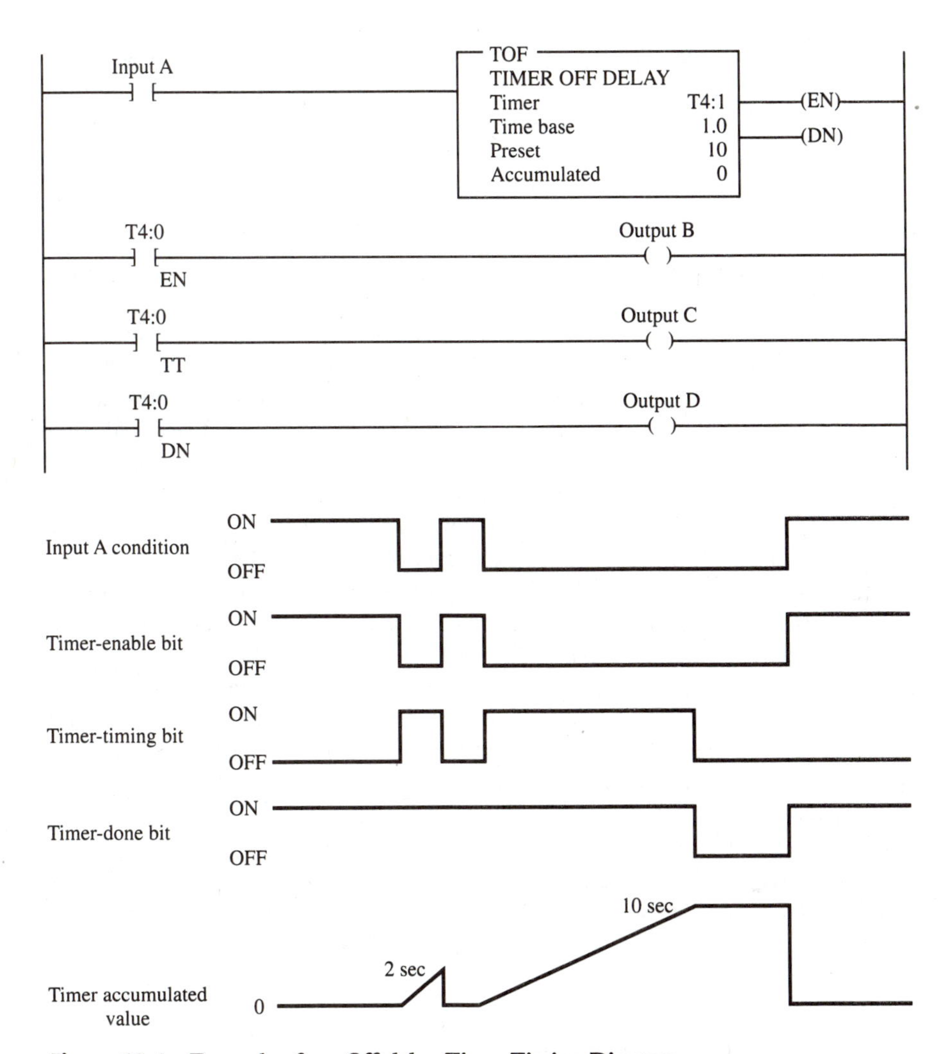

Figure 10-4 Example of an *Off-delay Timer* Timing Diagram

timing bit is false. When the timer rung goes false, the enable bit goes false and the timer-timing bit goes true. Let's follow the timing diagram from left to right.

When input A goes false, the timer starts timing and the accumulated value starts incrementing. The enable bit (EN) goes false, the timer-timing bit (TT) goes true, and the done bit (DN) remains true. Input A goes true again after being false for 2 seconds, which results in the timer's being reset. The enable bit (EN) goes true, the timer-timing bit (TT) goes false, and the accumulated value resets to 0. The TOF timer is nonretentive, for its accumulated value is not retained but instead is reset to 0 when the timer instruction goes true. Input A goes false again, and the TOF is allowed to time in excess of 10 seconds (its preset value). When the accumulated value reaches 10 seconds, the timer-timing bit goes false and the done bit goes false. The enable bit remains false, since it follows the timer instruction's condition. When Input A goes true, the timer instruction again resets, resetting the accumulated value to 0, and setting the enable and done bits. The *reset* instruction cannot be used to reset the TOF timer.

Retentive Timer On Instruction (RTO) and *Reset* Instruction (RES)

The retentive timer functions similar to the on-delay timer, except that it retains its accumulated value when the timer instruction goes false. A *reset* instruction is required to reset the retentive timer. Figure 10-5 shows the RTO instruction and the RES instruction. The information to be entered into the instruction is the same as for the previous timers. Again, the timer address must be unique for that timer and may not be used again for another timer. The *reset* instruction will reset the RTO timer when the RES instruction is true. When the timer resets, the accumulated value and the control bits are reset to 0. Note that the RTO instruction and the RES instruction have the same address.

Figure 10.6 shows the RTO instruction being controlled by input A, three different outputs being controlled by the control bits, and the *reset* instruction being controlled by input E. When input A goes true, the RTO timer begins timing and its accumulated value increments. The status of the enable bit (EN) follows the rung condition of the RTO instruction. The enable bit is true when the RTO instruction is true and false when the RTO instruction is false. The done bit (DN) goes true when the accumulated value reaches the preset value. The timer-timing bit is true whenever the timer accumulated value is changing (the timer is timing). If the RTO instruction goes false after the timer has timed out, the enable bit will go false, the done bit will remain true, and the accumulated value will be retained until the RES instruction goes true. The RTO timer is thus considered a retentive timer. When the RES instruction goes true, the done bit is reset and the accumulated value is reset to zero, whether or not the timer has timed out.

10-2 Counters

Counter elements each take three data table words: the control word, the preset word, and the accumulated word.

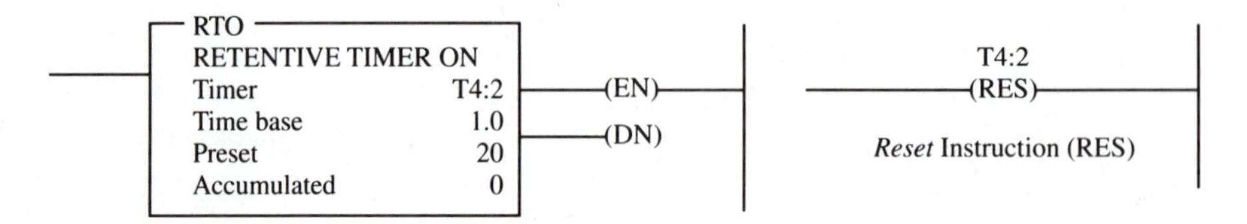

Retentive Timer On Instruction (RTO)

Figure 10-5 Example of the *Retentive Timer On* Instruction and the *Reset* Instruction

Bits	15 14 13 12 10 9 8 7 6 5 4 3 2 1 0
Control	CU CD DN OV UN
Preset	Range: −32,768 through +32,767
Accumulated	Range: −32,768 through +32,767

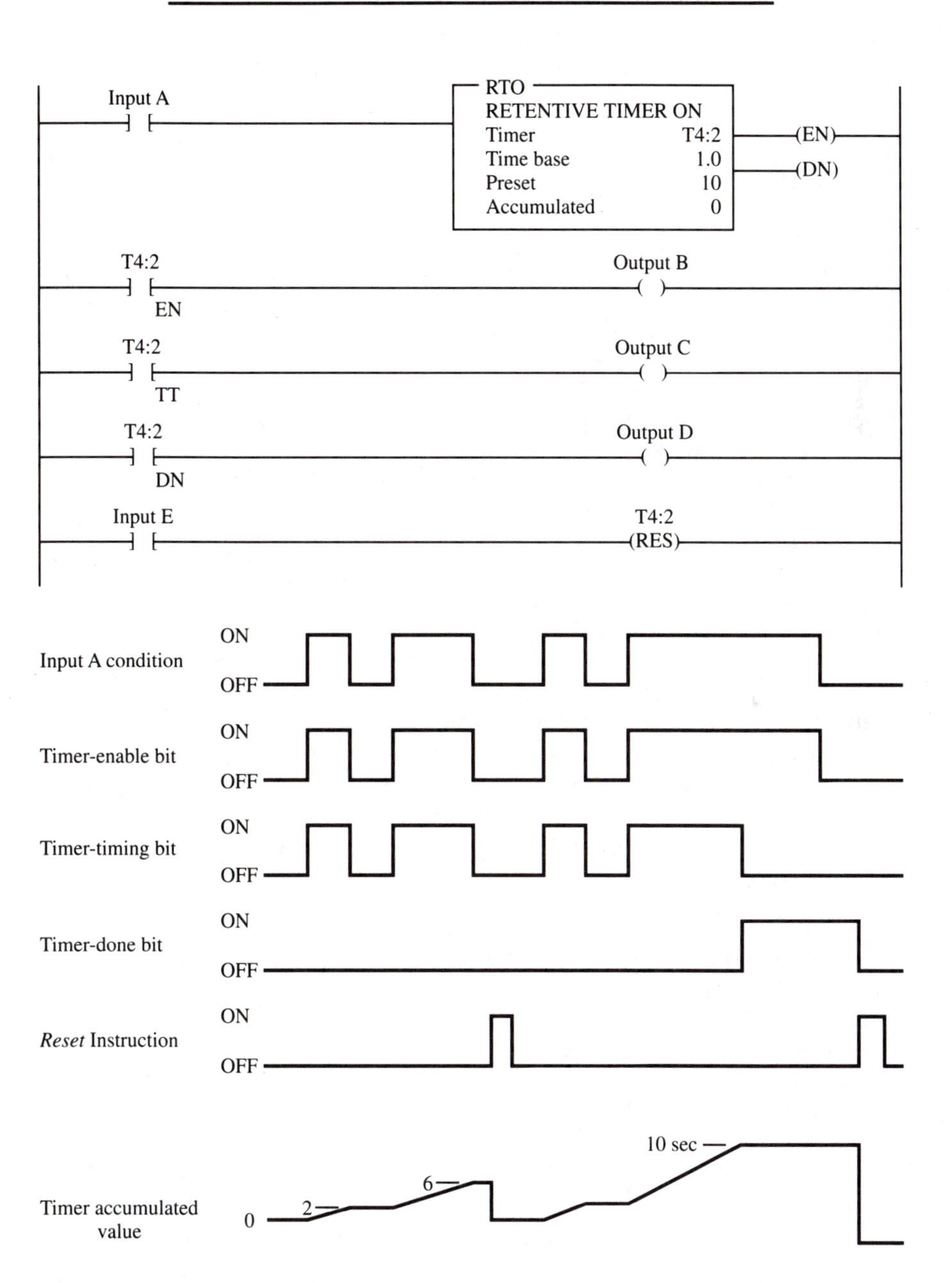

Figure 10-6 Example of a *Retentive Timer* Timing Diagram

The control word uses five control bits: the count-up enable bit (CU), the count-down enable bit (CD), the done bit (DN), the overflow bit (OV), and the underflow bit (UN). The count-up enable bit is used with the count-up counter and is true whenever the *count-up counter* instruction is true. If the *count-up counter* instruction is false, the (CU) bit is false. The count-down enable bit is used with the count-down counter and is true whenever the *count-down counter* instruction is true. If the *count-down counter* instruction is false, the (CD) bit is false. The done bit is true whenever the accumulated value is equal to or greater that the preset value of the counter, for either the count-up or the count-down counter. The overflow bit is true whenever the counter counts past its maximum value, which is 32,767. On the next count the counter will wrap around to −32,768 and will continue counting from there towards 0 on successive false-to-true rung transitions of the count-up counter. The underflow bit will go true when the counter counts below −32,768, and the counter will wrap around to +32,767 and continue counting down towards 0 on successive false-to-true rung transitions of the count-down counter.

The preset value is the set point of the counter, and ranges from −32,768 through +32,767. The number is stored in binary form, with any negative numbers being stored in 2's-complement binary.

The accumulated value either increments with a false-to-true transition of the *count-up counter* instruction or decrements with a false-to-true transition of the *count-down counter* instruction. It has the same range as the preset, −32,768 through +32,767. The accumulated value will continue to count past the preset value, instead of stopping at the preset like a timer does.

Count-up Counter (CTU)

The count-up counter is an output instruction whose function is to increment its accumulated value upon false-to-true transitions of its instruction. It thus can be used to count false-to-true transitions of an input instruction and then trigger an event after a required number of counts or transitions. Figure 10-7 shows the *count-up counter* instruction. The following information must be entered into the instruction:

The *counter number,* which must be from a counter file. The example shown in Figure 10-7 is C5:0, which represents counter file 5, counter 0 in that file. There may be up to 1000 counters in each counter file, numbered from 0 through 999. The address for this counter should not be used for any other count-up counter.

The *preset value,* ranging from −32,768 to +32,767.

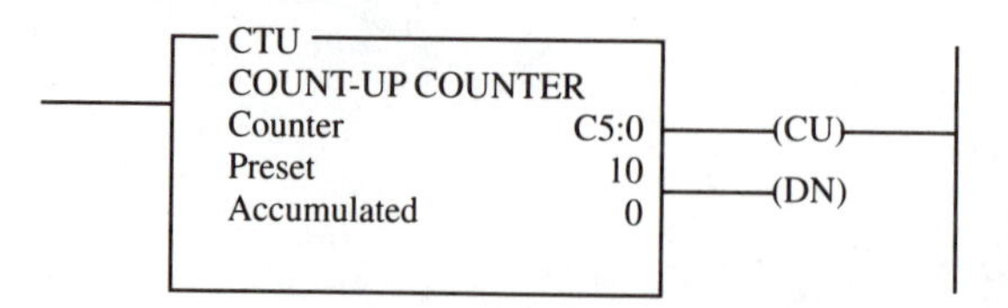

Figure 10-7 Example of the *Count-up Counter* Instruction

The *accumulated value*, which can also range from −32,768 through +32,767. Typically, the value entered in the accumulated word is 0. Regardless of what value is entered, the reset instruction will reset the accumulated value to 0.

The CU and DN bits shown to the right of the instruction are for display purposes and automatically appear when the instruction is inserted in the program.

Figure 10-8 shows the CTU instruction being conditioned by input A. The count-up enable bit (CU), the done bit (DN), and the overflow bit (OV) are each connected to an output. Input B controls the *reset* instruction. Whenever input A goes from false to true, and the *reset* instruction is not true, the accumulated value will increase by one count. When the CTU instruction is true, C5:0/CU will also be true, turning on output A. The condition C5:0/DN will be true when the accumulated value is equal to or greater than the preset value, turning on output B. If the accumulated value is incremented past 32,767, then C5:0/OV will become true, causing output C to become true. The accumulated value will wrap around to −32,768, one count past +32,767. Input B's going true will cause the RES instruction to go true, resetting the counter, at which time the accumulated value goes to 0 and the done bit and the overflow bit are reset to 0. Note that the address on the *reset* instruction is the same as the counter address that it is resetting.

Count-down Counter (CTD)

The count-down counter is an output instruction whose function is to decrement its accumulated value upon false-to-true transitions of its instruction. It then can also be used to trigger an event after a required number of counts or transitions. Figure 10-9 shows the *count-down counter* instruction. The information to be entered into the instruction is the same as for the count-up counter.

Figure 10-10 shows the CTD instruction being conditioned by input A. The count-down enable bit (CD), the done bit (DN), and the underflow bit (UN) each control an output. The *reset* instruction is controlled by input B. Whenever input A goes from false to true, and the *reset* instruction is not true, the accumulated value will decrease by one count. When the CTD instruction is true, C5:1/CD will also be true, turning on output C. The condition C5:1/DN will be true when the accumulated value is equal to or greater than the preset value, turning on output D. Note that this is the same way the done bit functions with the count-up counter. If the accumulated value is decremented past −32,768, then C5:1/UN will become true, making output E true, and the accumulated value will wrap around to +32,767. Input B's going true will cause the RES instruction to go true, resetting the counter, at which time the accumulated value goes to 0 and the done bit and the underflow bit are reset to 0.

Up–Down Counter

The count-down counter frequently is used with the count-up counter, as shown in Figure 10-11. This application is known as an *up–down counter.* Note that the

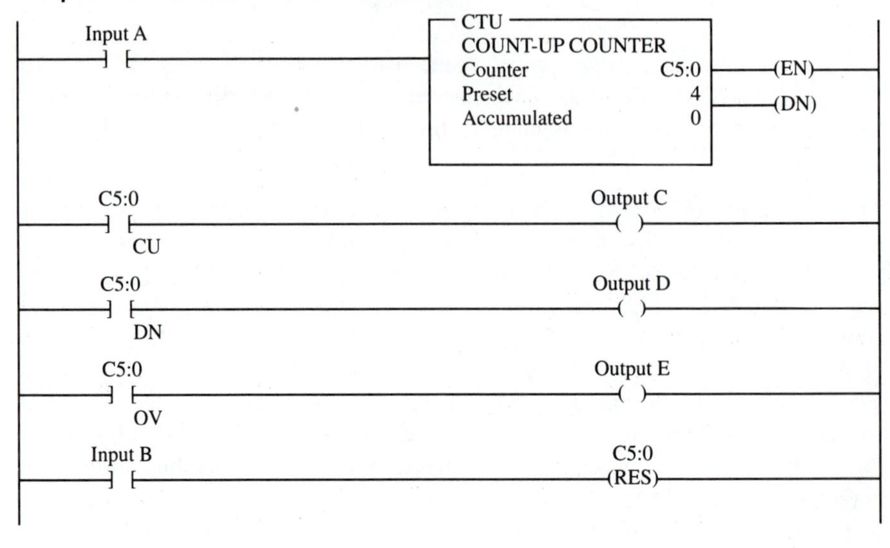

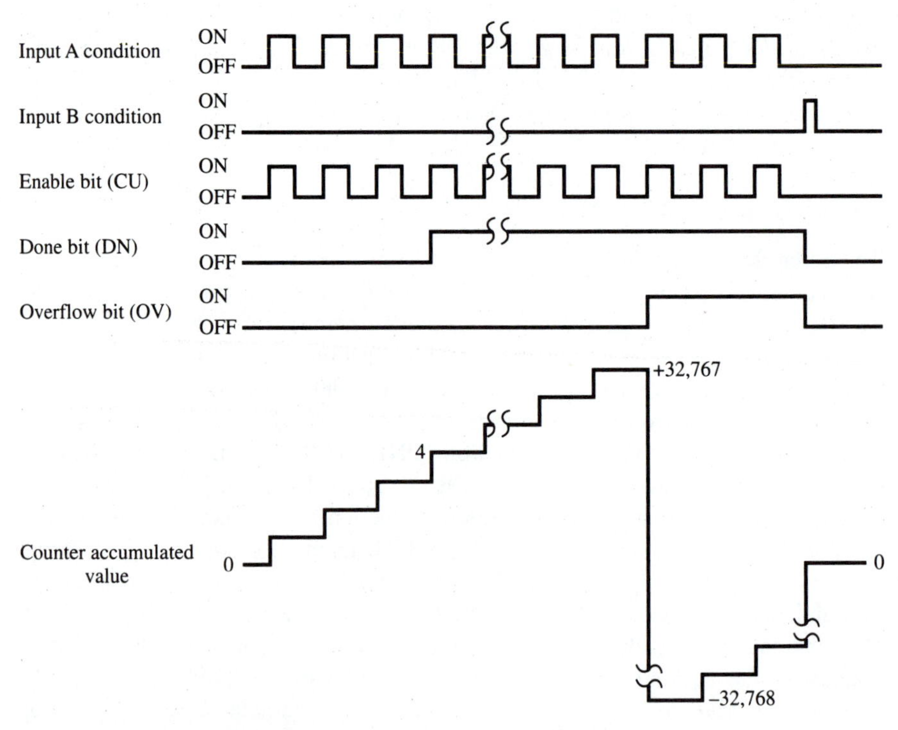

Figure 10-8 Example of a Count-up Counter

same address is given the *up-counter* instruction, the *down-counter* instruction, and the *reset* instruction. All three instructions will be looking at the same address in the counter file. When input A goes from false to true, one count is added to the accumulated value. When input B goes from false to true, one count

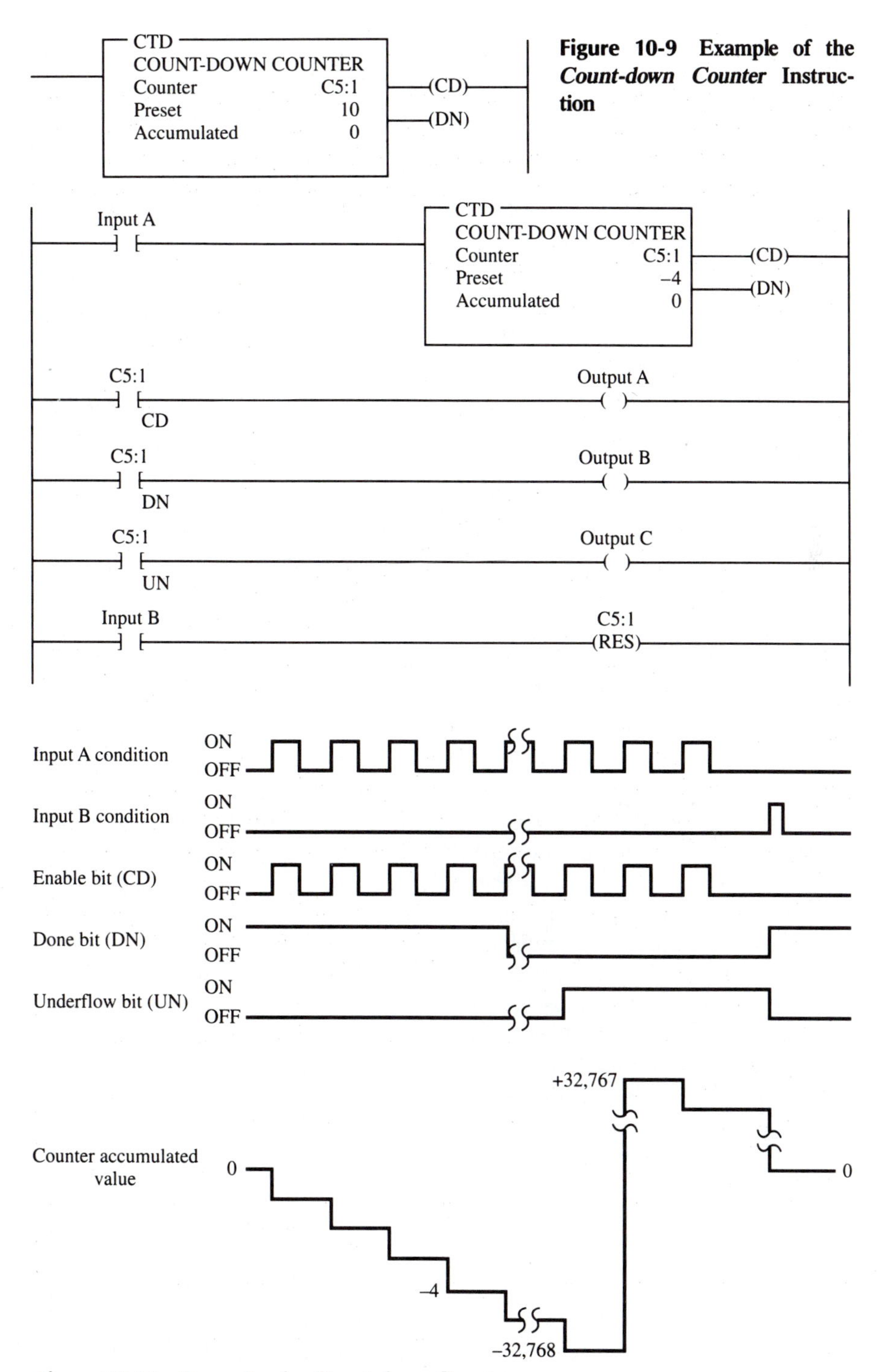

Figure 10-9 Example of the *Count-down Counter* Instruction

Figure 10-10 Example of a Count-down Counter

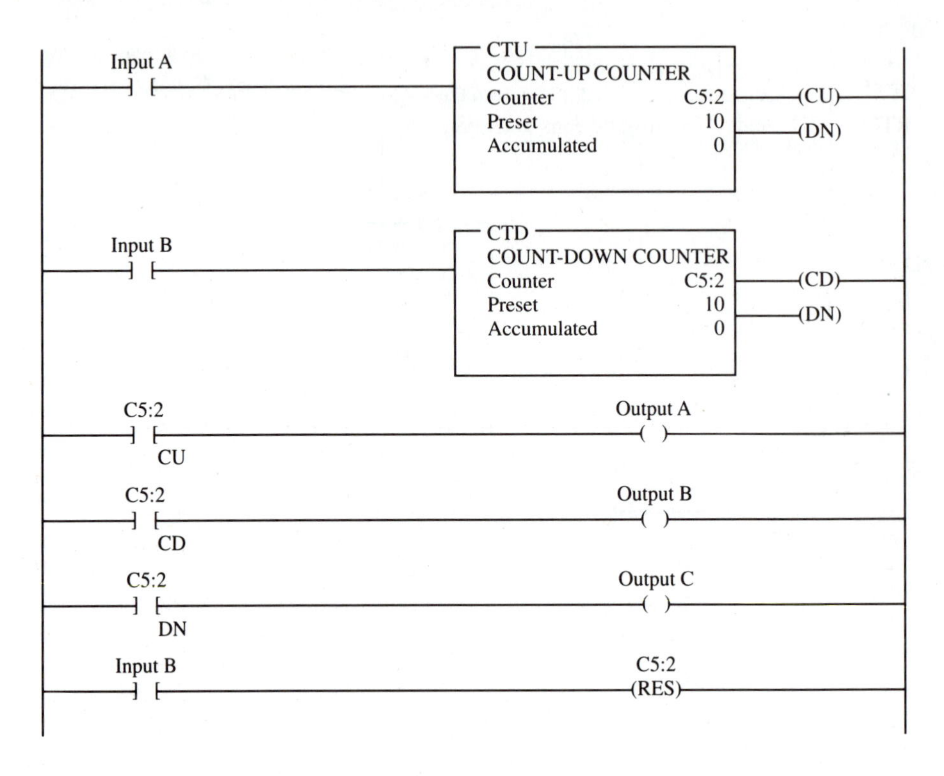

Figure 10-11 Example of an Up–Down Counter

is subtracted from the accumulated value. The accumulated value is the same for both counters, again because of their same address.

When the CTU instruction is true, C5:2/CU will be true, causing output A to be true. When the CTD instruction is true, C5:2/CD will be true, causing output B to be true. When the accumulated value is greater than or equal to the preset value, C5:2/DN will be true, causing output C to be true. Input B's going true will cause both counters to reset. When reset, by the *reset* instruction, the accumulated value will be reset to 0 and the done bit will be reset, along with either the underflow or the overflow bit, if it was set at the time. Though they are not shown, the overflow and underflow bits will function the same as they did on their individual instructions.

Summary

The processor's instruction set includes three timer instructions and two counter instructions. The timer instructions are the on-delay timer (TON), the off-delay timer (TOF), and the retentive timer on (RTO). The timer instructions allow for programming the occurrence of events after a specified time interval following other events. The counter instructions are the count-up counter (CTU) and the count-down counter (CTD). The counter instructions allow for the counting of events and the controlling of other events based on the accumulated count. The RTO, CTU, and CTD instructions require a *reset* instruction (RES) to reset the instruction's control bits and accumulated values back to 0.

Exercises

10-1. How many data table words are required per timer element?

10-2. Describe the bits used in the control word of the timer element.

10-3. What is the maximum amount of time a timer can time?

10-4. What timer requires the *reset* instruction?

10-5. Which timer instruction or counter instruction cannot be reset using the *reset* instruction?

10-6. You have a conveyor that cycles on and off during its operation. You need to keep a record of its total run time, for maintenance purposes. Which timer would accomplish this?

10-7. Develop the ladder logic that will turn on a light, PL1, 15 seconds after switch SS1 has been turned on.

10-8. Develop the ladder logic that will turn on a light when switch SS2 is turned on and will turn off the light 12 seconds after SS2 is turned off.

10-9. Develop the ladder logic that will turn on a light, PL3, after push button PB 1 has been pressed ten times. Push button PB2 will reset the counter.

10-10. When does the overflow bit on a counter become true? What value does the counter index to when the overflow bit comes on?

10-11. What are the minimum and maximum values that can be stored in the counter preset word?

10-12. Input A cycles on and off. Turn on an output when the total accumulated time that input A is on reaches 30 minutes. Input B will reset the timer.

10-13. Develop a program that will increase a counter's accumulated value when push button PB1 is pressed and will decrease the counter's accumulated value when push button PB2 is pressed. Pilot light PL2 will come on when the counter's accumulated value reaches 10. Push button PB3 will reset the counter.

10-14. Develop a program that will turn on pilot light PL1 10 seconds after switch SS1 is turned on. Pilot light PL2 will come on 5 seconds after PL1 comes on. Pilot light PL3 will come on 8 seconds after PL2 comes on. Pressing push button PB1 will reset all the timers but will not allow them to be reset unless PL3 is on.

10-15. Develop a program that will increment a counter's accumulated value one count every 60 seconds. A second counter's accumulated value will increment one count every time the first counter's accumulated value reaches 60. The first counter will reset when its accumulated value reaches 60, and the second counter will reset when its accumulated value reaches 12.

10-16. Develop a program that will latch on an output, PL1, after an input, PB1, has cycled on 20 times. When the count of 20 is reached, the counter will automatically reset itself. PB2 will unlatch PL1.

10-17. Develop a program that will latch on an output, PL5, 20 seconds after input SW2 has been turned on. The timer will continue to cycle up to 20 seconds, and reset itself, until SW2 has been turned off. After the third time the timer has timed to 20 seconds, PL5 will be unlatched.

Data Collecting

11

OUTLINE

OBJECTIVES

Upon completion of this chapter, the student will be able to:

- Describe the operation of the word-level instructions used to copy data from one memory location to another.
- Describe the operation of the word-level instructions used to do arithmetic operations on data stored in the processor.
- Describe the operation of word-level instructions that make mathematical comparisons on data stored in the processor.
- Describe the operation of word-level instructions that perform logical operations on data stored in the processor.
- Develop elementary programs that employ the word-level instructions.

Data manipulation involves: the transfer of data, operating on data with math functions, data conversions, data comparison, and logical operations on data. There are two basic classes of instructions to accomplish this: instructions that operate on word data, and those that operate on file data, which is multiple words. In this chapter we will cover the word-type instructions; Chapter 12 will cover the file instructions that do basically the same operation but on a file level.

11-1 Word-level Move Instructions

The following lists instructions that move data, along with their mnemonics:

Instruction	Mnemonic
Move	MOV
Masked move	MVM

These instructions copy data from a source word to a destination word. Data may be moved from one data-table type to another—for example, from a floating-point address to an integer address—while maintaining its correct value. The data is rounded at the destination, with 0.5 and above being rounded up, and less than 0.5 being rounded down. In the case of floating point, the data being operated on is an *element,* which is two words. In all other data types the instructions operate on single words.

Move Instruction (MOV)

The example of the *move* instruction presented in Figure 11-1 shows the value stored at the address indicated in the Source, N7:56, being copied into the address indicated in the Destination, N7:60. This value will be copied every time the instruction is scanned and the instruction is true. When the rung goes false, the Destination address will retain the value, unless it is changed elsewhere in the program. The instruction may be programmed with input conditions preceding it, or it may be programmed unconditionally.

Masked Move Instruction (MVM)

The example of the *masked move* instruction given in Figure 11-2 shows the data at the Source address, B3:0, being copied into the Destination address, B3:4, through the mask. Data will be transferred through the mask. The mask may be entered as an address or in hexadecimal format, and its value will be displayed in hexadecimal. Where there is a 1 in the mask, data will pass from the source to the destination. Where there is a 0 in the mask, data in the destination will remain in its last state. In Figure 11-2, the data in the destination in bits 04 through 07 remain in the same state as before the instruction became true. Status of the bits in the destination (bits 00 through 03 and 08 through 15) will match the status of the bits in the source, due to the transfer of data through the 1's in the mask. The instruction executes every time it is scanned and the instruction is true. It can be programmed either conditionally or unconditionally.

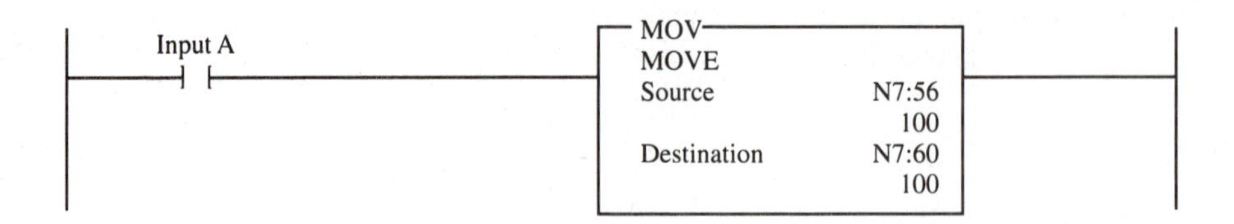

Figure 11-1 Example of the *Move* Instruction

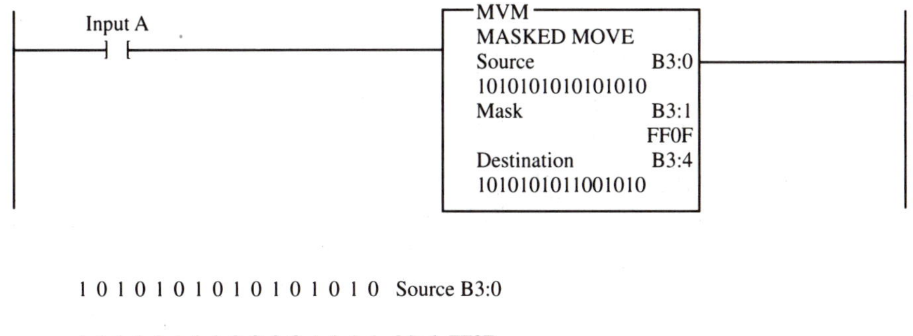

1 0 1 0 1 0 1 0 1 0 1 0 1 0 1 0 Source B3:0

1 1 1 1 1 1 1 1 0 0 0 0 1 1 1 1 Mask FF0F

1 1 0 0 1 1 0 0 1 1 0 0 1 1 0 0 Destination B3:4 before instruction went true

1 0 1 0 1 0 1 0 1 1 0 0 1 0 1 0 Destination B3:4 after instruction went true

Status unchanged due to zeroes in the mask (remained in last state)
Status in bits 0-3 and 8-15 copied from Source to Destination when instruction went true

Figure 11-2 Example of the *Masked Move* Instruction

11-2 Word-level Arithmetic Instructions

The following lists the arithmetic instructions and their mnemonics:

Instruction	Mnemonic
Add	ADD
Subtract	SUB
Multiply	MUL
Divide	DIV
Square root	SQR
Negate	NEG
Clear	CLR
Convert to BCD	TOD
Convert from BCD	FRD

These are output instructions that operate at the word level, meaning they work with the value stored in a 16-bit word, except for floating point, where the value is stored in 32 bits. The value will be stored in binary format for positive numbers and 2's complement for negative numbers. Different data-table types may be entered in an instruction, but care must be taken that values remain in the range the data-table type can store; for example, you cannot store a value larger than 32,767 at an integer address. The processor will round the result of the operation: In the integer file it will round to the nearest whole number, with 0.5 and above being rounded up, and less than 0.5 being rounded down.

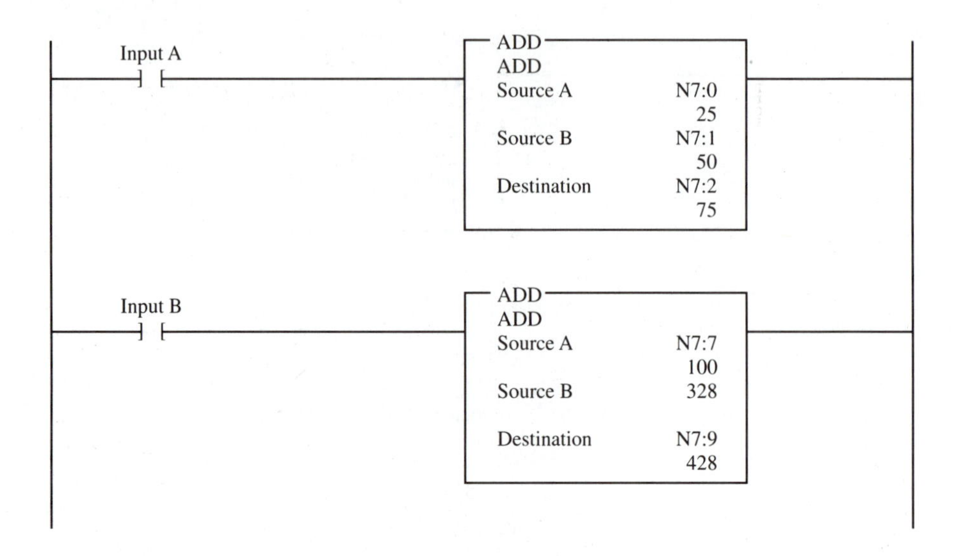

Figure 11-3 Two examples of the *Add* Instruction

Add Instruction (ADD)

Two examples of the *add* instruction are presented in Figure 11-3. The first example shows the value stored at the address indicated in Source A, N7:0, being added to the value stored at the address in Source B, N7:1. The answer will be stored at the address location indicated in the Destination, N7:2. The *add* operation will take place every time the instruction is scanned and the instruction is true. The instruction can be programmed unconditionally, without an input instruction preceding it, or conditionally, with an input instruction preceding it. The second example shows the instruction with a constant entered in Source B. A constant may be entered either in Source A or in Source B, and it is stored in the instruction, not in the data table.

Subtract Instruction (SUB)

Two examples of the *subtract* instruction are presented in Figure 11-4. The first example shows the value stored at the address indicated in Source B, N7:05, being subtracted from the value stored at the address in Source A, N7:10. The answer will be stored at the address location indicated in the Destination, N7:20. The *subtract* operation will take place every time the instruction is scanned and the instruction is true. The instruction can be programmed unconditionally or conditionally. The second example shows the instruction with a constant entered in Source B. A constant may be entered in either Source A or Source B, and it is stored in the instruction.

Multiply Instruction (MUL)

The example of the *multiply* instruction given in Figure 11-5 shows the value stored at the address indicated in Source A, F8:10, being multiplied by the value

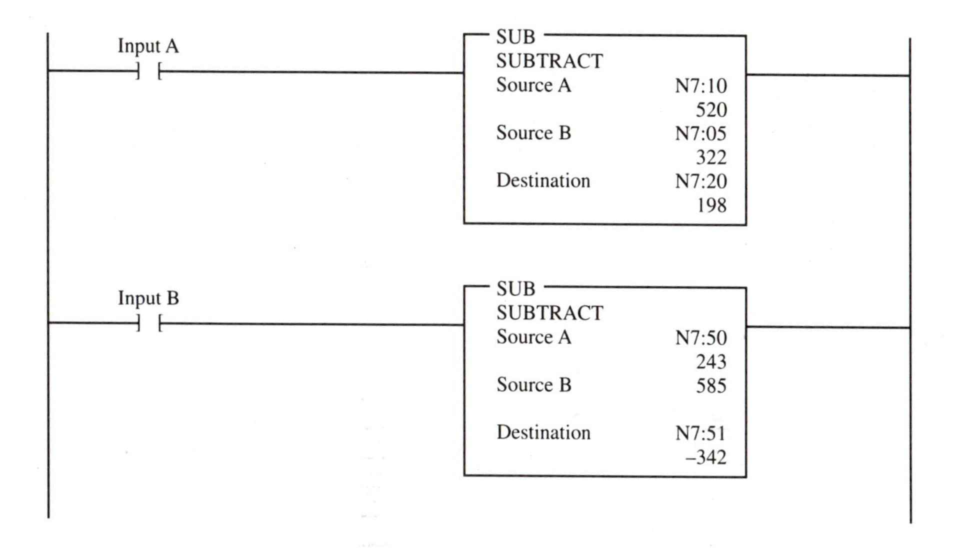

Figure 11-4 **Two examples of the *Subtract* Instruction**

stored at the address in Source B, N7:30, with the result being stored at the address in the Destination, F8:11. The multiplication will take place every time the instruction is scanned and the instruction is true. The instruction can be programmed unconditionally, without an input instruction preceding it, or conditionally, with an input instruction preceding it. A constant may also be entered in either Source A or Source B, and it will be stored in the instruction.

Divide Instruction (DIV)

The *divide* instruction in the example in Figure 11-6 divides the value stored at Source A's address (N7:50) by the value stored at Source B's address (N7:55), and stores the result at the Destination's address (N7:60). Note that the result stored at the Destination address is rounded off: If the remainder is 0.5 or above, the result is rounded up; if the remainder is less than 0.5, the answer is rounded down. The *divide* instruction will execute every time the instruction is scanned and the instruction is true. The instruction can be programmed conditionally or unconditionally.

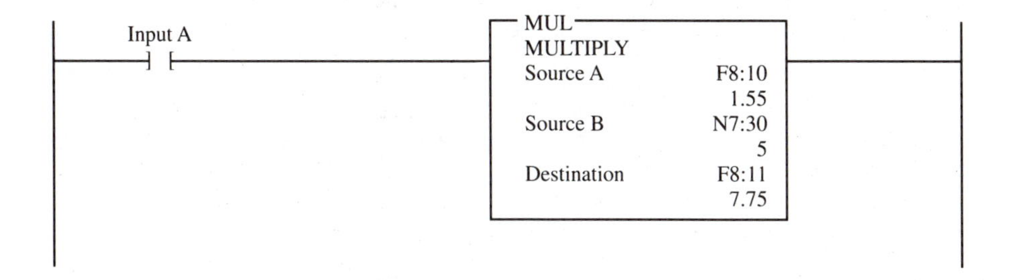

Figure 11-5 **Example of the *Multiply* Instruction**

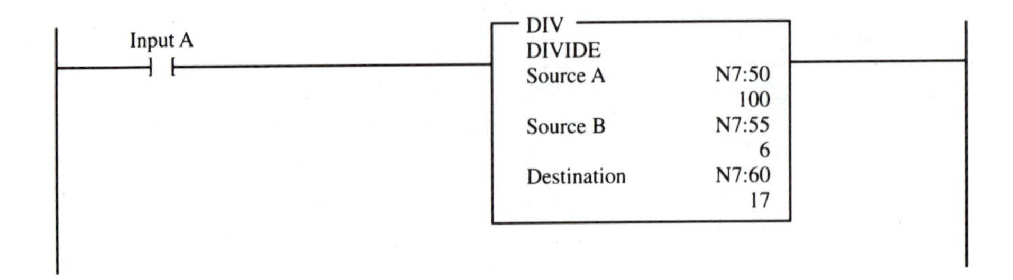

Figure 11-6 Example of the *Divide* Instruction

A constant can be entered in either Source A or Source B, and it will be stored in the instruction.

Square Root Instruction (SQR)

The example of the *square root* instruction presented in Figure 11-7 shows the instruction determining the square root of the value stored at the address of the Source, N7:101, and storing the result at the address of the Destination, N7:105. If the value of the source is negative, the instruction will store the square root of the absolute value of the source at the destination. The instruction will execute every time it is scanned and the instruction is true. The instruction can be programmed either conditionally or unconditionally.

Negate Instruction (NEG)

The *negate* instruction in the example in Figure 11-8 negates (changes the sign of) the value stored at the Source address, N7:52, and stores the result at the Destination address N7:53, when the instruction is true. Positive numbers will be stored in straight binary format, and negative numbers in two's complement. The instruction will be executed every time it is scanned and the instruction is true. The instruction can be programmed either conditionally or unconditionally.

Clear Instruction (CLR)

The *clear* instruction in the example in Figure 11-9 zeros the value stored in the Destination address, N7:22, when the instruction is true. The instruction executes

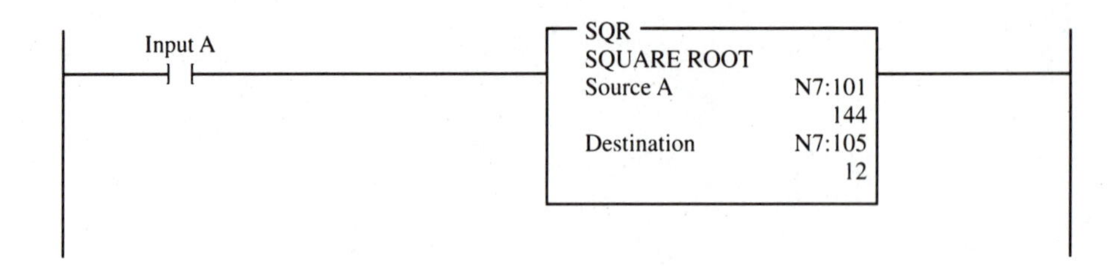

Figure 11-7 Example of the *Square Root* Instruction

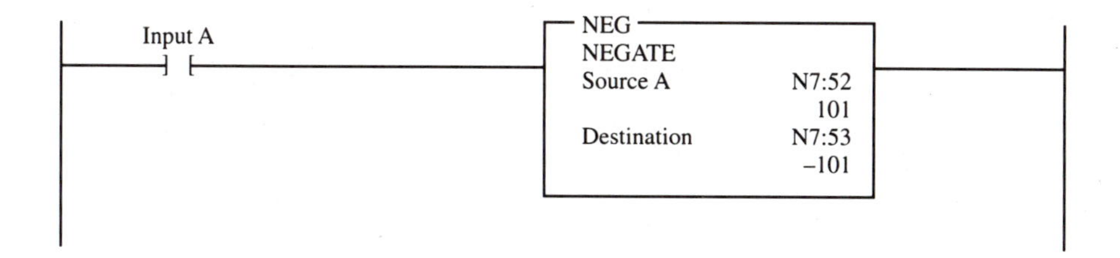

Figure 11-8 Example of the *Negate* Instruction

every time it is scanned and the instruction is true. The instruction can be programmed conditionally or unconditionally.

Convert to Decimal Instruction (TOD)

The *convert to decimal* instruction in the example in Figure 11-10 will convert the binary bit pattern at the Source address, N7:23, into a BCD bit pattern of the same decimal value at the Destination address, O:20. The conversion is necessary when transferring data from the processor (which stores data in a binary format) to an external device that functions in BCD format, such as an LED display. Note the values displayed in the instruction: The Source displays the value 10, which is the correct decimal value; however, the Destination displays the value 16. Since the processor interprets all bit patterns as binary, the value 16 displayed is the binary interpretation of the BCD bit pattern. The bit pattern for 10 BCD is the same as the bit pattern for 16 binary. The instruction executes every time it is scanned and the instruction is true. It can be programmed either conditionally or unconditionally.

Convert from Decimal Instruction (FRD)

The *convert from decimal* instruction in the example in Figure 11-11 will convert the BCD bit pattern from the source to a binary bit pattern of the same decimal value at the destination. The data stored at the Source address, I:30, would be in a BCD format (probably from a source external to the processor). The data stored in the Destination address, N7:24, will be in a binary format, recognizable by the processor. Note that the value displayed at the Source address is 16, which is the

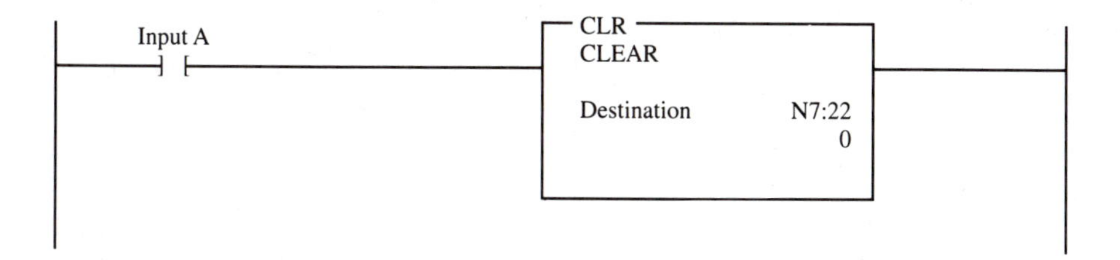

Figure 11-9 Example of the *Clear* Instruction

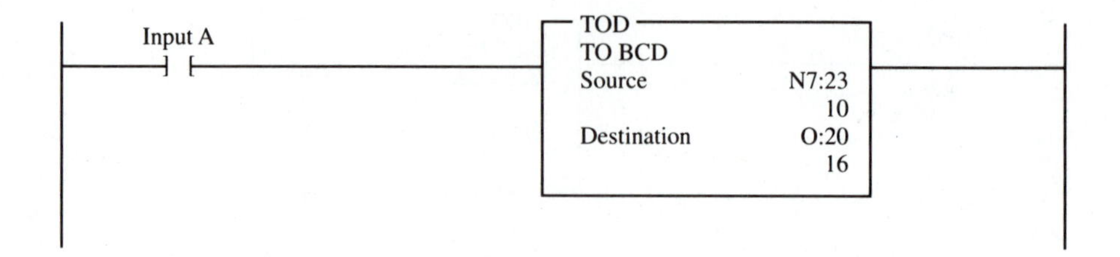

Figure 11-10 Example of the *Convert to Decimal* Instruction

binary interpretation of the BCD bit pattern, since the processor looks at all bit patterns as binary. The correct value in the instruction is 10, which is displayed at the Destination address. The bit pattern at the destination is in binary, which the processor recognizes. The instruction could be used to convert data from a BCD external source, such as BCD thumb wheels, to the binary format in which the processor operates. The instruction executes each time it is scanned and the instruction is true. The instruction may be programmed either conditionally or unconditionally.

11-3 Data-comparison Instructions

The following lists the data-comparison instructions and their mnemonics:

Instruction	Mnemonic
Equal	EQU
Not equal	NEQ
Less than	LES
Less than or equal	LEQ
Greater than	GRT
Greater than or equal	GEQ
Masked comparison for equal	MEQ
Limit test	LIM

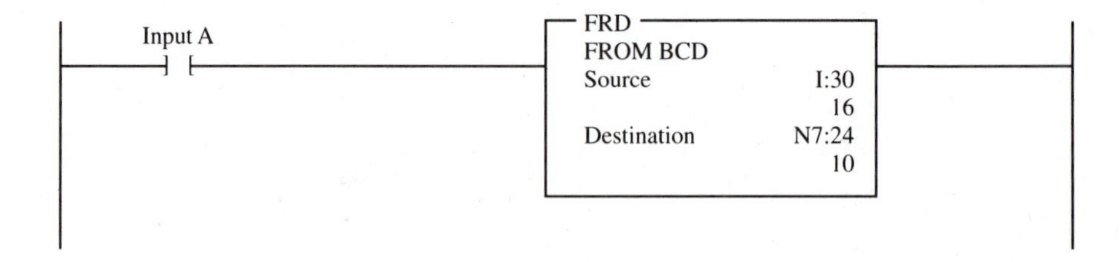

Figure 11-11 Example of the *Convert from Decimal* Instruction

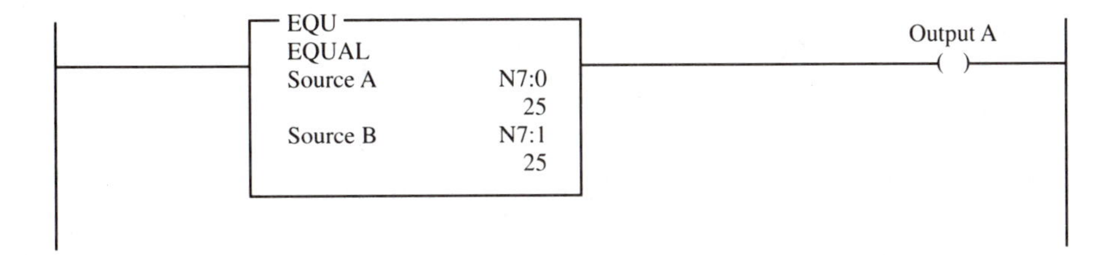

Figure 11-12 **Example of the *Equal* Instruction**

These are input instructions and thus are located to the left of the output instructions on the rung. Each instruction will have a logical status, either true or false, depending on whether or not the instruction parameters are met. The comparison instructions will compare unlike data-table types. For example, values from the integer file can be compared to values in the floating-point file. There can also be multiple-comparison instructions in a rung, which can also be used with other input instructions in the same rung.

Equal Instruction (EQU)

The *equal* instruction presented in the example in Figure 11-12 is an input comparison instruction that is either true or false, depending on the values being compared in the instruction. When the value in Source A equals the value in Source B, the instruction is true. In Figure 11-12, when the value stored in Source A's address, N7:0, equals the value stored in Source B's address, N7:1, the instruction is true, causing Output A to go true. Source A or Source B could also be a constant stored in the instruction.

With the *equal* instruction, the floating-point data type is not recommended, due to the exactness required. One of the other comparison instructions, such as the limit test, is preferable.

Not Equal Instruction (NEQ)

The *not equal* instruction, exemplified in Figure 11-13, is an input-comparison instruction that is either true or false, depending on the values being compared in the instruction. When the value in Source A is not equal to the value in Source B, the instruction is true. In Figure 11-13, when the value stored at Source A's address, N7:5, is not equal to 25, Output B will be true; otherwise, Output B will be false.

Less Than Instruction (LES)

The *less than* instruction presented in Figure 11-14 is an input-comparison instruction that is either true or false, depending on the values being compared in the instruction. When the value in Source A is less than the value stored in Source B, the instruction is true. In Figure 11-14, when the value stored at Source A's

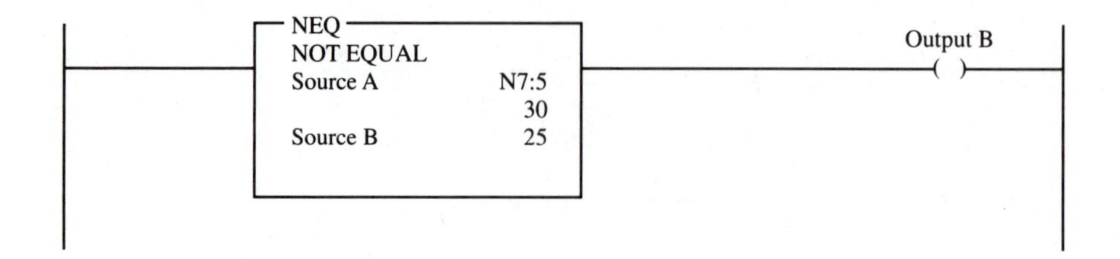

Figure 11-13 Example of the *Not Equal* Instruction

address, N7:13, is less than the value stored at Source B's address, N7:23, Output A will be true; otherwise, Output A will be false.

Less Than or Equal Instruction (LEQ)

Figure 11-15 displays the *less than or equal* instruction. This is an input-comparison instruction that is either true or false, depending on the values being compared. When the value at Source A is less than or equal to the value at Source B, the instruction is true; otherwise, it is false. In Figure 11-15, when the value stored at Source A, 55, is less than or equal to the value stored at Source B's address, N7:101, then Output C is true; otherwise, Output C is false.

Greater Than Instruction (GRT)

The *greater than* instruction, illustrated in Figure 11-16, is an input instruction that is either true or false, depending on the values being compared. If the value at Source A is greater than the value at Source B, the instruction is true; otherwise, it is false. In Figure 11-16, when the value stored at the address of Source A, N12:10, is greater than the value stored at the address of Source B, F8:115, Output E will be true; otherwise, it is false.

Greater Than or Equal Instruction (GEQ)

Figure 11-17 illustrates the *greater than or equal* instruction. When the value at Source A is greater than or equal to the value at Source B, the instruction is true; otherwise, the instruction is false. In Figure 11-17, if the value stored at the ad-

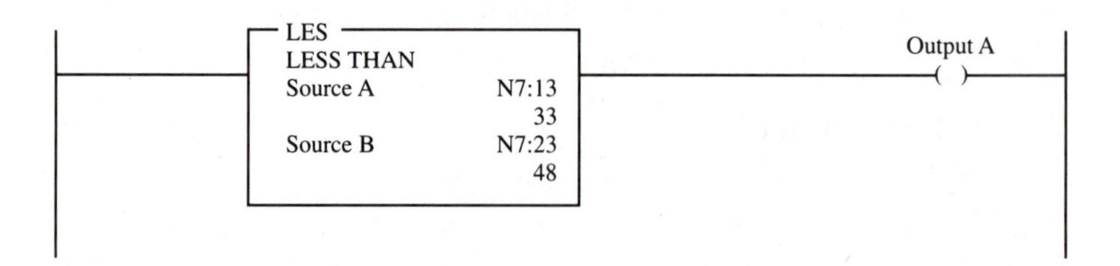

Figure 11-14 Example of the *Less Than* Instruction

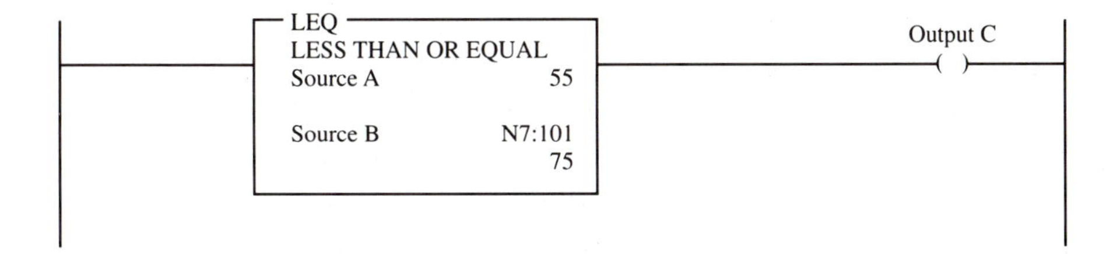

Figure 11-15 Example of the *Less Than or Equal* Instruction

dress of Source A, N7:555, is greater than or equal to the value stored at the address of Source B, N7:12, Output D will be true; otherwise, Output D will be false.

Masked Comparison for Equal Instruction (MEQ)

The *masked comparison for equal* instruction, shown in Figure 11-18, is an input instruction that is either true or false, depending on the status of the comparison being made. If the status of the bits in the Source match the status of the bits at the Compare address, wherever there is a 1 in the mask the instruction is true. In Figure 11-18, the bit pattern in the upper byte of the Source, B3:1, must match the bit pattern of the upper byte of the Compare, B3:3, for the instruction to be true. Since the upper byte of the mask is all 1's, it must be the upper byte that matches. Wherever there is a 0 in the mask (as in the mask's lower byte), that match is automatically true. Hence, the statuses of the bits in the lower bytes of the Source and the Compare do not affect the instruction. The data displayed in the mask will always be in hexadecimal format, regardless of the data type of the mask address. The mask may be an address, or data may be entered directly in the mask in hexadecimal format. The *masked comparison for equal* instruction in Figure 11-18 is true because the data in the upper byte of B3:1 matches the data in the upper byte of B3:3.

Limit Test (Circular) Instruction

The *limit test* instruction compares a test value to values in the low limit and in the high limit. The low limit, the test, and the high limit may be either an address

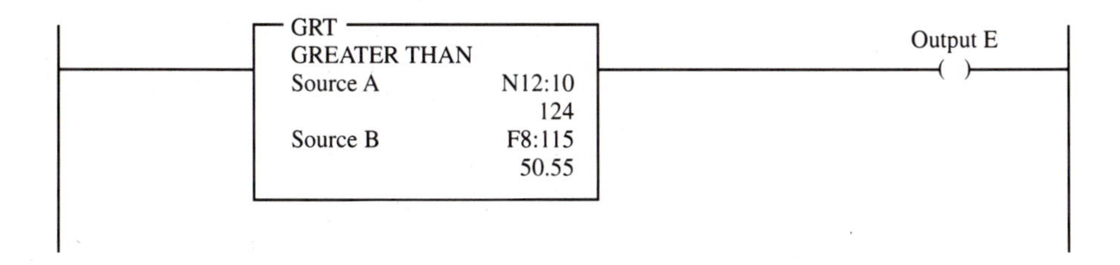

Figure 11-16 Example of the *Greater Than* Instruction

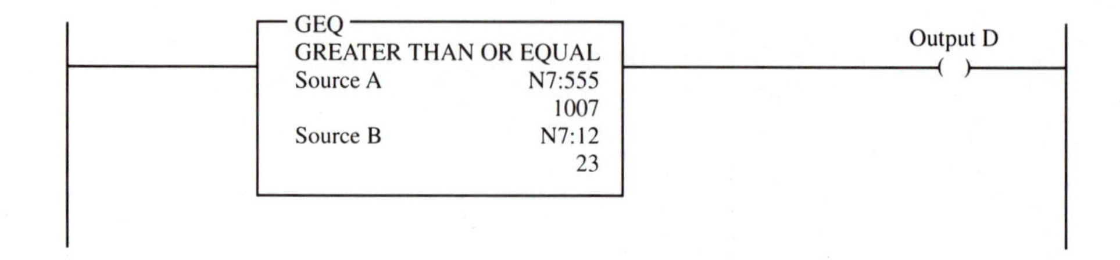

Figure 11-17 Example of the *Greater Than or Equal* Instruction

or a decimal value. The decimal value may be an integer value or a floating-point value.

The *limit test* instruction is said to be *circular,* which means it can function in either of two ways: (1) If the high limit has a greater value than the low limit, then the instruction is true if the value of the test is between or equal to the values of the high limit and the low limit. (2) If the value of the low limit is greater than

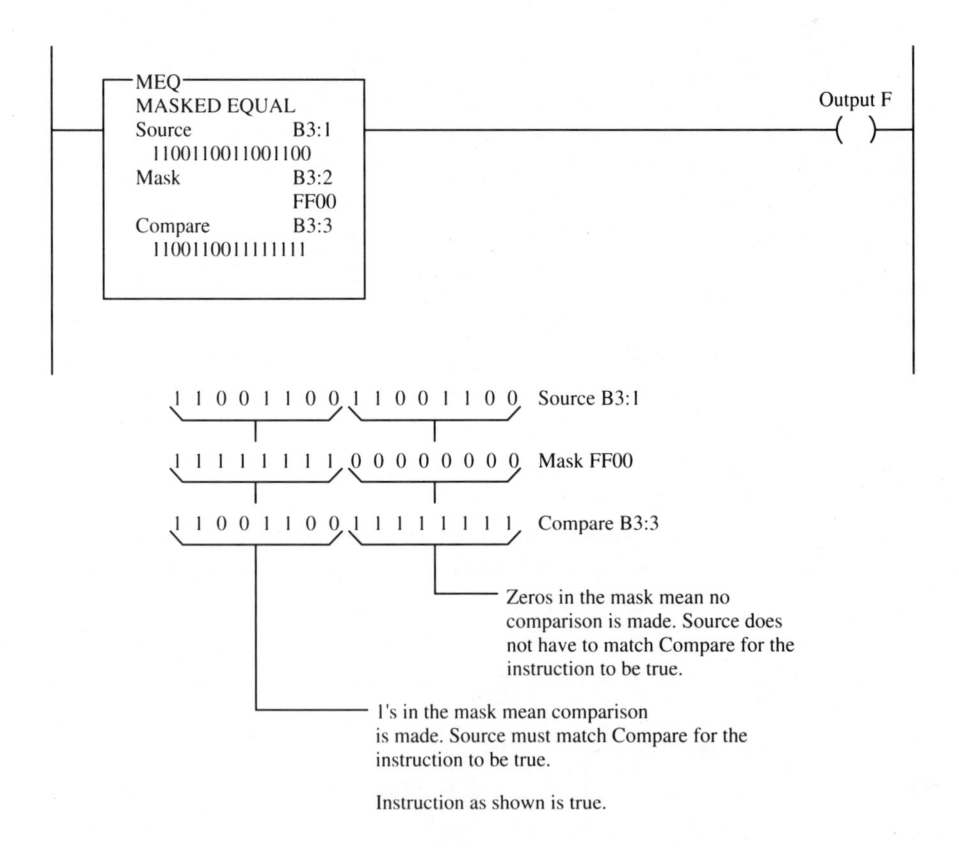

Figure 11-18 Example of the *Masked Comparison for Equal* Instruction

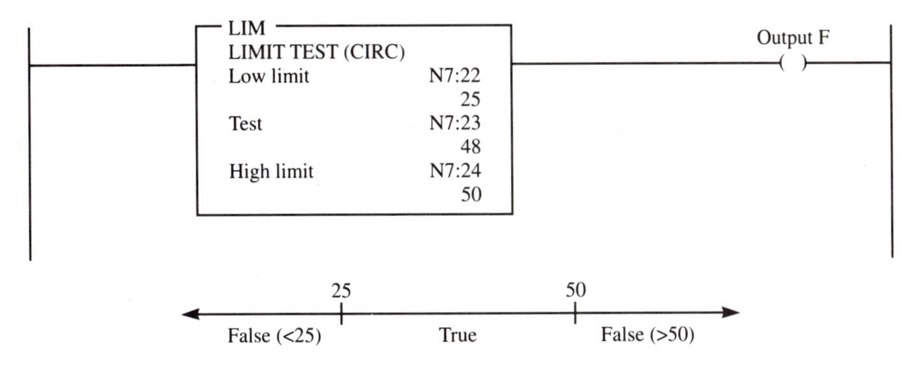

Figure 11-19 Example of the *Limit Test* Instruction

the value of the high limit, the instruction is true if the value of the test is equal to or less than the low limit or equal to or greater than the high limit.

In Figure 11-19, the high limit has a value of 50 and the low limit a value of 25. The instruction is true then for values of the test from 25 through 50. The instruction as shown in the figure is true because the value of the test is 48.

In Figure 11-20, the high limit has a value of 50 and the low limit a value of 100. The instruction is true then for test values of 50 and less than 50, and 100 and greater than 100. The instruction as shown in the figure is true because the test value is 125.

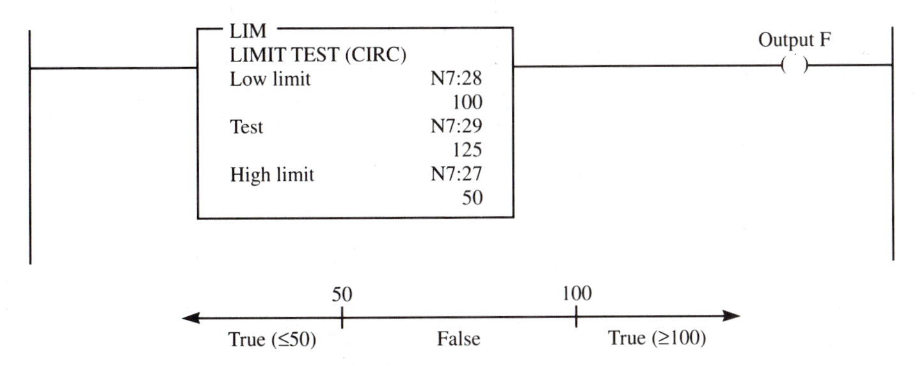

Figure 11-20 Example of the *Limit Test* Instruction

11-4 Word-level Logical Instructions

The following lists the logical instructions and their mnemonics:

Instruction	Mnemonic
And	AND
Or	OR
EXclusive OR	XOR
Not	

AND Instruction (AND)

The logical AND instruction ANDs Source A with Source B, bit by bit, and stores the result at the Destination address. The truth table for the logical AND operation is:

Source A	Source B	Destination
0	0	0
1	0	0
0	1	0
1	1	1

As you can see from the table, the only time you get a 1 in the Destination is when both A and B have a 1.

The AND instruction is exemplified in Figure 11-21. The source A address is B3:5, the source B address is B3:7, and the destination address is B3:10. The results at the destination are due to the instruction's being true. As you can see, the bits are the result of the logical AND operation. The instruction executes every scan while the instruction is true. The instruction may be programmed with input conditions preceding it, or it may be programmed unconditionally.

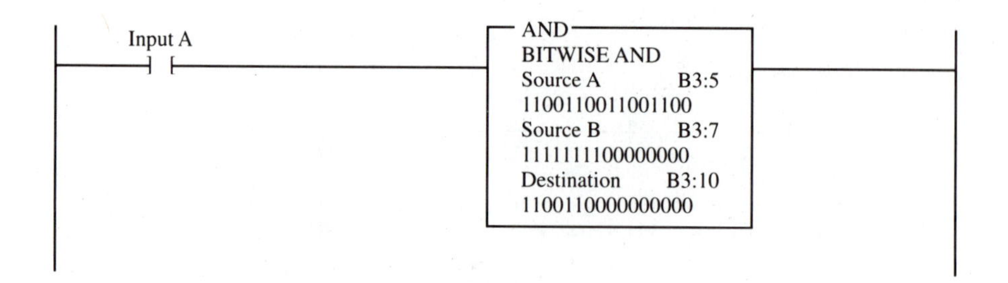

Figure 11-21 Example of the *AND* Instruction

OR Instruction (OR)

The logical OR instruction ORs the data in Source A, bit by bit, with the data in Source B and stores the result at the Destination address. The truth table for the logical OR operation is:

Source A	Source B	Destination
0	0	0
1	0	1
0	1	1
1	1	1

As you can see from the table, the only time you do not get a 1 at the destination is when both Source A and Source B are 0.

In the example of the logical OR instruction shown in Figure 11-22, the address of Source A is B3:1, the address of Source B is B3:2, and the Destination address is B3:20. The results at the Destination are due to the instruction's being true. The instruction executes during every scan in which it is true. The instruction may be programmed conditionally, with input instruction(s) preceding it, or unconditionally, without any input instructions preceding it.

EXclusive OR Instruction (XOR)

The logical XOR instruction EXclusive-OR's the data in Source A, bit by bit, with the data in Source B and stores the result at the Destination address. The truth table for the XOR operation is:

Source A	Source B	Destination
0	0	0
1	0	1
0	1	1
1	1	0

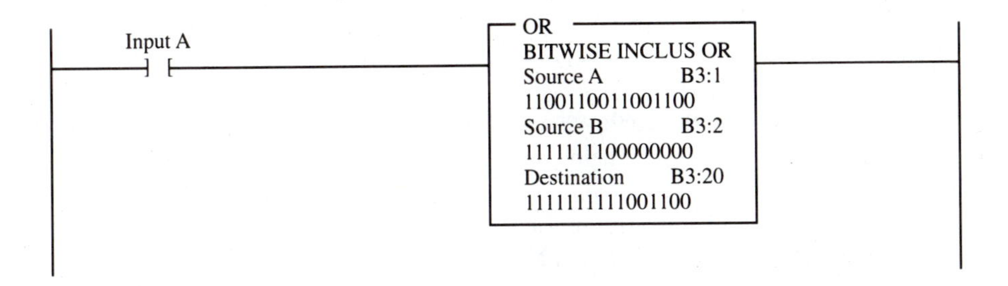

Figure 11-22 Example of the *OR* Instruction

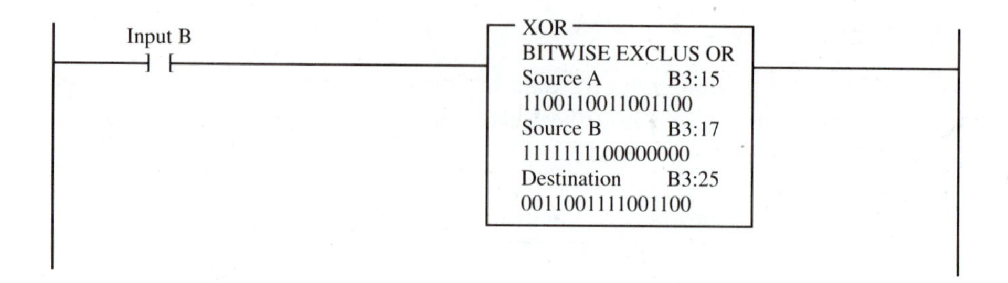

Figure 11-23 Example of the *XOR* Instruction

As you can see from the table, you get a 1 at the destination when Source A or Source B have a 1, but not when both have a 1.

In the example of the logical XOR instruction shown in Figure 11-23, the address of Source A is B3:15, the address of Source B is B3:17, and the address of the Destination is B3:25. The results at the Destination are due to the instruction's being true. As can be seen, there is a 1 in every bit location in the destination corresponding to the bit locations where Source A and Source B are different, and a 0 in the Destination where Source A and Source B are the same. The instruction executes during every scan in which it is true. The instruction may be programmed conditionally, with input instruction(s) preceding it, or unconditionally, without any input instructions preceding it.

The XOR is often used in diagnostics, such as comparing real-world inputs with their desired states. If Source A were a real-world input address representing sixteen input devices, such as limit switches, and the Source B address stored the desired status of those inputs, the Destination address would store the location where the actual inputs did not match their desired state.

NOT Instruction (NOT)

The logical NOT instruction inverts the bits from the Source A word to the Destination word, bit by bit. The truth table for the NOT operation is as follows:

Source A	Destination
0	1
1	0

As you can see from the table, the bits are inverted between Source A and the Destination, with the 0's being changed to 1's and the 1's being changed to O's.

In the NOT instruction exemplified in Figure 11-24, Source A's address is B3:25 and the Destination address is B3:27. The bit pattern in B3:27 is the result of the instruction's being true and is the inverse of the bit pattern in B3:25.

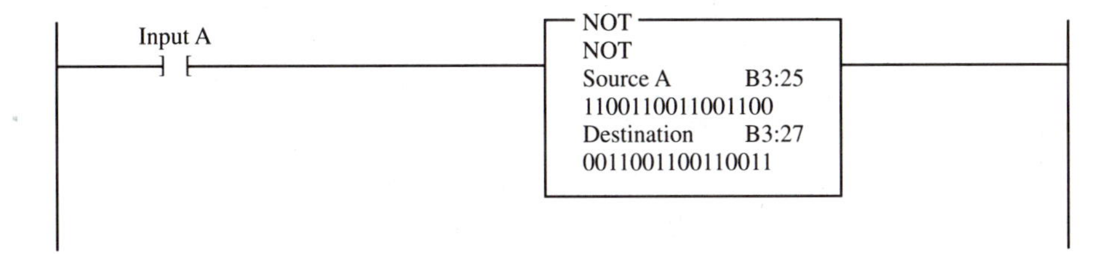

Figure 11-24 **Example of the *NOT* Instruction**

Summary

This chapter covered the word-level data-manipulation instructions. These include the moving of data, mathematical operation on data, data conversion, data comparison, and logical operation on data. The moving of data, math operations, data conversion, and logical operations are all output instructions, whereas the data-comparison instructions are input instructions.

Exercises

11-1. Develop a program that will add the values stored at N7:23 and N7:24 and store the result in N7:30 whenever input A is true, and then, when input B is true, will copy the data from N7:30 to N7:31.

11-2. Develop a program that will multiply the value stored in N7:21 by the constant 32 whenever input C is true. Store the result in N7:25.

11-3. Develop a program that will divide the value stored in N7:48 by 50 and store the result in F8:0 whenever input C is true.

11-4. Develop a program that will take the accumulated value from TON timer T4:1 and display it on a four-digit, BCD format set of LED's. Use address O:023 for the LED's.

11-5. Add to the program for Exercise 11-4 the ability to change the preset value of the timer from a set of four-digit BCD thumb wheels when input D is true. Use address I:012 for the thumb wheels.

11-6. Set up a program that will: (a) count the number of times PB1 is pushed and will reset the counter when the button has been pushed 15 times; (b) turn on pilot light PL1 when the button has been pushed five times; (c) turn on a second pilot light, PL2, when the counter accumulated value is greater than 0; and, (d) turn on a third light, PL3, when the counter accumulated value is between 8 and 12.

11-7. Is the *masked equal* instruction in Figure 11-25 true?

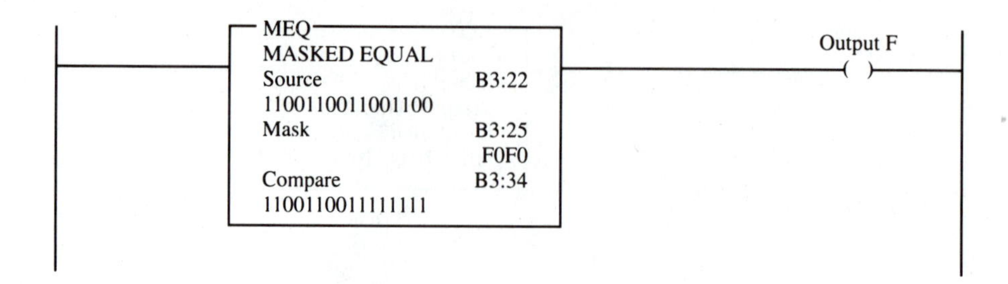

Figure 11-25 ***Masked Equal* Instruction, for Exercise 11-7**

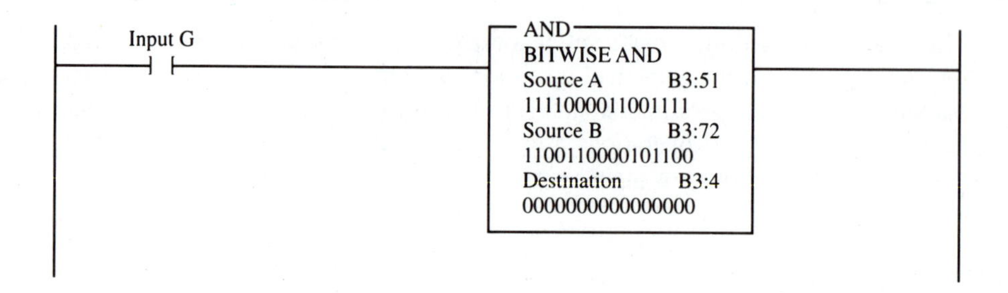

Figure 11-26 ***Bitwise AND* Instruction, for Exercise 11-8**

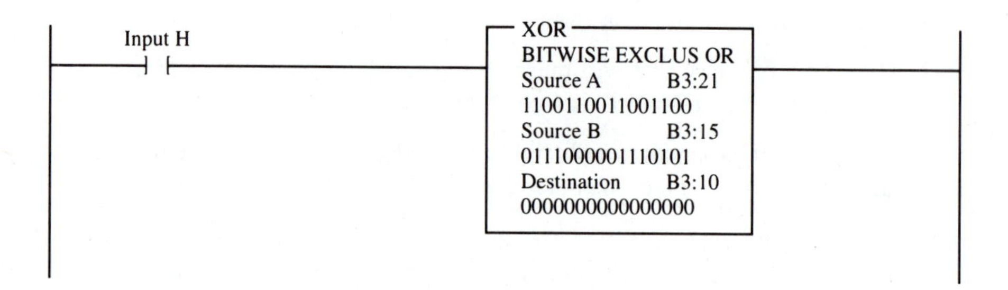

Figure 11-27 ***Bitwise XOR* Instruction, for Exercise 11-9**

11-8. In Figure 11-26, what will be the data stored in the destination address, B3:4, when input G is true?

11-9 In Figure 11-27, what will be the data stored in the destination address, B3:10, when input H is true?

File-Data Manipulation

12

OUTLINE

OBJECTIVES

Upon completion of this chapter, the student will be able to:

- Explain the concepts of *file* and *file addressing*, and describe the difference between a file address and a word address.
- Enter the appropriate parameters in the file instructions so as to accomplish desired functions.
- Use the mode of operation for the FAL instruction that is appropriate to a particular application.
- Describe the difference between file-to-file, word-to-file, and file-to-word operations of the FAL instruction, and enter the appropriate addresses in the instruction.
- Explain how the *file copy* instruction and *file fill* instruction differ from the FAL instruction.

FILE instructions deal with groups of words in the data files. First we will cover file concepts, and then we will go on to the basic file instructions. The main file instruction covered in this chapter will be the *file arithmetic and logic instruction* (FAL). With this single instruction, the processor copies data from one data file to another and performs the file arithmetic functions and the file logical functions. Two high-speed file instructions, the *file copy* instruction (COP) and the *fill file* instruction (FLL), also will be discussed.

12-1 File Concepts

As used in this section, *file* means a group of consecutive words in the data table with a beginning address and of a specified length. This use of the word *file* easily can cause confusion, for we have already discussed data files and program files. What we have defined here would better be described as a subfile within a data file. But convention has us defining this subfile as a file, and so we will follow that convention. The file may be a subfile of any one of the eleven data-table file types.

The starting address that defines the beginning of a file starts with a pound sign (#), whereas an address that starts without a pound sign is considered a word address. Figure 12-1 illustrates the difference between a file address and a word address. Address N7:20 is a word address, which represents a single word: word number 20 in integer file 7. Address #N7:30 represents the starting address of a group of consecutive words in integer file 7. The length shown is eight words, which is determined by the instruction in which the file address is used. This will be discussed later. The maximum file length is 1000 elements.

It is possible to overlap, or *nest,* files. For example, if you have a file #N7:20 that is 30 words long and another file #N7:30 that is 10 words long, then the file starting with address #N7:30 would be nested within the file starting with address #N7:20. A 20-word file starting with address #N7:15 would overlap file #N7:20. What isn't permitted is a file that overlaps a data-table file boundary. For example, you cannot have a 100-word file with a starting address of #N7:950, for it would extend beyond the maximum length of the integer file 7, which is word N7:999.

The exceptions to this rule that file addresses must take consecutive words in the data table are in the timer, counter, and control data files for the FAL instruction. In these three data files, if you designate a file address, the FAL instruction will take every third word in that file and make a file of preset, accumulated, length, or position data within the corresponding file type. For example, in Figure

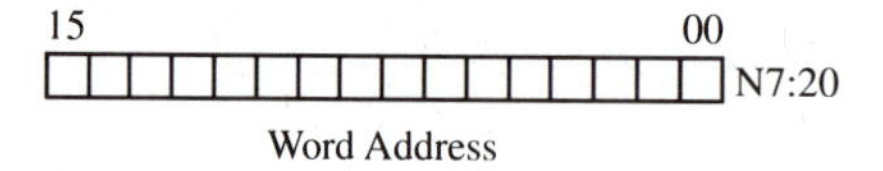

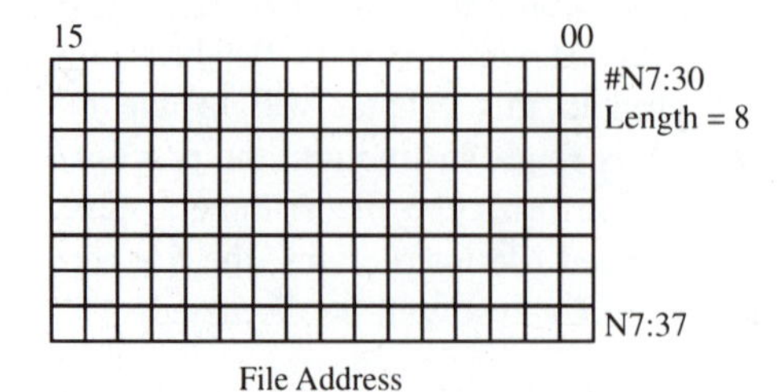

Figure 12-1 Comparison of Word and File Addresses

T4:0 Preset value	#T4:0.PRE Length = 4
T4:1 Preset value	
T4:2 Preset value	
T4:3 Preset value	

File of Timer Preset Values

C5:5 Accumulated value	#C5:5.ACC Length = 4
C5:6 Accumulated value	
C5:7 Accumulated value	
C5:8 Accumulated value	

File of Counter Accumulated Values

R6:3 Position value	#R6:3.POS Length = 4
R6:4 Position value	
R6:5 Position value	
R6:7 Position value	

File of Control Position Values

Figure 12-2 Files of Nonconsecutive Words

12-2, which presents a file of timer presets, a file of counter accumulated values, and a file of control positions, if the file address #T4:0.PRE is used, then every third word is accessed in the timer file, meaning the next word accessed in the file is T4:1.PRE. This might be done, for example, so that recipes storing values for timer presets can be moved into the timer presets (see Figure 12-3). Again, this holds true *only* for the FAL instruction. When used in other instructions, the file address will take consecutive words in the data table.

12-2 *File Arithmetic and Logic* (FAL) Instruction

The *file arithmetic and logic* instruction is used to copy data from one file to another and to do file math and file logic. The basic operation of the instruction is similar in all functions and requires certain parameters and addresses to be entered in the instruction. We will cover the instruction's basic functions in this section using a file-to-file copy as an example (this copies data from one file into another). In the following sections we will cover the other file operations using the FAL instruction but without repeating the basic operation each time.

An example of the FAL instruction is shown in Figure 12-4. The information that must be entered into the instruction when it is programmed is: the control element, file length and position, the mode of operation, the destination, and the expression. The function of each of these entries will be explained in the following.

Control Element

The first entry of the FAL instruction is the control element address, which will always be from a control data file. The default file for the control file is data file

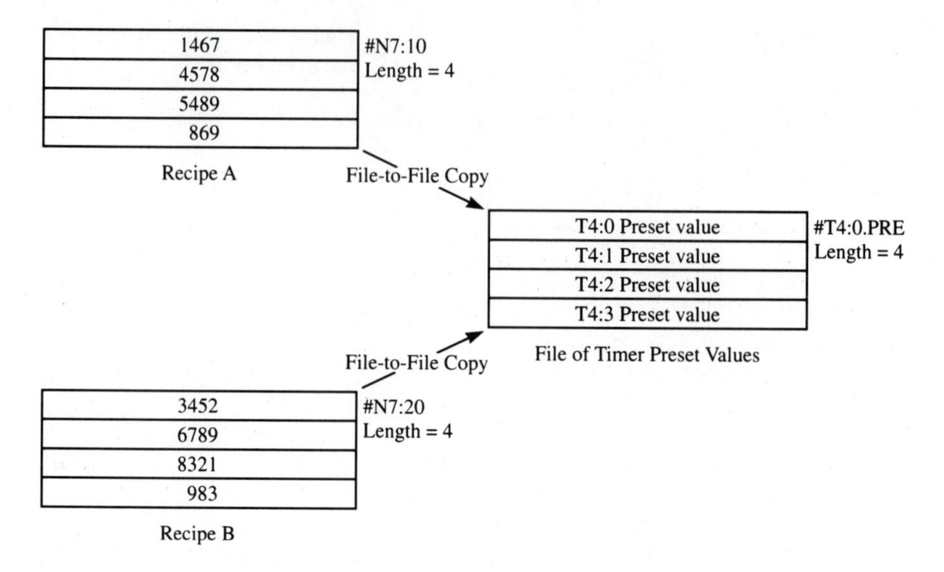

Figure 12-3 Copying Recipe Data into a File of Timer Presets

6. Other control files may be assigned from data files 9 through 999. The control element for the FAL instruction must be unique for that instruction and may not be used to control any other instruction.

The control element is made up of three words, as follows (listed here with an example for data file 6):

Bit:	15	13	11	10
Control word: R6:0	EN	DN	ER	UL
Length word: R6:0.LEN	Stores file length, 1–1000			
Position word: R6:0.POS	Position in the file			

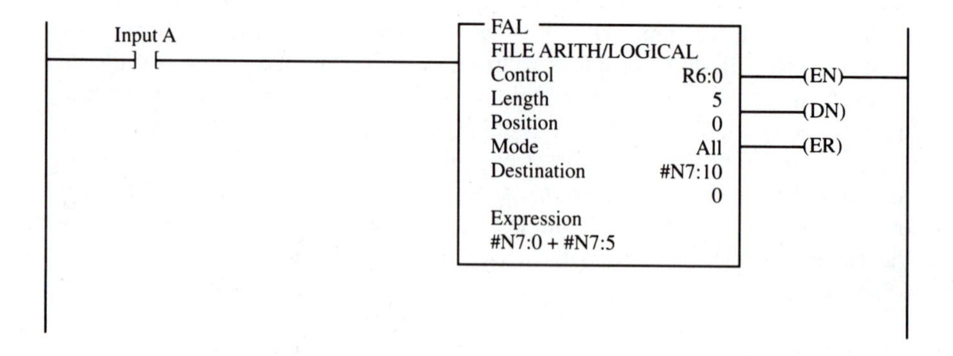

Figure 12-4 Example of the FAL Instruction

Control word

The control word in the FAL instruction uses four control bits: bit 15 (the enable bit), bit 13 (the done bit), bit 11 (the error bit), and bit 10 (the unload bit). The status of each of these bits during the operation of the instruction will be covered shortly, when the mode of operation of the instruction is discussed.

File length

The second entry (the second word of the control element) is the file length. This will be in words, except for the floating-point file, in which the length is in elements. (A floating-point element consists of two words.) The maximum length possible is 1000 elements.

File position

The third entry (the third word of the control element) is the position, that is the current word or element being operated on in the file. It functions to point to the word being operated on. The position starts with position 0 and indexes to 1 less than the file length. You may also enter another position at which you want the FAL to start its operation. However, when the instruction resets, it will reset the position to 0. You can manipulate the position from the program.

Mode of operation

Next, you enter the mode of operation. There are three choices: the *all mode,* for which you enter an *A;* the *numeric mode,* for which you enter the number of elements you want transferred per scan (anywhere from 1 to the file length); and the *incremental mode,* for which you enter an *I*. Descriptions of each mode follow. Examples of the different modes of operation, and when one is preferable over another, will be covered later as we go through the different FAL functions.

All mode

In the all mode, the instruction will transfer the complete file of data in one scan. The enable bit (EN) will go true when the instruction goes true and will follow the rung condition. When all of the data has been transferred, the done bit (DN) will go true. This will be on the same scan as when the instruction goes true. If the instruction does not go to completion, due to an error in the transfer of data (such as trying to store too large or too small a number for the data-table type), the instruction will stop at that point and set the error bit (ER). The scan will continue, but the instruction will not continue until the error bit is reset. If the instruction goes to completion, the enable bit and the done bit will remain set until the instruction goes false, at which point the position, the enable bit, and the done bit all will be reset to 0.

Numeric mode

If it is not necessary to have all of the data transferred in one scan, you may want to use the numeric mode. In the numeric mode, you enter a value from 1 to the file length that sets the number of elements to be transferred per scan. The numeric mode can decrease the time it takes to complete a program scan: Instead of waiting for the total file length to be transferred in one scan, the numeric mode breaks up the transfer of the file data into multiple scans, thereby cutting down on the instruction execution time per scan.

There are a number of ways in which the execution of the instruction may occur. When the instruction goes true, the EN bit will go true and the number of elements indicated in the mode will be operated on in successive scans. There are three basic possibilities: (1) The instruction remains true until all of the data is operated on. In this case, the DN bit will come on when all of the file data has been operated on, and the EN and the DN bits will remain set until the instruction goes false, at which time the EN and the DN bits will reset and the position will reset to 0.

(2) The second possibility: The instruction goes false before it reaches completion and remains false. In this case, the EN bit will remain true and the DN bit will go true on the scan when the last data in the file is operated on. The EN bit will go false on the same scan as the DN bit goes true. The DN will remain true for one scan and then go false, and the position will reset to zero.

(3) The third possibility is that the instruction goes false and then back to true when the instruction goes to completion. In this case, the EN bit will remain true throughout the operation, the DN bit will become true when the operation on the last data is complete, and the EN bit and the DN bit will remain true until the instruction goes false, at which time the EN bit, the DN bit, and the position will all be reset.

Incremental mode

The third mode of operation is the incremental mode. In this mode, one element of data is operated on for every false-to-true transition of the instruction. The first time the instruction sees a false-to-true transition, and the position is at 0, the data in the first element of the file is operated on. The position will remain at 0 and the UL bit will be set. The EN bit will follow the instruction's condition. On the second false-to-true transition, the position will index to 1 and data in the second word of the file will be operated on. Actually, the UL bit controls whether the instruction will just operate on data in the current position, or whether it will index the position and then transfer data. If the UL bit is reset, the instruction—on a false-to-true transition of the instruction—will operate on the data in the current position and set the UL bit. If the UL bit is set, the instruction—on a false-to-true transition of the instruction—will index the position by 1 and operate on the data in its new position.

Destination

The destination entry is the data file location where the data will be stored when the FAL instruction does its operation. It may be either a file address or an element address.

Expression

The expression entered determines the function of the FAL instruction. The expression may consist of file addresses, element addresses, or a constant, and may contain only one function, since the FAL instruction may perform only one function. The following sections will cover the various functions of the FAL instruction, grouped into file copy functions, arithmetic functions, and file logical functions.

12-3 File Copy Functions

There are three file copy functions: the *file-to-file copy,* the *element-to-file copy,* and the *file-to-element copy.*

File-to-File Copy

The example of the *file-to-file copy* function in the FAL instruction is shown in Figure 12-5. The destination and the expression must contain file addresses. In this example, data from the expression file, #N7:20, will be copied into the destination file, #N7:50. The length of the two files is set by the value entered in the control element word R6:1.LEN. Since this controls the length of both expression and destination files, they must have the same length. In this instruction we have also used the all mode, which means all of the data will be transferred in the first

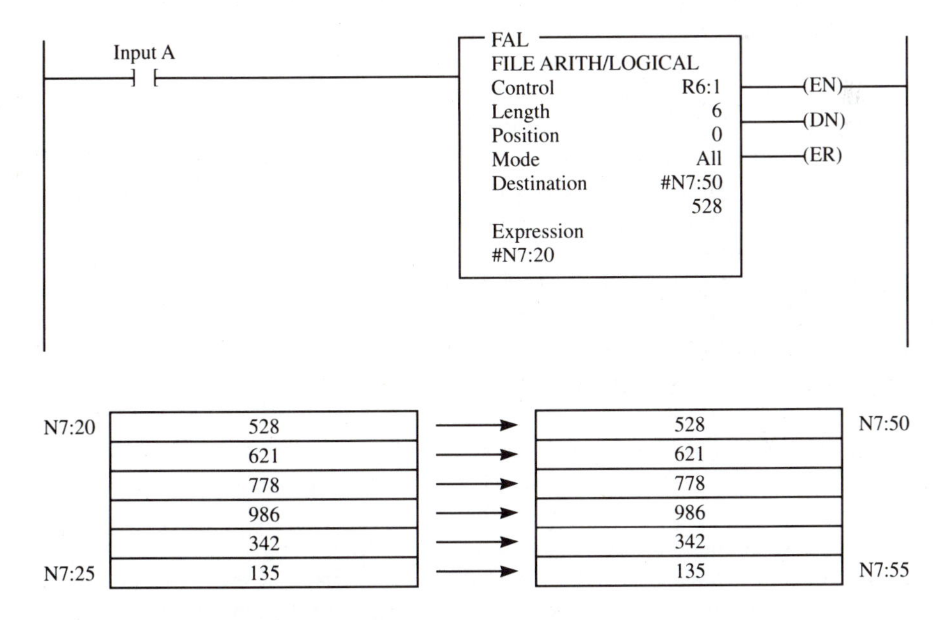

Figure 12-5 Example of *File-to-File Copy* (FAL)

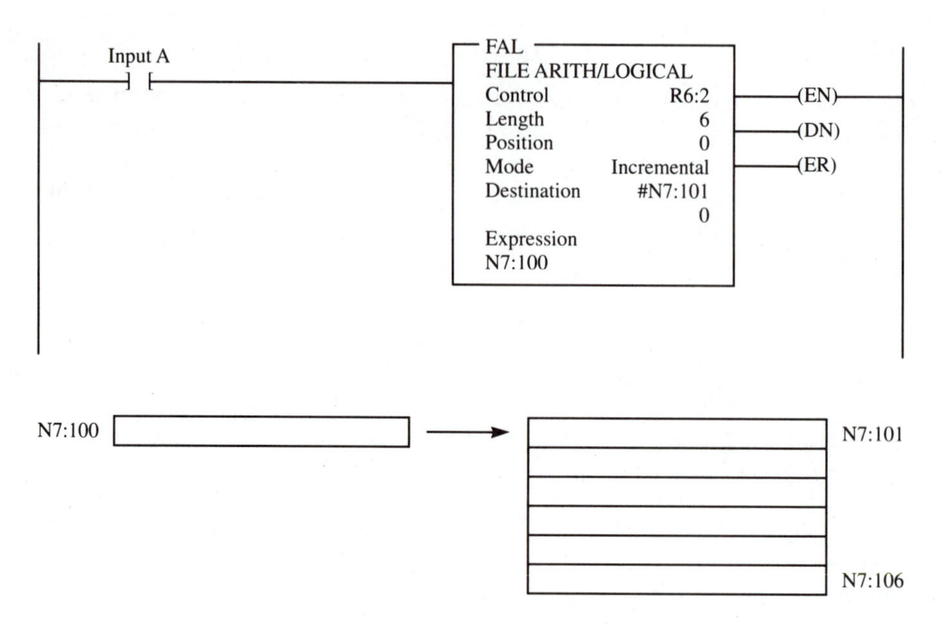

Figure 12-6 Example of *Element-to-File Copy* (FAL)

scan in which the FAL instruction sees a false-to-true transition. The DN bit will also come on in that scan, unless an error occurs in the transfer of data, in which case the ER bit will be set, the instruction will stop operation at that position, and the scan will continue at the next instruction.

Element-to-File Copy

The *element-to-file copy* function in the FAL instruction is exemplified in Figure 12-6. Note that now the expression is an element address, N7:100, and the destination is a file address, #N7:101. Also, the mode of operation is incremental. The data from N7:100 will be copied into a particular file location, which depends on the position and the status of the UL bit. If we start with position 0 and the UL bit reset, the data from N7:100 will be copied into N7:101 on the first false-to-true transition of instruction. The second false-to-true transition of the instruction will copy the data from N7:100 into N7:102. On successive false-to-true transitions of the instruction, the data will be copied into the next position in the file until the end of the file, N7:106, is reached. The DN bit will be set when the instruction is true and the position is 1 less than the length, which means all of the file data has been copied; and when the instruction goes false, the DN bit will be reset and the position will reset to 0.

In this example, the element address, N7:100, could be storing the current temperature in a tank. If the FAL instruction were indexed every minute, then the file would store the tank temperature for the last 6 minutes, at 1-minute intervals.

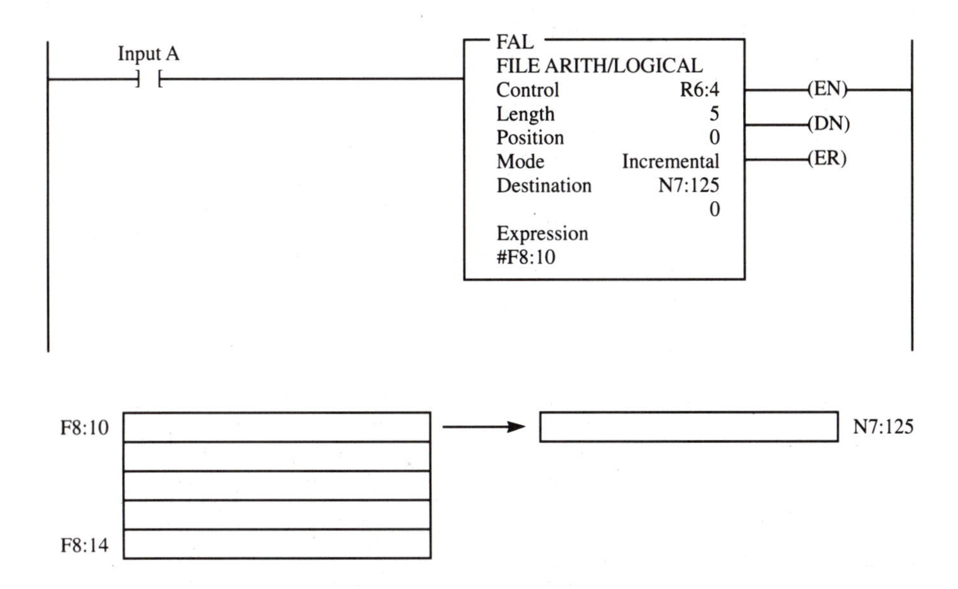

Figure 12-7 Example of *File-to-Element Copy* (FAL)

File-to-Element-Copy

In the example of the *file-to-element copy* function in the FAL instruction presented in Figure 12-7, the expression is a file address, #F8:10, and the destination address is an element address, N7:125. Note that two different file types are being used here. The data from the floating-point file will be copied into the integer element. When transferring data from a floating-point file type to an integer file type, the data will be rounded: 0.5 or greater will be rounded up, and less than 0.5 will be rounded down. Care must be taken not to exceed the ranges of the integer file. Also, the mode of operation of the instruction here is incremental, which means that one element of data will be copied from the file address of the expression to the element address of the destination for every false-to-true transition of the FAL instruction. If we start with position 0 and the UL bit reset, it will take five false-to-true transitions to transfer the data from the file address in the expression to the element address in the destination.

For the *file-to-element copy* instruction, the practical entries in the mode of operation would be either incremental or numeric with the rate of 1 per scan. In the all mode, the only data seen at the destination would be the last word of the file address in the expression, and a value other than 1 in the numeric mode would limit the data that could be accessed at the destination address.

12-4 File Arithmetic Functions

File arithmetic functions include *file add, file subtract, file multiply, file divide, file square root, file convert from BCD,* and *file convert to BCD.*

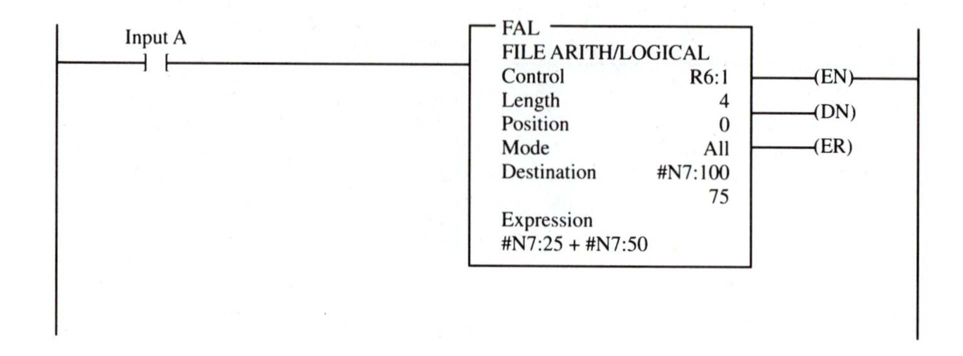

N7:25	25	+ N7:50	50	= N7:100	75
	234	+	22		256
	1256	+	456		1712
N7:28	77	+ N7:53	100	= N7:103	177

Figure 12-8 Example of *File Add* (FAL)

File Add

The *file add* function of the FAL instruction is given in Figure 12-8. The expression in this example adds the data in file address #N7:25 to the data stored in file address #N7:50 and stores the result in file address #N7:100. The expression will be in the format $A + B$, where A may be a file address, an element address, or a constant, and where the same is true for B, although it isn't practical for both A and B to be constants. The destination may be either an element address or a file address.

File Subtract

The *file subtract* function of the FAL instruction is shown in Figure 12-9. The expression in this example is a file address, #N10:0, that is having the constant 255 subtracted from it, with the result being stored at the destination file address, #N7:255. The rate per scan is set at 2, so it will take 2 scans from the moment the instruction goes true to complete its operation.

File Multiply

The *file multiply* function of the FAL instruction is exemplified in Figure 12-10. In the expression, the data in file address #N7:330 is multiplied by the data in element address N7:23, with the result being stored in file address #N7:500. The rate per scan is set at all, so the instruction goes to completion in one scan.

File Divide

An example of the *file divide* function of the FAL instruction is given in Figure 12-11. In the expression, the data in file address #F8:20 is divided by the data in file address #F8:100, with the result being stored in element address F8:200. The

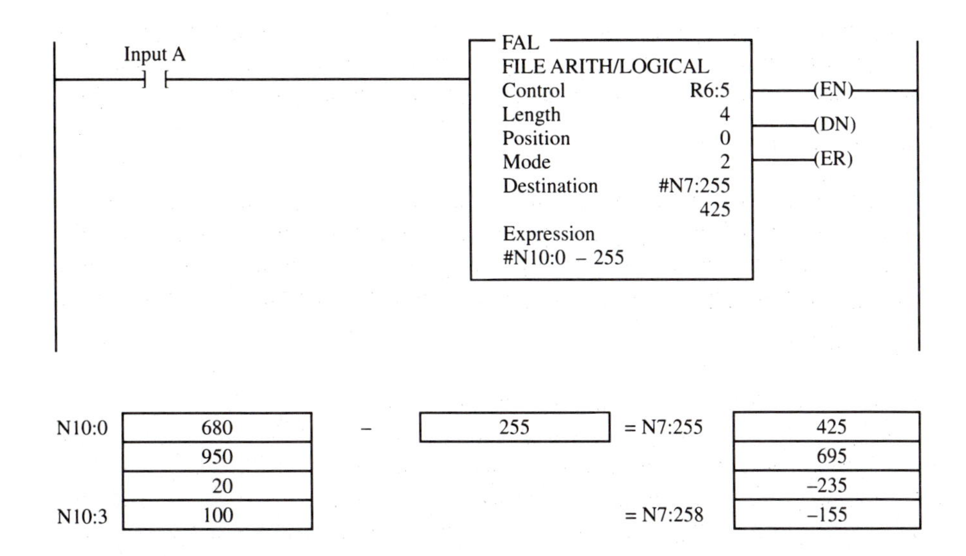

Figure 12-9 Example of *File Subtract* (FAL)

mode is incremental, so the instruction operates on one set of elements for each false-to-true transition of the instruction.

File Square Root

The *file square root* function of the FAL instruction is shown in Figure 12-12. The expression in the example is now just a single term, as indicated by the file address, #F8:100. The destination in the example is file address #F8:200. The

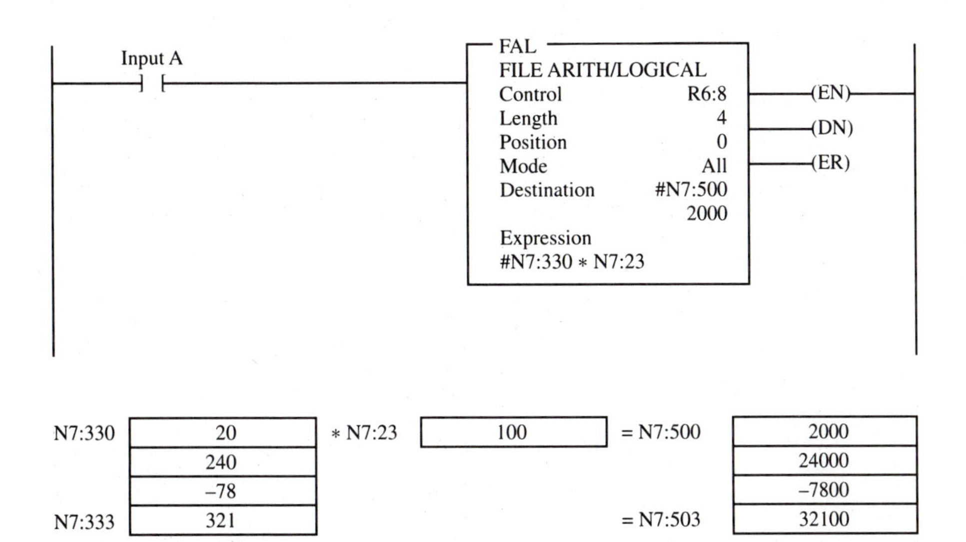

Figure 12-10 Example of *File Multiply* (FAL)

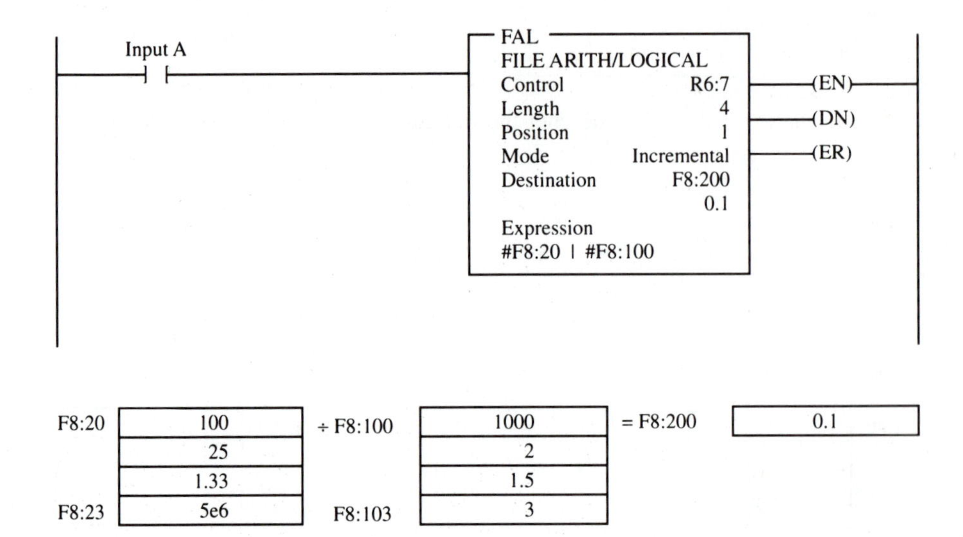

Figure 12-11 Example of *File Divide* (FAL)

expression may be a file address and the destination a file address, as illustrated, or the expression may be an element address and the destination a file address, or the expression may be a file address and the destination an element address.

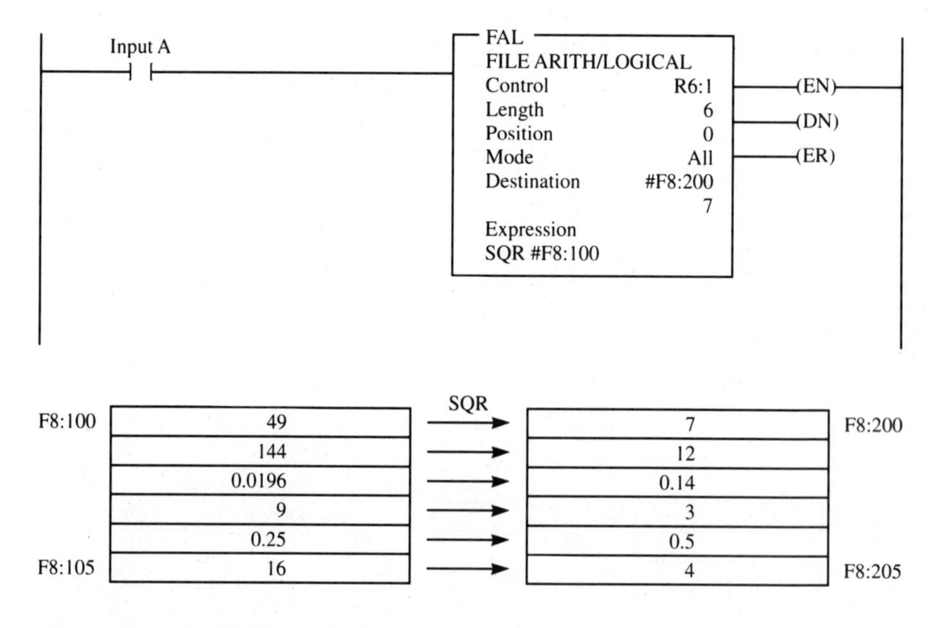

Figure 12-12 Example of *File Square Root* (FAL)

File Convert from BCD

The *file convert from BCD* function of the FAL instruction is shown in Figure 12-13. In the example, the data from the file address, #I:000 in the expression, is being converted from a BCD bit pattern to a binary bit pattern stored in the destination file address, #N7:0. This is often used to transfer data from an external source, which is a BCD bit pattern, to the processor data files, which are read by the processor in binary format. This transformation is necessary to retain the correct data value. The expression may be a file address or an element address, as may be the destination address. Both, however, should not be element addresses.

File Convert to BCD

The *file convert to BCD* function of the FAL instruction is exemplified in Figure 12-14. The data from the expression file address, #N7:150, will be converted from a binary bit pattern to a BCD bit pattern that will be stored at the destination file address, #O:05. This transformation of bit pattern is often necessary when transferring from the binary bit pattern the processor uses to an external address that uses a BCD bit pattern. The expression address may be a file address or an element address, as may be the destination address. Both, however, should not be element addresses.

12-5 File Logical Functions

File Logical AND

The *file logical AND* function of the FAL instruction is shown in Figure 12-15. The following is a truth table for the logical AND function:

0 AND 0 = 0

1 AND 0 = 0

0 AND 1 = 0

1 AND 1 = 1

For the AND logical function, as you can see, the only time you get a 1 in the result is when you AND two 1's.

The file logical AND will AND the data in file address #B3:100 with the data in file address #B3:200, bit by bit and element to element through the file, and store the result in the destination file address, #B3:255. The binary data-file type is most common, since its display is in a bit format. As in the previous functions, file addresses, element addresses, and constants may be used.

File Logical OR

The *file logical OR* function of the FAL instruction is shown in Figure 12-16. The truth table for the logical OR function is:

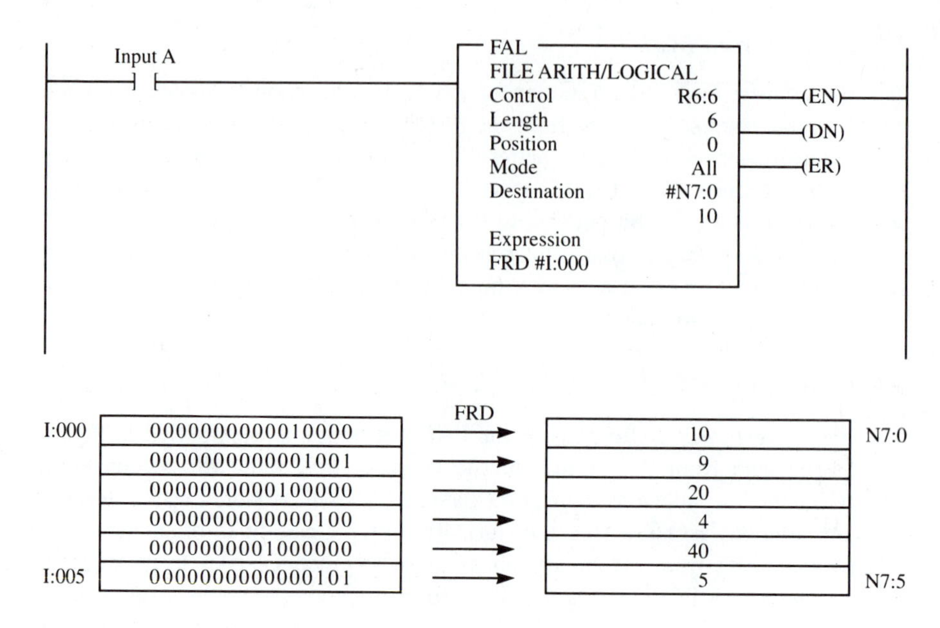

Figure 12-13 Example of *File Convert from BCD* (FAL)

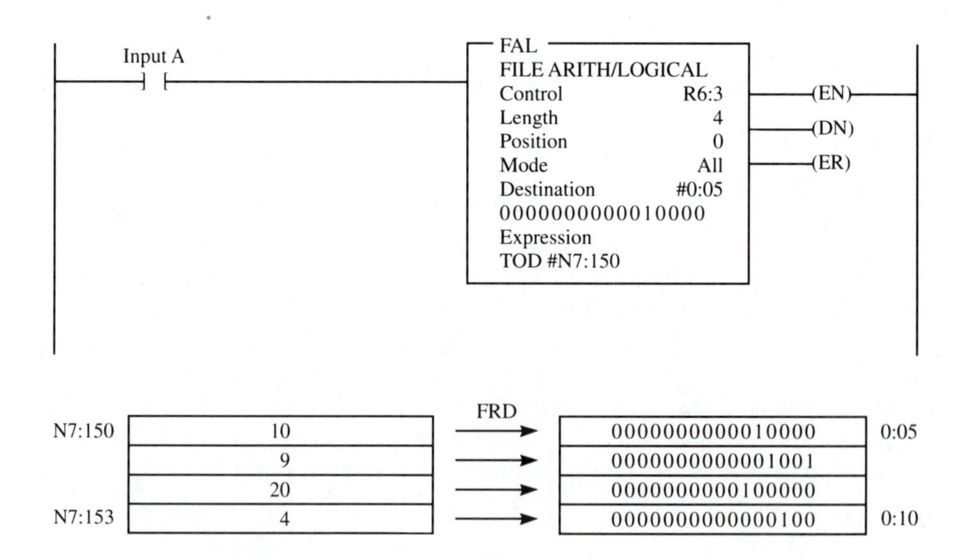

Figure 12-14 Example of *File Convert to BCD* (FAL)

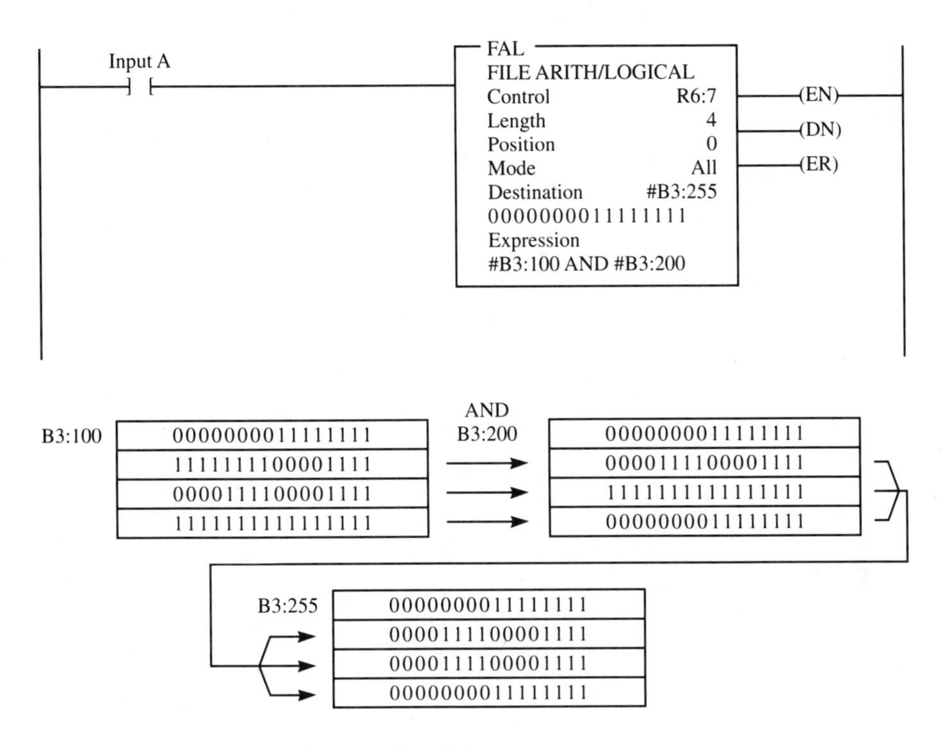

Figure 12-15 Example of *File AND* (FAL)

0 OR 0 = 0

1 OR 0 = 1

0 OR 1 = 1

1 OR 1 = 1

As you can see, the only time you do not get a 1 in the result is when you OR a 0 with a 0. In the example, the data in file address #B3:260 is OR'd with the data in file address #B3:280, with the result being stored in file address #B3:500.

File XOR

The *file XOR* function of the FAL instruction is presented in Figure 12-17. The truth table for the XOR instruction is:

0 XOR 0 = 0

1 XOR 0 = 1

0 XOR 1 = 1

1 XOR 1 = 0

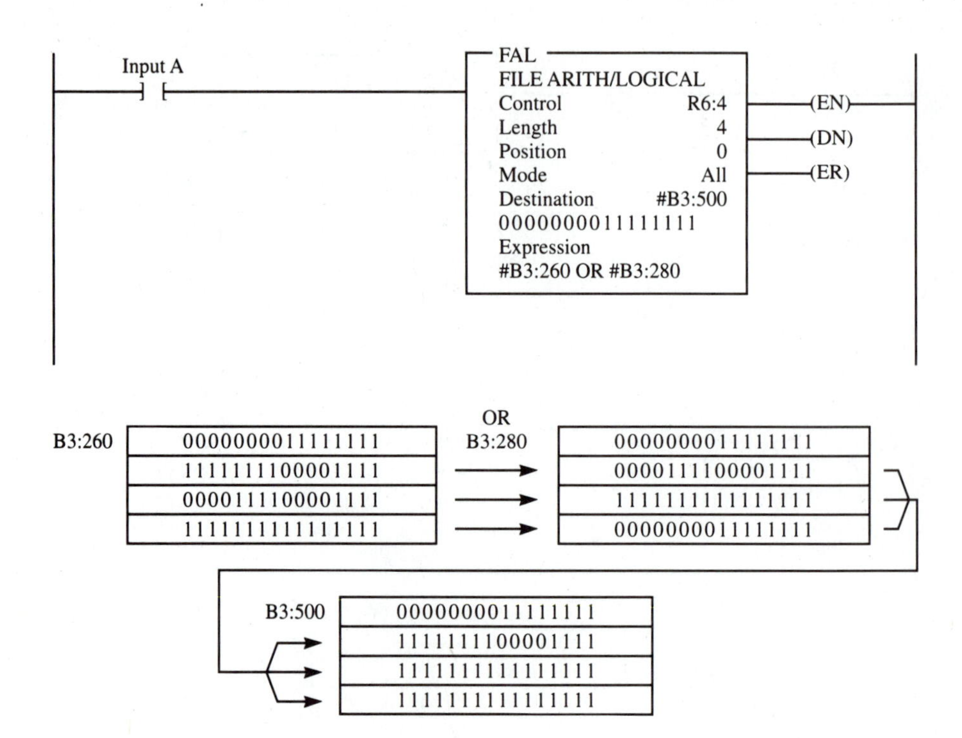

Figure 12-16 Example of *File OR* (FAL)

The only time you get a 1 in the result is when two different values are XOR'd. In the example, the data in file address #I:10 is XOR'd with the data in file address #B3:980, and the result is stored in data file address #B3:520.

File Logical *NOT*

The *file NOT* function of the FAL instruction is given in Figure 12-18. The truth table for the logical NOT function is:

NOT 0 = 1

NOT 1 = 0

As you can see, the NOT function inverts the bits. In the example, the data in file address #B3:120 is NOT'd, and the result is stored in the destination file address, #B3:140.

12-6 *File Copy* Instruction (COP) and *Fill File* Instruction (FLL)

The *file copy* instruction (COP) and the *fill file* instruction (FLL) are high-speed instructions that both use a similar format. Their advantage is that they operate at a higher speed than the same operation with the FAL instruction. Their dis-

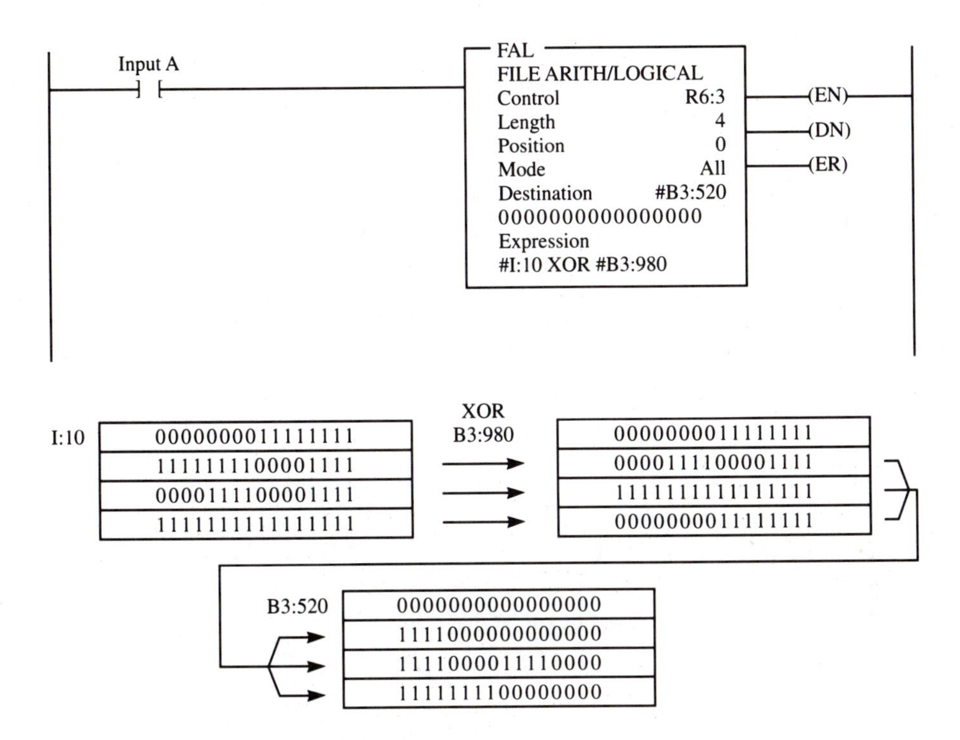

Figure 12-17 Example of *File XOR* (FAL)

advantages are that there is no control element to monitor or manipulate, data conversion does not take place so the source and destination should be the same file types, and they will not make a file of preset or accumulated values for timers, counters, or control elements like the FAL instruction does. The destination file type determines the length of the element, in words, that the instruction transfers.

File Copy (COP)

An example of the *file copy* instruction is shown in Figure 12-19. The *file copy* copies data from the source file address to the destination file address. In the COP instruction, both source and destination are file addresses. The length entered is the number of elements that will be copied from the source to the destination, with the destination data type determining the length in words. For example, if the destination is a counter or timer element, then the file length in words will be three times the length entered in the instruction. If the destination were an integer data type, the file length in words would be the same as the length in elements entered in the instruction. The instruction copies the file length each scan during which the instruction is true. It operates in a manner similar to the *file-to-file copy* of the FAL instruction.

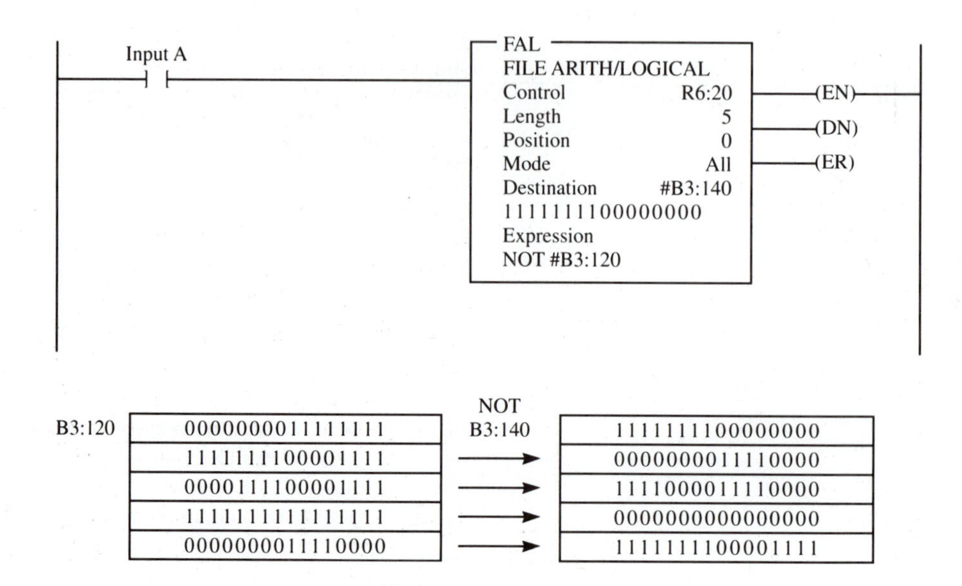

Figure 12-18 Example of *File Logical NOT* (FAL)

Fill File (FLL)

The *fill file* instruction (FLL), also a high-speed instruction, operates in a manner similar to the FAL instruction that performs the *element-to-file copy* in the all mode, except that it transfers the file length in every scan during which the instruction is true. See Figure 12-20. The source will be either an element address or a constant, and the destination will be a file address. Since the instruction transfers to the end of the file, the file will be filled with the same data value in each word. Thus, this instruction is frequently used to 0 all of the data in a file.

Summary

Files allow you to work with groups of words and make the program more efficient. The basic file instruction is the *file arithmetic and logic* instruction, which allows the copy, arithmetic, and logic functions to be accomplished with a single instruction, by changing the expression. Two additional instructions operate

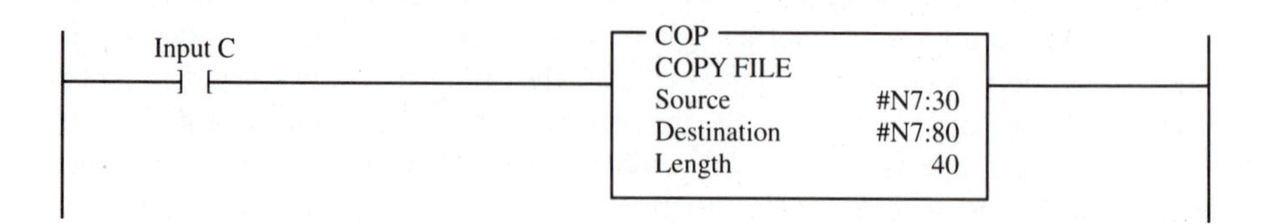

Figure 12-19 Example of the *File Copy* Instruction (COP)

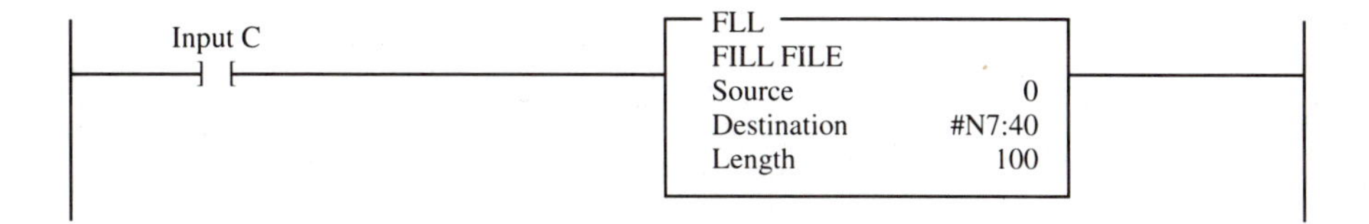

Figure 12-20 Example of the *Fill File* Instruction (FLL)

on file data at a faster rate than the FAL instruction: the *file copy* instruction (COP) and the *fill file* instruction (FLL). The COP instruction operates in a manner similar to the *file-to-file copy* function of the FAL instruction; the FLL instruction operates in a manner similar to the *element-to-file copy* function of the FAL instruction.

Exercises

12-1. Construct a program that will copy twenty elements of data from the integer data file starting with element N7:40 into the integer data file starting with element N7:80. The data will be copied whenever input A goes from false to true.

12-2. There are three timers that control the bake time in three separate ovens, one timer for each oven. The baked product will remain in each oven a specified time, controlled by the appropriate timer. There are three separate recipes to run through the ovens. The following table gives the bake times for each recipe.

Bake Times (seconds)

	Recipe		
	A	B	C
Oven A	10	23	33
Oven B	8	34	15
Oven C	22	9	17

The objective is to have an operator press push button PB1 for recipe A, push button PB2 for recipe B, and push button PB3 for recipe C. Each timer will be controlled by a separate input. Input A will control the timer for oven A, input B will control the timer for oven B, and input C will control the timer for oven C. You will not be able to change the presets if the input to any timer is true. Output A will control the heat to oven A, output B will control the heat to oven B, and output C will control the heat to oven C.

12-3. The temperature in an oven is stored in the processor's data table file address, N7:150. The temperature reading is brought in continuously, but it is desired to have a record of the temperature at 5-minute intervals for the previous 4 hours. Construct a program that will accomplish this.

12-4. Construct a program that will determine the average value of the accumulated value from four counters.

12-5. What is the maximum number of elements that can be transferred via the FAL instruction?

12-6. What will be the range of the position in the FAL instruction when compared to the length?

12-7. Why might you want to use the numeric mode in the FAL instruction rather than the all mode?

12-8. How many false-to-true transitions of the FAL instruction would it take to transfer 100 elements of data if the mode were incremental? How many false-to-true transitions would it take to do the same thing if the mode were numeric with the value 1 entered in the mode?

12-9. Construct a program that will compare 64 consecutive input bit addresses starting with I:013/00 to their desired states. Store their desired states in a binary file type starting with B3:100/00. Store the location of the differences in a binary file starting with B3:010/00.

12-10. To what FAL function is the COP instruction similar? What is the main advantage of using the COP instruction?

12-11. Construct a program that will bring in data from four sets of BCD four-digit thumb wheels—addresses I:001, I:002, I:003, and I:004—into the processor and change the data format to binary.

Shift Register and Sequencer Instructions

13

OUTLINE

OBJECTIVES

Upon completion of this chapter, the student will be able to:

- Describe the operation of the *bit shift left* and *bit shift right* instructions.
- Develop a program using the bit shift instructions to track parts on a manufacturing line.
- Describe the operation of the *FIFO load* and *FIFO unload* instructions.
- Develop a program using the *FIFO load* and *FIFO unload* instructions to track parts.
- Describe the operation of the *sequencer input, sequencer output,* and *sequencer load* instructions.
- Develop a program using the *sequencer output* instruction to control outputs in a sequential operation.
- Develop a program using the *sequencer input* instruction to make comparisons between real-world inputs and desired states, and then use the *sequencer input* instruction to control the *sequencer output* instruction.
- Develop a program using the *sequencer load* instruction to load the necessary data in a *sequencer input* file.

SHIFT register instructions are often used to track parts on automated manufacturing lines. This is accomplished by shifting either status or values through data files. When tracking parts on a status basis, bit shift registers are used. This chapter will cover the *bit shift left* instruction (BSL) and the *bit shift right* instruction (BSR). When tracking on a value basis, the "first in–first out" instructions are employed: the *FIFO load* instruction (FFL) and the *FIFO unload* instruction (FFU). Also used to track on a value basis are the *shift file* functions, which do not exist as processor instructions but may be created via programming. *Sequencer* instructions can program a sequence of operation and thereby greatly reduce the program length (compared to traditional programming methods). There are three sequencer instructions: sequencer output (SQO), sequencer input (SQI), and sequencer load (SQL).

13-1 Bit Shift Instructions

Bit shift instructions will shift bit status from a source bit address, through a data file, and out to an unload bit, one bit at a time. There are two bit shift instructions: *bit shift left* (BSL), which shifts bit status from a lower address number to a higher address number, through a data file; and the *bit shift right* (BSR), which shifts data from a higher address number to a lower address number, through a data file. The data file used for a shift register usually is the bit file, since its data is displayed in binary format, making it easier to read.

Bit Shift Left Instruction (BSL)

An example of the *bit shift left* instruction (BSL) is displayed in Figure 13-1. The first entry in the instruction is the file address through which the bit status is going to be shifted. In the *bit shift left* instruction, the data always enters the file at bit 00 of the first word of the file address. In the example, the file address is #B3:010, so the data enters the file at B3:010/00. At times this address is also shown as B3/160, which is just another way of displaying B3:010/00.

Next, the *control element* is entered. This controls the operation of the instruction. It is reserved for the instruction and cannot be used to control any other instruction. The three words that make up the control element are:

Bit:	15	13	11	10
Control word	EN	DN	ER	UL
Length word	Stores the length of the file, in bits			
Position word	Points to current bit			

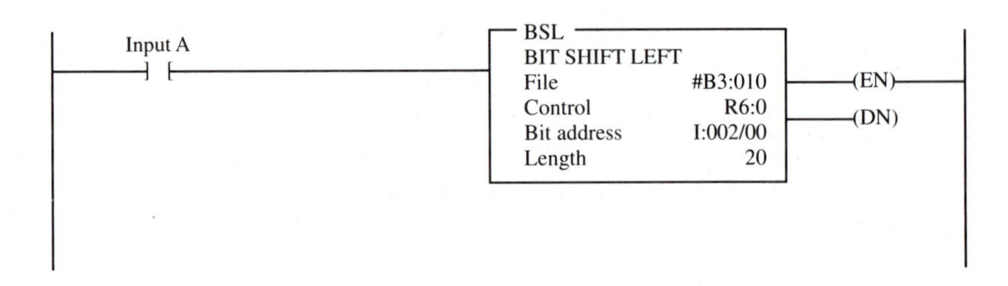

Figure 13-1 Example of the *Bit Shift Left* Instruction (BSL)

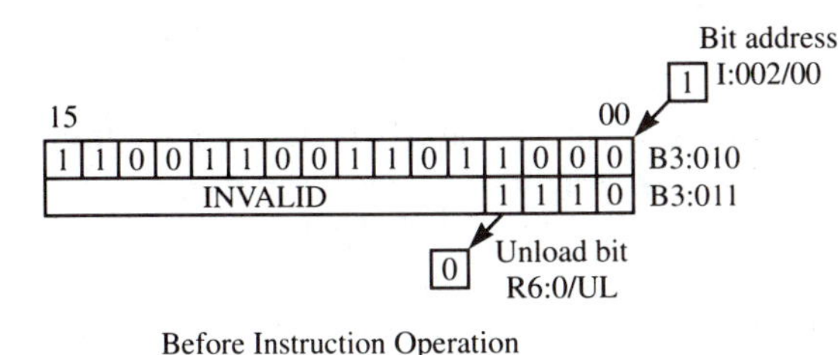

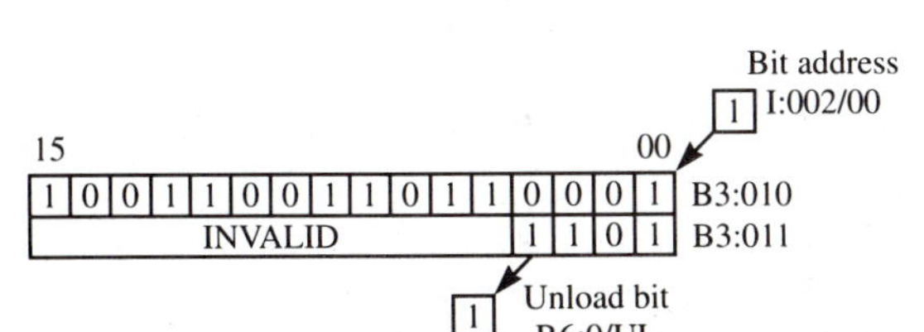

Figure 13-2 ***Bit Shift Left*** **Data File for Figure 13-1**

The enable bit, bit 15, is set when the instruction is true. The done bit, bit 13, is set when the instruction has shifted all of the bits in the file one position. It resets when the instruction goes false. The error bit, bit 11, is set when the instruction has detected an error. This can happen when a negative number is entered in the length. The unload bit, bit 10, is the bit location into which the status from the last bit in the file shifts when the instruction goes from false to true. When the next shift occurs, this data is lost, unless additional programming is done to retain it.

The next entry in the instruction is the *bit address,* which is the source bit from which status shifts into the file. The status of this bit is copied into the first bit in the file at each false-to-true transition of the BSL instruction.

The last entry is the *length,* that is, the file length, in bits. This may be up to 16,000 bits.

Figure 13-2 shows the data file layout for the *bit shift left* instruction in Figure 13-1. The file data is shown before the instruction goes true and also after the instruction has had a false-to-true transition. Before the instruction goes true, the status of the bits in words B3:010 and B3:011 is as shown. The status of the bit address, I:002/00, is a 1, and the status of the unload bit, R6:0/UL, is 0. When the instruction goes true, the status of the bit address, I:002/00, is shifted into bit 00 of word B3:010. The status of the bits in the file are all shifted one position to the left through the length of twenty bits. The status of bit 20, B3:11/03, is shifted to the unload bit, R6:0/UL. The status that was previously in the unload bit is lost. Note that all of the bits in the unused portion of the last word of the file are invalid and should not be used elsewhere in the program. The *bit shift left* instruction does affect the status of these bits.

Bit Shift Right Instruction (BSR)

The *bit shift right* instruction (BSR) operates in a manner similar to the *bit shift left* instruction, except it shifts the bit status from the highest address in the file

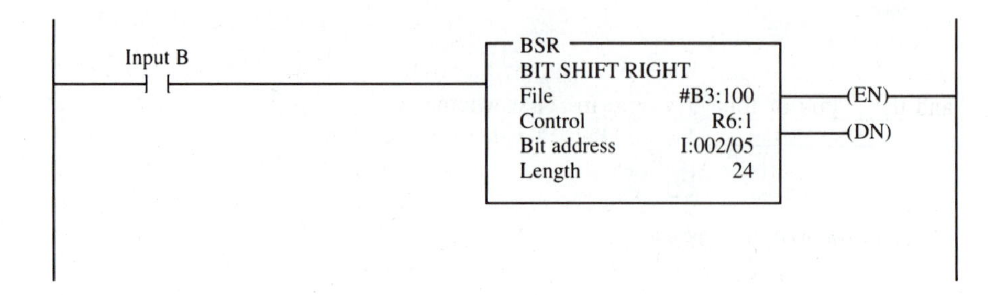

Figure 13-3 Example of *Bit Shift Right* Instruction (BSR)

towards the lowest address in the file. Figure 13-3 shows an example of the *bit shift right* instruction. The file address, the control element, the bit address, and the length are entered as in the *bit shift left*.

Figure 13-4 shows the data file status of the *bit shift right* instruction in Figure 13-3. The file data is shown before the instruction goes true and also after the instruction has had a false-to-true transition. Before the instruction goes true, the status of bits in words B3:100 and B3:101 are as shown. The status of the bit address, I:002/05, is a 0, and the status of the unload bit, R6:1/UL, is a 1. When the instruction goes true, the status of the bit address, I:002/05, is shifted into B3:101/07, which is the twenty-fourth bit in the file. The status of all of the bits in the file are shifted one position to the right, through the length of twenty-four bits. The status of B3:100/00 is shifted to the unload bit, R6:1/UL. The status that was previously in the unload bit is lost. Note that all of the bits in the unused portion of the last word of the file are invalid and should not be used elsewhere in the program.

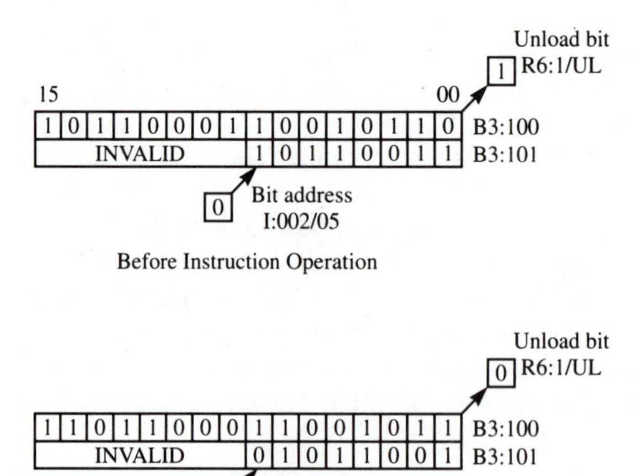

Figure 13-4 *Bit Shift Right* Data File for Figure 13-2

The *bit shift left* and *bit shift right* instructions can be used for tracking parts on a manufacturing line. Each bit location would represent a station on the line, and the status of the bit would indicate whether or not a part was present at that station. The bit address would detect whether or not a part had come on the line. The *bit shift left* normally would be used to indicate a forward motion of the line. If the line were reversed because of a fault, the *bit shift right* instruction could be used to indicate reverse motion. In this application, both the *bit shift left* and the *bit shift right* would have the same file address.

13-2 *FIFO Load* Instruction (FFL) and *FIFO Unload* Instruction (FFU)

FIFO stands for "first in–first out." Both of the FIFO instructions are output instructions, and they are used as a pair. The *FIFO load* instruction (FFL) loads data into a file from a source element; the *FIFO unload* instruction (FFU) unloads data from a file to a destination word.

Examples of the *FIFO load* and *FIFO unload* instructions are presented in Figure 13-5. The source in the FFL instruction is the word address location from which the data that is entered into the FIFO file comes. The FIFO file is the file address that stores the data entered from the source. The address for the FIFO file may not be from the timer, counter, control, or floating-point section of the data files. The FIFO file address for both the FFL and FFU instructions is to be the same. The destination is the word address to which the data is copied when the FFU instruction is executed.

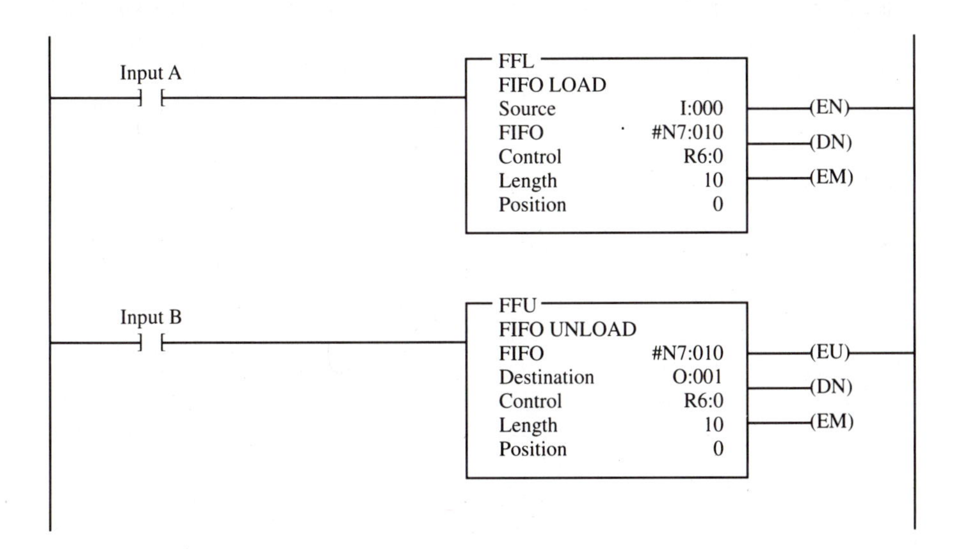

Figure 13-5 Example of the *FIFO Load* Instruction (FFL) and the *FIFO Unload* Instruction (FFU)

The *control element* stores the control bits, the FIFO length, and the position. Note that the *FIFO load* and *FIFO unload* instructions share the same control element, which may not be used to control any other instructions. The control bits in the FIFO control element are:

Bit:	15	14	13	12
	EN	EU	DN	EM

The EN bit is the *FIFO load* enable bit and follows the status of the FFL instruction. The EU bit is the *FIFO unload* enable bit and follows the status of the FFU instruction. The DN bit, the done bit, indicates that the position has reached the FIFO length, that is, that the FIFO file is full. When the DN bit is set, it inhibits the transfer of any additional data from the source to the FIFO file. The EM bit is set when the last piece of data entered from the source has been transferred to the destination and the position is 0. If the FFU has a false-to-true transition after the EM bit is set, zeroes will be loaded into the destination.

The *control length* is the length of the FIFO file, in words. The control position is the pointer in the FIFO file. It indicates where the next piece of data from the source will be entered and also how many pieces of data are currently entered in the FIFO.

The *destination* is the word address where the data goes as it is indexed from the FIFO file upon a false-to-true transition of the FFU instruction. Any data currently in the destination is written over by the new data when the FFU is indexed.

Figure 13-6 shows how data is indexed in and out of the FIFO file. Data enters the FIFO file from the source address upon a false-to-true transition of the FFL instruction. Data is placed at the position indicated in the instruction upon a false-to-true transition of the FFL instruction, after which the position will indicate the current number of data entries in the FIFO file. The FIFO file fills from the beginning address of the FIFO file and indexes to one higher address for each false-to-true transition of the FFL instruction. A false-to-true transition of the

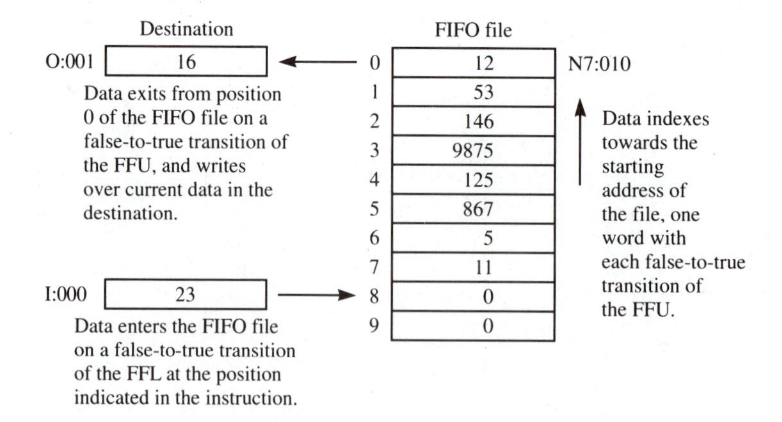

Figure 13-6 Data Transfer Using the FFL and FFU Instructions

FFU instruction causes all data in the FIFO file to shift one position toward the starting address of the file, with the data from the starting address of the file shifting to the destination address.

13-3 Sequencer Instructions

There are three sequencer instructions: the *sequencer output* instruction (SQO), the *sequencer input* instruction (SQI), and the *sequencer load* instruction (SQL). The *sequencer output* and *sequencer load* instructions are output instructions, and the *sequencer input* is an input instruction.

The *sequencer output* instruction can be used to control output devices in a sequential manner. The desired sequence of operation is stored in a data file, and this information then is transferred sequentially to the outputs. The *sequencer output* instruction functions in a manner similar to a mechanical drum switch, which is used to control output devices sequentially.

The *sequencer input* instruction makes comparisons between the states of input devices and their desired states: If the conditions match, the instruction is true. The *sequencer input* instruction is much like the *masked equal* instruction, except it will compare data at different locations in a file rather than on just a word basis like the *masked equal*. The *sequencer input* instruction can be programmed with additional input instructions in the rung.

The *sequencer load* instruction can serve as a teaching tool, and functions like a word-to-file move. A machine may be manually jogged through its sequence of operation, with its input devices being read at each step. And at each step the status of the input devices is written to the data file in the *sequencer input* instruction. As a result, the file is loaded with the desired input status at each step and this data is then used for comparison with the input devices when the machine is run in automatic mode. The *sequencer load* instruction is used to load the file, and does not function during the machine's normal operation. It replaces the manual loading of data into the file with the programming terminal.

Sequencer Output Instruction (SQO)

An example of the *sequencer output* instruction (SQO) is given in Figure 13-7. The *file address* is the first entry in the instruction. The file contains the data that will be transferred to the destination address when the instruction undergoes a false-to-true transition. Each word in the file represents a position, starting with position 0 and continuing to the file length. The actual file length will be one word longer than that indicated by the length in the instruction.

The *mask* is the next entry in the instruction. It may be entered as a hexadecimal value, a word address, or a file address. If a file address, the instruction will track the position of the file to the corresponding position of the mask. The mask serves to filter the data that passes from the file to the destination address: A 1 at a bit position in the mask allows data to flow through to the destination; a 0 in the mask blocks the flow of data from the file to the destination, and leaves the data in the corresponding destination bit position in its last state.

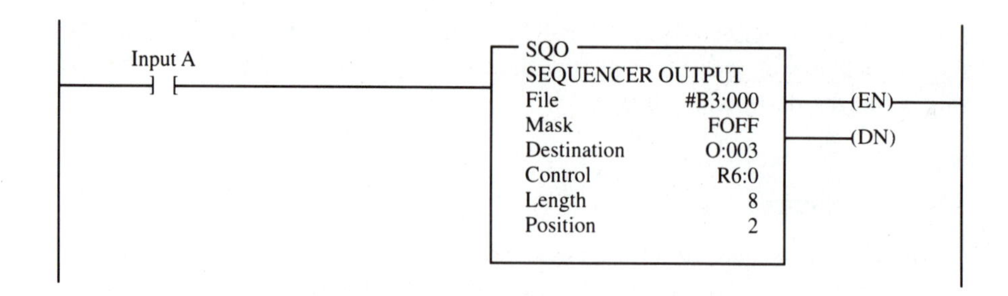

Figure 13-7 Example of the *Sequencer Output* Instruction (SQO)

The next entry in the instruction, the *destination,* is the address location where the data is written to when it is copied from the file. The destination may be a word address or a file address. If a file address, the position of the file and the position of the destination will be the same automatically. In most applications, the destination will be a word address.

The *control element,* the next entry, will be from an R data file. The control bits used will be the enable bit (EN), which is bit 15 and which will follow the instruction condition; the done bit (DN), which is bit 13 and which will be set on the false-to-true transition when the sequencer has operated on the last position; and the error bit (ER), which is bit 11 and which is set when the position has a negative value or the length has a negative or 0 value. The done bit will reset on the true-to-false transition once the instruction has operated on the last position. The length of the file is stored in the second word of the control element, and the position is stored in the third word of the control element.

The *length* is the next entry. The actual file length will be 1 plus the file length entered in the instruction.

The last entry is the *position.* Any value up to the file length may be entered, but the instruction will always reset to 1 on the true-to-false transition after the instruction has operated on the last position.

The addresses in the destination, mask, or file may only be from the data file types with a single word per element.

Figure 13-8 shows the flow of data in the operation of the *sequencer output* instruction. The example uses the instruction as given in Figure 13-7. Data is copied from the file, #B3:000, though the mask, at the bit locations where there is a 1 in the mask, to the destination, O:003, upon a false-to-true transition of the instruction. The position indexes one position, and the data is then copied. Once the position reaches the last position, then upon the true-to-false transition of the instruction, the position will reset to 1. Position 0 is executed under the following conditions: The position is at 0, the instruction is true, and the processor goes from the program to the run mode. Position 0 often is used as a home or starting position, with a 0 loaded in the position through program. Then when the instruction sees a false-to-true transition, it indexes to position 1 and copies the data from the first position in the file to the destination. On subsequent sequences, it will reset to position 1.

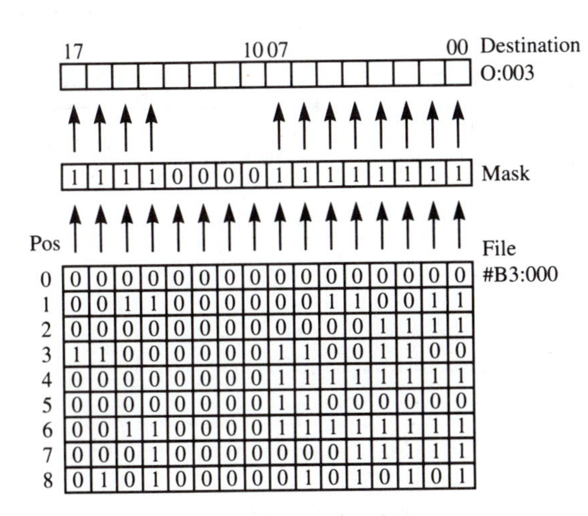

Figure 13-8 SQO Data Flow for Figure 13-7

Figure 13-9 shows the data from Figure 13-8 at the destination address, before and after the sequencer has indexed to the next position. Note that the data in O:003 matches the data in position 2 in the file, except for the data in bits 10 through 13. These bits may be controlled from elsewhere in the program, since they are not affected by the sequencer instruction due to the 0 in these bit positions in the mask. When the instruction is indexed to position 3, the data is copied from the file to the destination. The data in the destination then matches the data in the file at position 3, except for the data in bits 10 through 13, which has not changed.

Sequencer Input Instruction (SQI)

The *sequencer input* instruction (SQI) is exemplified in Figure 13-10. The entries in the instruction are similar to those in the *sequencer output* instruction, except the destination is replaced by the source. The *file address* is where the data that will be compared to the data at the source address is stored. The *source address* may be a word address or a file address. The comparison will be done through the *mask,* which may be entered as a hexadecimal value, a word address, or a file address. Wherever there is a 1 in the mask, the data in the file at that position must match the data in the source for the instruction to be true.

The *control element* functions somewhat differently in the *sequencer input* instruction. Since it is in an input instruction, it cannot be internally indexed, but

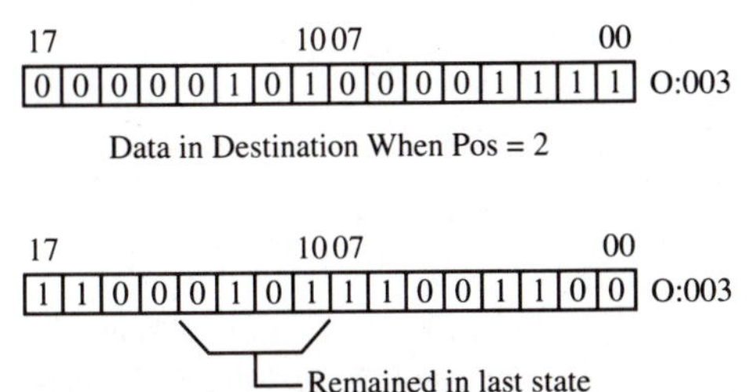

Figure 13-9 Destination Data (from Figure 13-8) Before and After the Instruction Has Indexed

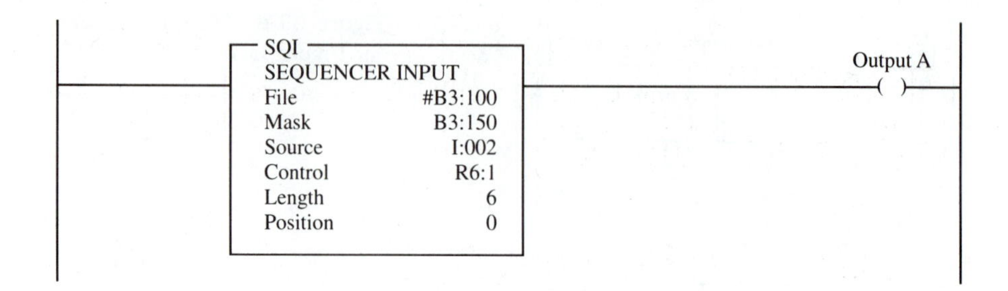

Figure 13-10 Example of the *Sequencer Input* Instruction (SQI)

must be indexed by other programming. Either data must be moved into the position of the control element, or another instruction—with the same control element—will have to index the position. This is frequently done via the *sequencer output* instruction.

Figure 13-11 presents the comparison between the data at the source address and the data in the file, through the mask. The example uses the instruction as given in Figure 13-10. The data in the source word is compared through the mask—at each bit position with a 1 in the mask—to the data in the current position in the file. For the instruction to be true, the data in the source must match the data in each bit position, wherever there is a 1 in the mask. In the example shown, the instruction would be true if it were in position 2, since the data in I:002 matches the data in position 2 except for bits 04 through 07 (the bit locations where there are 0's in the mask). Note that the instruction does not change any data, but just makes the comparison.

An example of a *sequencer input* instruction used in conjunction with a *sequencer output* instruction is given in Figure 13-12. Note that the control element for the *sequencer ouput* instruction has the same address as the control element for the *sequencer input* instruction. The *sequencer output* instruction is thus driving the *sequencer input* instruction. When the *sequencer input* instruction goes true (a true comparison is made between the source and the file word at the cur-

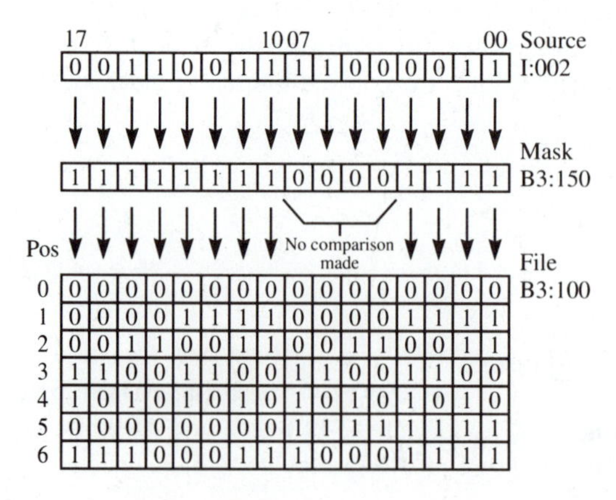

Figure 13-11 Data Comparison for the SQI Instruction in Figure 13-10

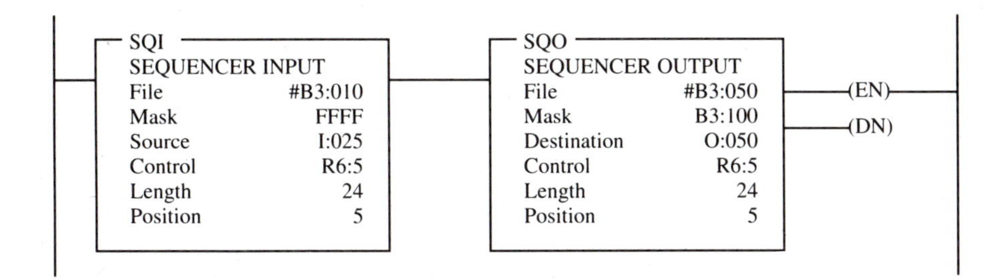

Figure 13-12 Example of a Sequencer Output Driving a Sequencer Input

rent position), the resultant false-to-true transition of the *sequencer output* instruction causes the sequencer output to index to the next step and to copy data from the file address to the destination address. The *sequencer input* position also changes as a result, so now it is making a comparison at the new position. It can be seen that a match at a position of the sequencer input triggers the next step at the sequencer output. This must be taken into consideration when loading data in the *sequencer input* and *sequencer output* files.

Sequencer Load Instruction (SQL)

An example of the *sequencer load* instruction (SQL) is given in Figure 13-13. The parameters entered in the instruction are similar to those entered in the SQI and SQO instructions. The *sequencer load* instruction does not use a mask. It copies data from the source address to the file. When the instruction goes from false to true, the instruction indexes to the next position and copies the data. When the instruction has operated on the last position, and has a true-to-false transition, it resets to position 1. It only transfers data in position 0, if it is at position 0 and the instruction is true, and the processor goes from the program to run mode.

The *sequencer load* instruction can become a teaching tool. Note that the address of the file and the address of the source of the *sequencer input* instruction in Figure 13-12 match those addresses in the *sequencer load* instruction in Figure 13-13. By manually jogging the machine through its cycle, the switches connected

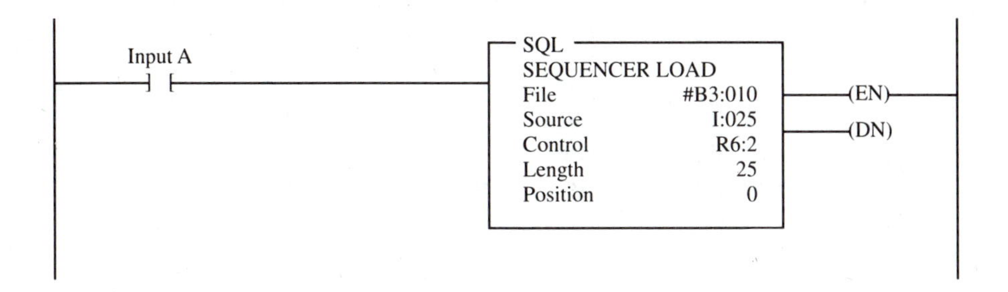

Figure 13-13 Example of the *Sequencer Load* Instruction (SQL)

to the input addresses of the source can be read and written into the file by the *sequencer load* instruction at each position. Thus, the file data for the *sequencer input* instruction is loaded by use of the *sequencer load* instruction.

Summary

In this chapter we covered the shift register and sequencer instructions. Bit shift registers index bit status through a data file and can be used for tracking parts on a manufacturing line. The FIFO instructions shift data through a file, indexing data out of the file in the order in which it is put in (first in–first out). The *sequencer output* instruction can be used to control outputs, in a sequential manner, by copying data from a file to a destination. The *sequencer input* instruction makes comparisons, in a sequential manner, from a source to data stored in a file. The *sequencer load* instruction copies data from a source to a file and can be used to load data in *sequencer input* or *sequencer output* instruction files.

Exercises

13-1. The length of the bit shift instructions is given in what units?

13-2. What is the bit address of the unload bit?

13-3. Construct a program that will keep track of the presence of parts on a 23-station conveyor line. If a part is placed on the line, then a limit switch connected to input address I:002/15 will close. The conveyor will be indexed by pushing a push button connected to input I:013/12. An indicator light connected to output address O:003/15 will turn on when a part comes off the line.

13-4. Add to the program in Exercise 13-3 the ability to track the parts if a jam-up occurs and the conveyor must be reversed. Load the file with a 0 stored in B3:000/00. A push button connected to input I:014/16 will index the conveyor in reverse.

13-5. What must be true for the *FIFO load* and the *FIFO unload* control element addresses?

13-6. Construct a program that will keep track of the order in which alarm codes are received. The alarm codes will be written into word address N7:20, and input B3:003/10 will index the alarm codes into the file. Input B3:003/11 will index the alarm codes out of the file and display them on a set of LED's connected to output address O:015. The file will be able to store twenty alarms.

13-7. When the *FIFO unload* instruction unloads the last data that has been entered by the *FIFO load* instruction, the next false-to-true transition will cause a 0 to be unloaded to the destination address. Program the *FIFO unload* instruction in Exercise 13-6 so as to prevent this.

13-8. Construct a program that will turn on the outputs listed, in the following sequence. The output sequence will allow 10 seconds between steps. Mask out the unused bits in the output word. Construct a table showing the data that has to be entered in the data file.

Step	Outputs Energized
Step 1	O:015/10, O:015/14, O:015/17
Step 2	O:015/00, O:015/07
Step 3	O:015/03
Step 4	O:015/05, O:015/13, O:015/16, O:015/17

13-9. Construct a program using a *sequencer output* instruction driving a *sequencer input* instruction that meets the following criteria. Mask out all unused input and output bits. Construct a chart to show the data that must be entered in the *sequencer input* and *sequencer output* files.

Inputs *True* to Cause Outputs to Index Indicated Output Step	Output Step	Outputs *True* at Indicated Output Step
I:002/00, I:002/10	1	O:015/15, O:015/17
I:002/11, I:002/15	2	O:015/04
I:002/11	3	O:015/03, O:015/13
I:002/05, I:002/07	4	O:015/10
I:002/04	5	O:015/11, O:015/16

13-10. Program a *sequencer load* instruction that will load the data into a *sequencer input* instruction file, #B3:100, with a length of 20. The source address for the *sequencer input* instruction is I:016. Index the *sequencer load* instruction with input I:017/10.

14 Program Control Instructions

OUTLINE

OBJECTIVES

Upon completion of this chapter, the student will be able to:

- State the purpose of program control instructions.
- Describe the operation of the *master control reset* instruction (MCR), and develop an elementary program illustrating its use.
- Describe the operation and limitations of the *jump* instruction (JMP) and the *label* instruction (LBL), and develop an elementary program illustrating their use.
- Explain the functions of subroutines, and design an elementary program involving subroutines.
- Describe the passing of parameters when using subroutines, and incorporate them in a program.
- Describe the function of the selectable time-interrupt file, and use such a file in a program.
- Describe the function of the fault routine file, and explain how such files are programmed.
- Explain the use of *immediate input* instruction (IIN) and the *immediate output* instruction (IOT), and implement them in a program.

THE program control instructions allow for greater program flexibility and greater efficiency in the program scan. Portions of the program that are not being utilized at any particular time can be jumped over, and outputs

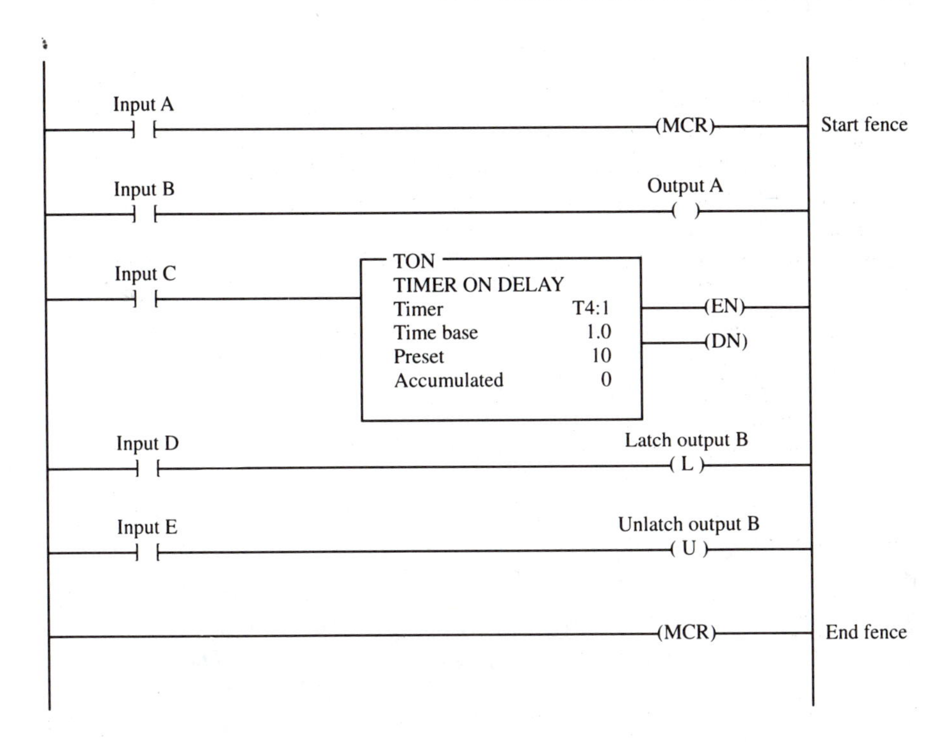

Figure 14-1 Example of the *Master Control Reset* Instruction (MCR)

in specific zones in the program can be left in their desired states. The instructions allowing for program control are the *master control reset* (MCR), the *jump* (JMP), and the *jump to subroutine* (JSR). Two special subroutine files are the *selectable timed-interrupt* program file and the *fault routine* program file. There are two additional instructions that alter the way in which the program is scanned: the *immediate input* instruction (IIN) and the *immediate output* instruction (IOT).

14-1 *Master Control Reset* Instruction (MCR)

The *master control reset* (MCR) instruction sets up a zone or multiple zones in a program. It is an output instruction, is used in pairs, and has no address.

Figure 14-1 shows the programming of a typical MCR zone. The MCR zone is enclosed by a *start fence,* which is a rung with a conditional MCR, and an *end fence,* which is a rung with an unconditional MCR. When the MCR in the start fence rung is true, the scan proceeds through the MCR zone (as normally), with the states of the outputs being dependent on the rung logic of the program. When the MCR in the start fence is false, all rungs within the zone are treated as false. The scan ignores the inputs and deenergizes all nonretentive outputs (that is, the *output energize* instruction, the on-delay timer, and the off-delay timer). All retentive devices, such as latches, retentive timers, and counters remain in their last state.

In Figure 14-1, input A controls the state of the MCR zone. When input A is true, the states of output A, T4:1, and output B will depend on the input conditions in their respective rungs. When input A is false, output A and T4:1 will be false and output B will remain in its last state. The input conditions in each rung will have no effect on the output conditions.

Multiple MCR zones may exist in a program, or the entire program could be enclosed in an MCR zone. MCR zones are not to be nested, nor are they to be overlapped.

A typical application of an MCR zone is when a fault bit or multiple fault bits are examined in the start fence, and a portion of the program containing outputs you want deenergized in case of a fault are enclosed in the MCR zone. In case of a fault, the outputs in that zone would be deenergized.

14-2 *Jump* Instruction (JMP) and *Label* Instruction (LBL)

The *jump* instruction (JMP) and the *label* instruction (LBL) are employed together so the scan can jump over a portion of the program. The *label* is a target for the *jump*, is the first instruction in the rung, and is always true. A *jump* jumps to a label with the same number. There can be up to 32 labels in each program file, numbered from 0 through 31. Each label in the program file must have its own unique label number. You can jump to the same label from multiple jump locations; in other words, there can be multiple jumps with the same number. The *jump* instruction will not jump between program files. It is possible to jump backwards in the program, but care must be taken that the scan does not remain in a loop too long, for the processor has a watchdog timer that sets the maximum allowable time for a total program scan, and if this time is exceeded, which is easy with a loop, the processor will have a major fault and shut down.

Figure 14-2 shows a program example using the *jump* and *label* instructions. There are two *jump* instructions numbered 1. There is a single label numbered 1. The scan can then jump from either *jump* instruction to label 1. If input A is true, jump 1 will also be true, causing the scan to jump to label 1. Since output A, T-4:1, and output C are unscanned if they are jumped over, they will be left in their last state and will not be affected by the input conditions in their rungs. Output D will be controlled by input F, since the *label* instruction is always true. If input A is false and input D is true, then the scan will jump from the *jump* instruction in the same rung as input D to label 1, leaving output C in its last state. When this is true, output C will not be controlled by input E.

14-3 Subroutines

A *subroutine* is a section of program that is not scanned unless there is a specific request for the scan to go to that particular subroutine. Subroutines are located in different program files from the main program: The main program is located in

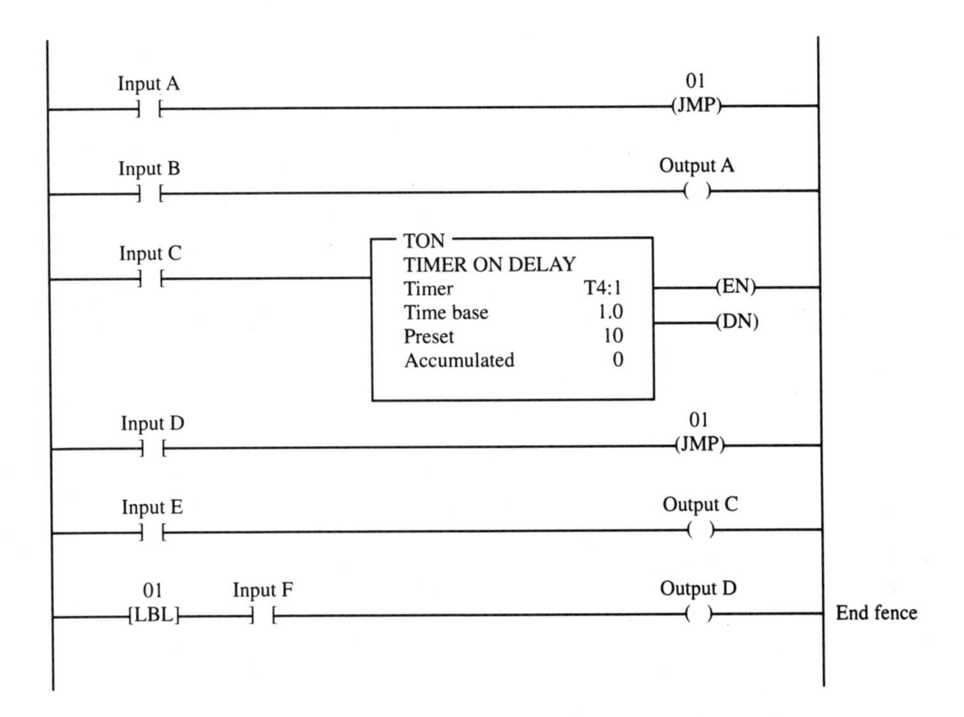

Figure 14-2 Example of the joint use of the *Jump* (JMP) and *Label* (LBL) Instructions

program file 2, whereas subroutines are assigned to program files 3 through 999. Subroutines may be accessed either on an event basis or on a time basis.

The *jump to subroutine* instruction (JSR), when true, tells the scan to jump to the subroutine file indicated in that instruction. The scan then jumps from the JSR instruction to the first rung in the subroutine program file and scans that file until it reaches a true *temporary end* instruction, a true *return instruction* (RET), or the end of file. The *return* instruction is an output instruction, either conditional or unconditional, that causes the scan to return from the subroutine to the instruction after the JSR instruction from which it jumped, with the scan continuing from there. The JSR instruction can be located in the main program file or in any subroutine file, where one subroutine can call another subroutine. Subroutines can be nested up to eight levels deep before returning to the main program.

Figure 14-3 shows an example of jumping to a subroutine from the main program. When the JSR instruction is true, the program scan will go from the JSR instruction to the program file indicated in that instruction. It will jump to the first instruction in the first rung in that file, which in Figure 14-3 would be input D. The scan would then continue in the subroutine until it reached the end of file or the first true *return* instruction, after which it would return to the instruction following the JSR in the main program (input B), and continue to scan the main program file. On the next scan of the main program, it would check input A, and its condition would determine whether or not it would jump to program file 3.

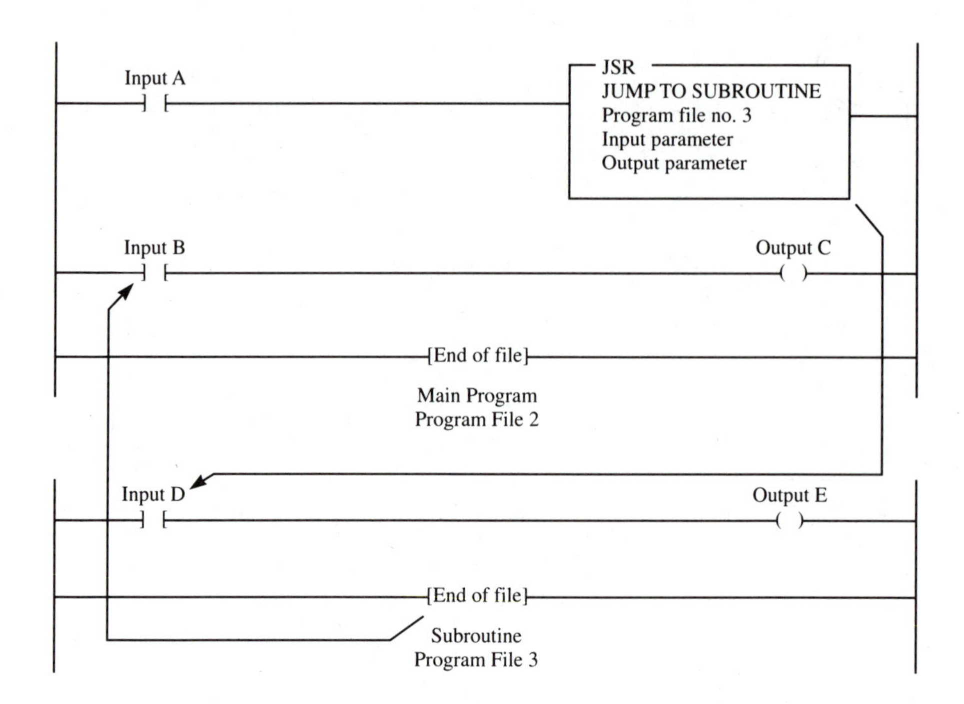

Figure 14-3 Subroutine Example

When the JSR instruction is false, the subroutine is not scanned, and its outputs are left in their last state, unless they are controlled from elsewhere in the program being scanned.

Figure 14-4 illustrates the use of the *subroutine* instruction (SBR) and the *return* instruction (RET). The SBR instruction is only required when used to pass parameters, and the RET instruction is only necessary where there is either a conditional return or the passing of return parameters.

In Figure 14-4, our objective is to find the average value of N7:5 and N7:20 and store the result in N7:30, and to find the average value of N7:7 and N7:22 and store the result in N7:35. We will accomplish this by passing parameters to the subroutine and doing the math in the subroutine, and then returning the answer to the main program through the RET instruction.

When input A is true, the data from the input parameter, N7:5, is copied into the first input parameter in the SBR instruction, N7:50. The data from the second input parameter, N7:20, is copied into the second input parameter in the SBR, N7:51. In the subroutine, N7:50 and N7:51 are then added together, with the result being stored in N7:52. The value in N7:52 is then divided by 2, which gives the average of N7:50 and N7:51, with the result being stored in N7:53. The RET instruction then returns the average value through the return parameter, N7:53, to N7:30 in the JSR instruction in the first rung in the main program.

When input B is true, we will pass the values stored in N7:7 and N7:22 to the subroutine, and the values will be copied into N7:50 and N7:51, respectively. The

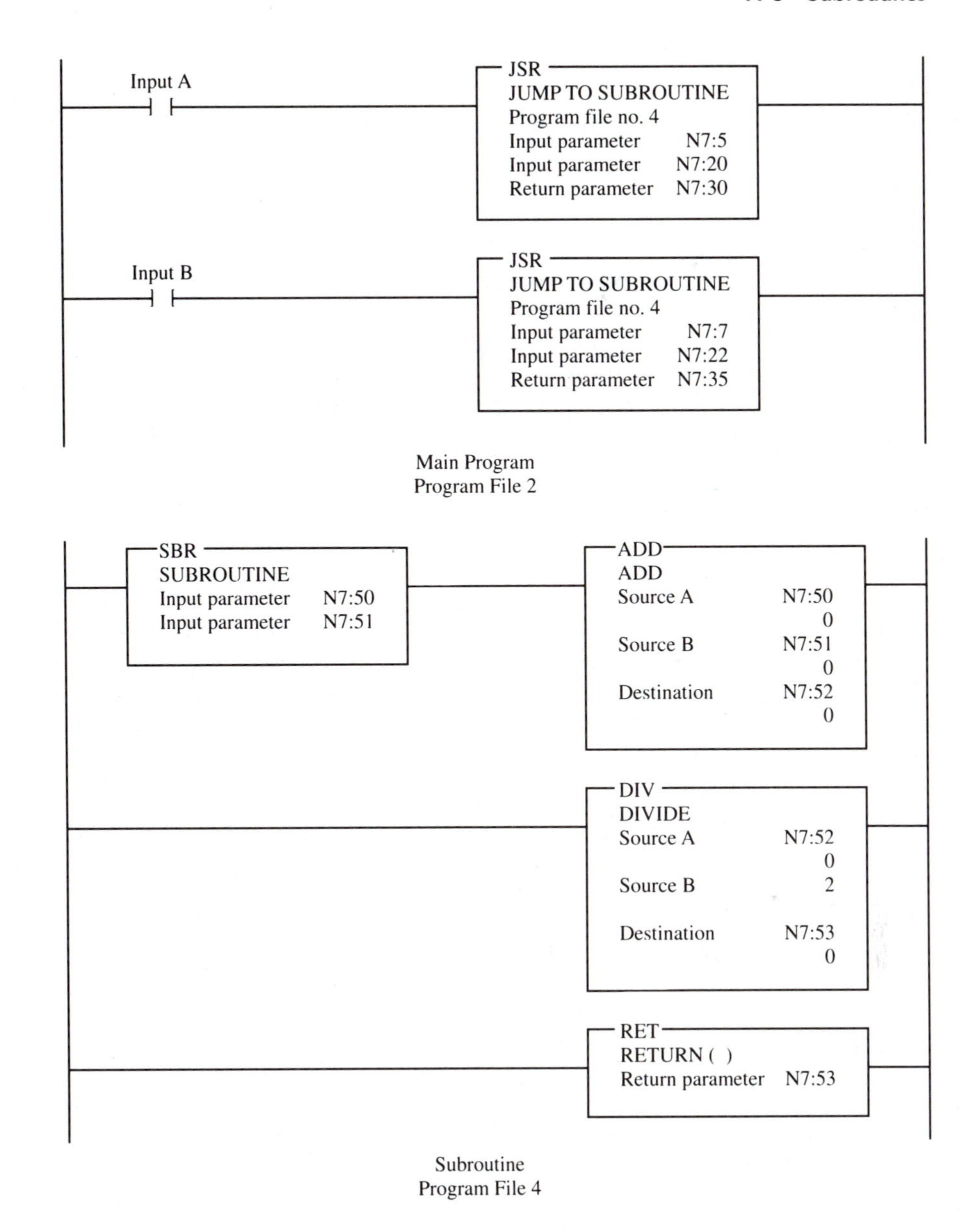

Figure 14-4 Passing Subroutine Parameters

average will be calculated in the subroutine, as in the previous example, and will be returned to N7:35 through the return parameter.

In this short example, it can be seen that we do not have to repeat the math operation programming twice in the main program, but instead can do it just once in the subroutine area. By jumping to the subroutine and passing parameters, we can condense the program.

14-4 Selectable Timed-Interrupt Program File

The selectable timed-interrupt program file is executed on a time basis. The time base at which the program file is executed and the program file assigned as the selectable timed-interrupt file are determined by the values stored in words 30 and 31 in the status section of the data files. The value in word 30 stores the time base, which may be from 1 through 32,767 milliseconds, at 1-msec intervals. Word 31 stores the program file assigned as the selectable interrupt file, which may be any program file from 3 through 999. The value stored in either of these two words may be manipulated from the program. Entering a 0 in the time-base word disables the selectable timed interrupt.

The normal program scan is interrupted at the interval determined by the value stored in status word 30. The scan then jumps to the file determined by the value stored in status word 31. The selectable timed-interrupt file is then scanned one time, after which the scan returns to the instruction following the instruction it was on when it jumped to the selectable timed-interrupt file.

The scan time of the selectable timed-interrupt file must be no longer than the time base stored in status word 31. If that limit is exceeded, then the processor will set a minor-fault bit in the status file and will continue operating the interrupt file.

Figure 14-5 presents an example of the program scan's being interrupted by the selectable timed interrupt. The program scan was at output A when the selectable timed interrupt occurred. The time set was 20 msec. The scan then jumped to the first rung in program file 6, which is determined by the value 6 stored in status word 30. After scanning file 6, the program scan returned to input B, which is the next instruction after output A. The next selectable timed interrupt occurred when the scan was at output E. It took 20 msec for the scan to go from output D to input H. The selectable timed-interrupt file was then executed again, and the scan returned to input F. Thus, every 20 msec the program scan is interrupted and the selectable timed-interrupt file scanned one time.

Programming the selectable timed interrupt is done when a section of program needs to be executed on a time basis rather then on an event basis. Programming that performs a calculation based on a delta time, such as the PID instruction, may require the calculation to be executed at a repeatable time interval, for accuracy, which can be accomplished by placing this programming in the selectable timed-interrupt file.

14-5 Fault Routine

The fault routine is a program file that, if used, will determine how the processor responds to a programming error. The program file assigned as the fault routine is determined by the value stored in status word 29. Entering a 0 in word 29 disables the fault routine. The value stored in word 29 may be manipulated from the program.

There are two kinds of major faults that result in a processor fault: recoverable faults and nonrecoverable faults. Bits 00 through 07 in the major-fault word,

Input C Output A
Input B Output D
Input H Output E
Input F Output G

Main Program
Program File 2

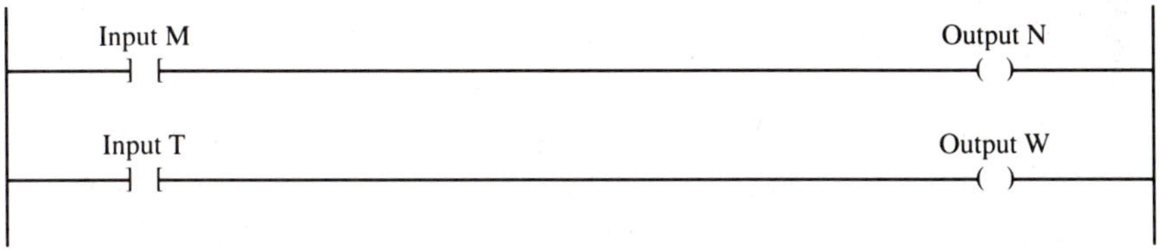

Input M Output N
Input T Output W

Selectable Timed-Interrupt File
Program File 6

Figure 14-5 Example of the Selected Timed-Interrupt File

word 11 in the status file, indicate recoverable faults; bits 08 through 15 indicate nonrecoverable faults.

When the processor detects a major fault, it looks for a fault routine. If a fault routine exists, it is executed; if one does not exist, the processor shuts down. When there is a fault routine, and the fault is *recoverable*, the fault routine is executed. The processor then would be shut down, or the major-fault word, status word 11, might be zeroed in the fault routine, and the program scan would return to the program at the instruction following the faulted instruction. When the scan again reached the point where the fault occurred, then if the fault existed the processor would again fault and go to the fault routine. This sequence would repeat until the fault was cleared. It may be desired to place a scan counter in the fault routine in order to limit the number of times the fault routine is called: When the scan counter reaches a predetermined count, the program would not zero the fault word, and the processor would shut down.

If the fault is *nonrecoverable,* the fault routine is scanned once and shuts down. Either way, the fault routine allows for an orderly shutdown.

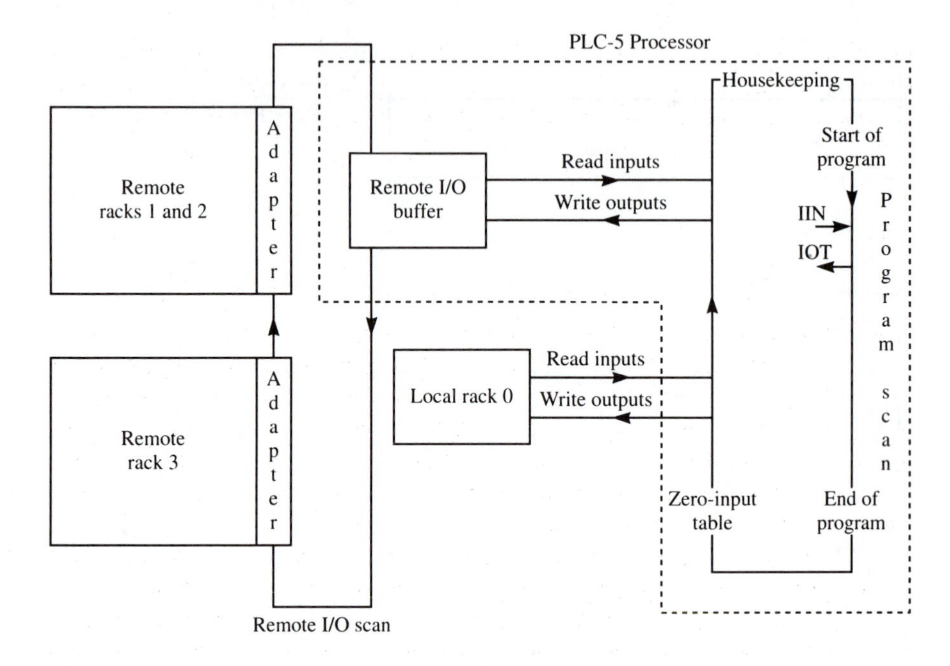

Figure 14-6 PLC-5 Processor Scan: Local I/O, Remote I/O, *Immediate Input* and *Immediate Output* Transfer of Discrete Data

14-6 *Immediate Input* Instruction (IIN) and *Immediate Output* Instruction (IOT)

The *immediate input* instruction (IIN) and the *immediate output* instruction (IOT) help to alleviate the problem of input conditions being frozen through the program scan and output data not being transferred to the real world until the end of the program scan.

Fast-acting inputs, such as those connected to electronic input devices, may have very short time periods in either an ON or an OFF state. Since the inputs are not updated during the program scan, change of state may not be detected by the program. Also, fast-acting devices should be wired into the local chassis, since its I/O scan is synchronous with the program scan and communication is in parallel with the processor, whereas the remote I/O scan is asynchronous with the program scan and communication with remote I/O is serial. Communication with the local chassis is thus many times faster than communication with the remote chassis, and all high-speed inputs and outputs that require fast updating should be wired to the local chassis. Figure 14-6 illustrates communication between processor and input/output.

The *immediate input* instruction is an output instruction. An example of its use is given in Figure 14-7. When the program scan reaches a true IIN instruction, the scan is interrupted and the processor updates sixteen bits in the input image table at the location indicated on the IIN instruction. The two-digit address on the IIN instruction is comprised of the rack number (first digit) and the I/O

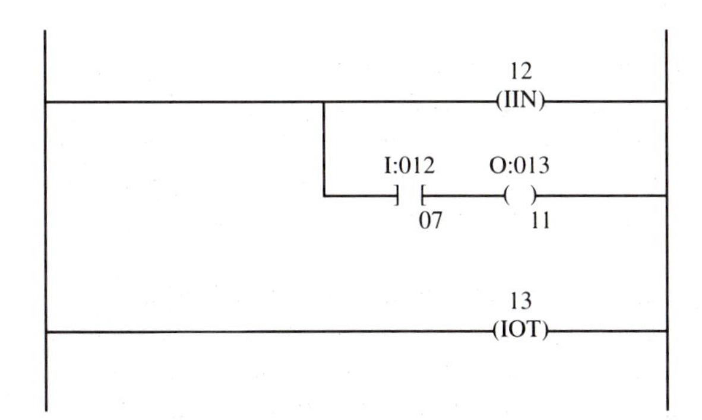

Figure 14-7 Example of the *Immediate Input* (IIN) and *Immediate Output* (IOT) Instructions

group number containing the input or inputs needing immediate updating (second digit). Do *not* put an *I* in front of the two-digit number, for unexpected results may occur when the scan reaches the true IIN instruction addressed in this manner. The same holds true for the IOT instruction. It is also addressed with a two-digit number, the first digit being the rack number and the second being the I/O group number. Do *not* put an *O* in front of the two-digit number, for unexpected results may occur when the scan reaches a true IOT instruction.

In Figure 14-7, note that the *examine on* instruction using a bit address from the same word address as on the IIN instruction follows the IIN instruction in the scan. In this way the input word in the data table is updated, and then the bit in the data table is examined by the *examine on* instruction.

The IOT instruction operates in a similar manner. If a real-world output requires updating prior to the end of the program, the IOT instruction can be used: When the scan reaches a true IOT instruction, the scan is interrupted at that point, and the data in the output image table at the word address on the instruction is transferred to the real-world outputs. All sixteen bits in the word are transferred. In Figure 14-7, the IOT instruction follows the *output energize* instruction. In this way, the output image table word is first updated, and then the data is transferred to the real-world outputs.

During programming, the *immediate input* and *immediate output* instructions are frequently located in a selectable timed-interrupt file, so that section of program is updated on a time basis, set by the selectable timed-interrupt set point. This could be done on a high-speed line, when items on the line are being examined and the rate at which they pass the sensor is faster than the scan time of the program. In this way, the item can be scanned multiple times during the program scan, and the appropriate action may be taken before the end of the scan.

Summary

The instructions covered in this chapter are those that alter the program scan from its normal sequence. They can shorten the time required to complete a program scan, for example, the *jump* and the *label* instructions and the use of subroutines.

The MCR instruction allows zones to be created so nonretentive outputs can be deenergized as a group within that zone. These instructions allow for unique programming techniques.

Exercises

14-1. How is an MCR zone created? What restrictions are placed on the creating of MCR zones?

14-2. If a *latch* instruction and an *unlatch* instruction with the same address are placed in an MCR zone, how will the bit status of that address be affected by the condition of the MCR zone?

14-3. How many MCR zones may exist in a program?

14-4. What happens to the status of the output bits located in a section of the program that is jumped over when a true JMP instruction causes the scan to jump to a LBL instruction?

14-5. How many JMP instructions can you use in a program file? How many LBL instructions?

14-6. Can the JMP instruction be used to jump to a LBL instruction located in a different program file?

14-7. What is the danger of jumping to a LBL instruction located above the JMP instruction in the ladder logic?

14-8. In what program files may subroutines be located?

14-9. Construct a program that will cause the scan to jump from the main program file to program file 4 when input A is true. When the scan jumps to program file 4, data will also be passed from N7:30 to N7:40. When the scan returns to the main program from program file 4, data will be passed from N7:50 to N7:60.

14-10. What determines which program file is the selectable timed-interrupt file? How is its time interval set?

14-11. What is the function of the fault routine? How is it determined which program file is the fault routine?

14-12. Should the *immediate input* instruction come before or come after the instruction examining the desired bit that is to be updated?

14-13. Should the *immediate output* instruction come before or come after the instruction controlling the output bit whose status is to be transferred to the real world?

14-14. How is the IIN instruction addressed?

14-15. How is the IOT instruction addressed?

Structured Programming on a Programmable Logic Controller

15

OUTLINE

OBJECTIVES

Upon completion of this chapter, the student will be able to:

- Discuss three structured methods for implementing a control scheme.
- Generate a state diagram for any given control problem.
- Design a Petri network for any given control problem.
- Convert a Petri network into a sequential function chart.
- Cite the pros and cons of structured vs. nonstructured programming.

THIS chapter explains how to develop control using structured methods rather than trial and error. Adding structure through a state diagram or a Petri network aids in analyzing the overall functioning of a control and breaks up complex controls into smaller, more manageable segments. Troubleshooting structured programs is much easier, because they are organized, and the state diagram and Petri network provide visual aid.

15-1 State Diagrams

Relay ladder logic often is written in a nonstructured way that makes it hard to analyze and troubleshoot. But this can be remedied via a simple, structured approach, called the *state diagram*. A **state diagram** is a pictorial representation, that is, a special type of flow chart, of a sequential control process that shows the possible paths the process can take and the Boolean conditions necessary to go from one state to another.

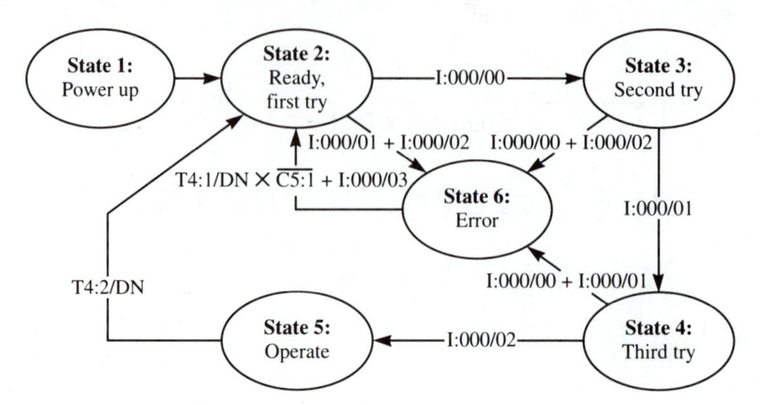

Designation	Device	Function
I:000/00	Push button 1	First combination
I:000/01	Push button 2	Second combination
I:000/02	Push button 3	Third combination
I:000/03	Push button 4	Reset from latch
T4:1	Timer 1	10-sec delay
C5:1	Counter	Error counter
T4:2	Timer 2	15-sec delay

Figure 15-1 State Diagram for Example 15-1 (Sequential-Lock Design)

The first step in this approach to structured programming is to prepare a verbal description of what you want to do. From this description, a state diagram is made by determining the unique states through which the process will go. Any time there is a change in output conditions, a new state needs to be shown. The third step is to connect these states with lines and arrows to indicate the desired directions of change. Finally, add the logical conditions, written in Boolean form, that will cause a change from one state to the other. Example 15-1 presents a sample verbal description and corresponding state diagram.

E X A M P L E 1 5 - 1
SEQUENTIAL-LOCK DESIGN

Design a sequential lock that will operate a device for 15 seconds after three push buttons have been pressed in the correct sequence. If the incorrect sequence is pressed at any time, an error state should be activated that will prevent further tries for 10 seconds and then permit another try. If the error state is activated more than twice during a try, then the lock is to be disabled or locked out. A push button is to be provided to reset the lock if it is locked out.

The state diagram for this sequential lock is given in Figure 15-1.

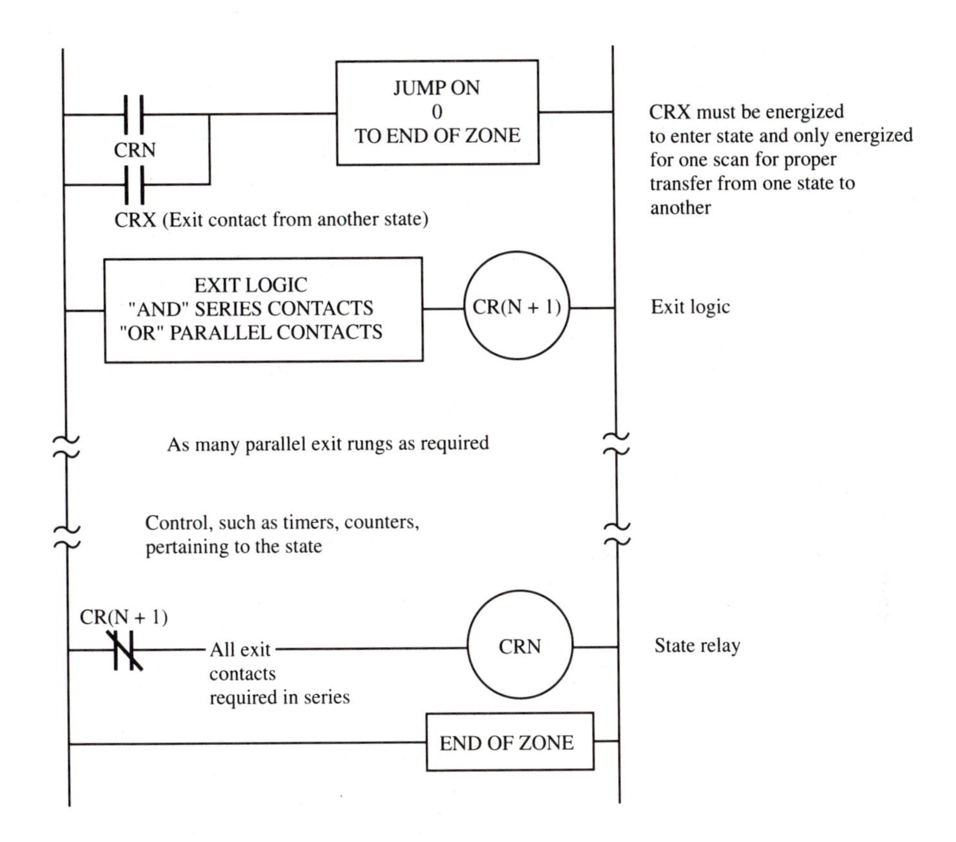

Figure 15-2 General Form for Implementing a State

All states will be implemented, one state at a time, by following the general format of Figure 15-2. The first rung contains the necessary condition (contact CRN) to enter a state. The second rung consists of the conditions required to get out of the state. The logic conditions involved are also listed on the lines connecting states on the state diagram. The area between the second and last rungs contains the control necessary for accomplishing what needs to be done in that particular state, and can range from a single rung of control to however many are required. The last rung acts to disable the state as we leave, by picking up the exit relay, CR(N + 1), leaving its normally closed contact open, and thereby deenergizing the state relay. This is necessary so we do not enter the state again on the next scan. If this were not done, the output would stay on until the zone control turned it off on the next scan, and this we cannot allow, for if we exit this state and enter another state, two states might be on at the same time.

Rule for Making State Diagrams Only one state may be active at a time.

This rule is ensured by having the last rung in a state disable the state relay for that state. Otherwise, a serious conflict between states could result that might prove unsafe to personnel and machines, such as trying to turn on a motor that was drilling a part while simultaneously turning on a motor that would move the part.

The sequential control for the lock described in Example 15-1 is shown implemented via this state diagram method in Figures 15-3 through 15-8. Table 15-1 gives, for each state, the rungs in which each of the three general tasks—enter, exit, and disable—is carried out. The PLC employed is an Allen-Bradley PLC-5. The states are programmed as zones, so the boundaries of the zones must be defined. This is accomplished in this control scheme via zone control. Note the pair of MCR's around the state in each of the figures.

Table 15-1 Rungs Used to Implement the General Form

State	Enter	Exit	State/Disable
1. Power up	2:0	2:1	Not required
2. Ready	2:3	2:4 and 6	2:7
3. Second try	2:9	2:10 and 2:12	2:13
4. Third try	2:15	2:16 and 2:18	2:19
5. Operate	2:21	2:22	2:25
6. Error	2:27	2:28	2:29

The obvious structure in this control makes it easy to follow and troubleshoot. The hardest part of the design is generating the state diagram. But once that is done, the rest is fairly straightforward. You could implement this same control with fewer control relays. But since the PLC's memory is not expensive, the cost for structure is minimal. The advantages of structured over nonstructured programming make it an excellent way to implement control.

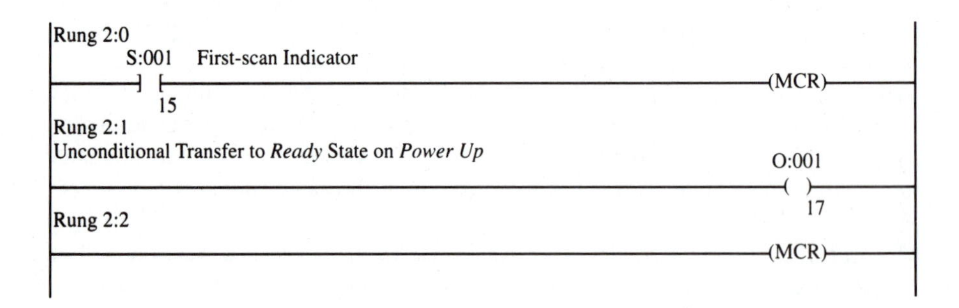

Figure 15-3 ***Power Up*** **State**

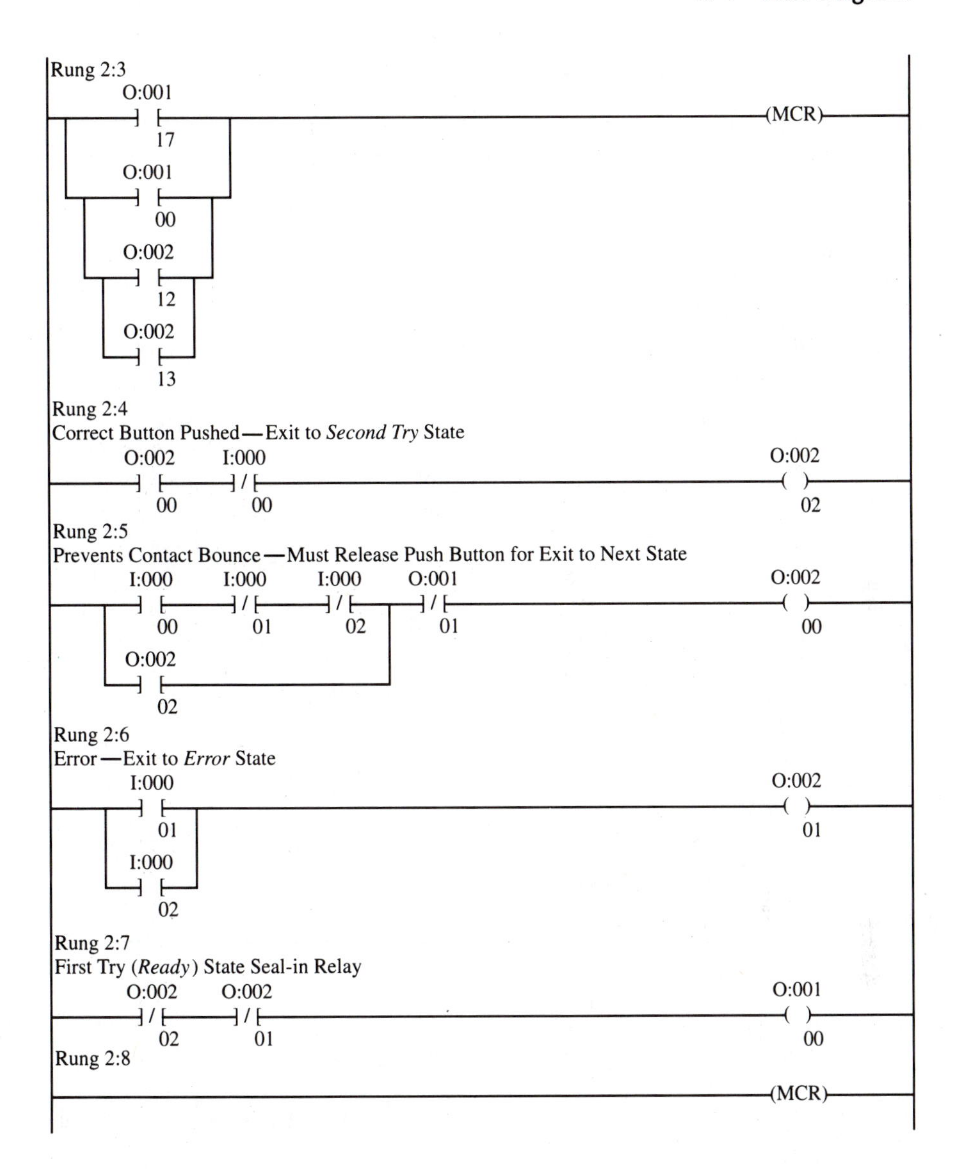

Figure 15-4 ***Ready* State**

EXAMPLE 15-2
TWO-PUMP PROBLEM

A water tank has three sensors to detect three different water levels: F—full, L—low, and E—empty. There are two pumps to fill the tank: P1 and P2. When the automatic control is on, the following should happen:

1. An initially empty tank should result in both pumps being turned on until the tank is full.

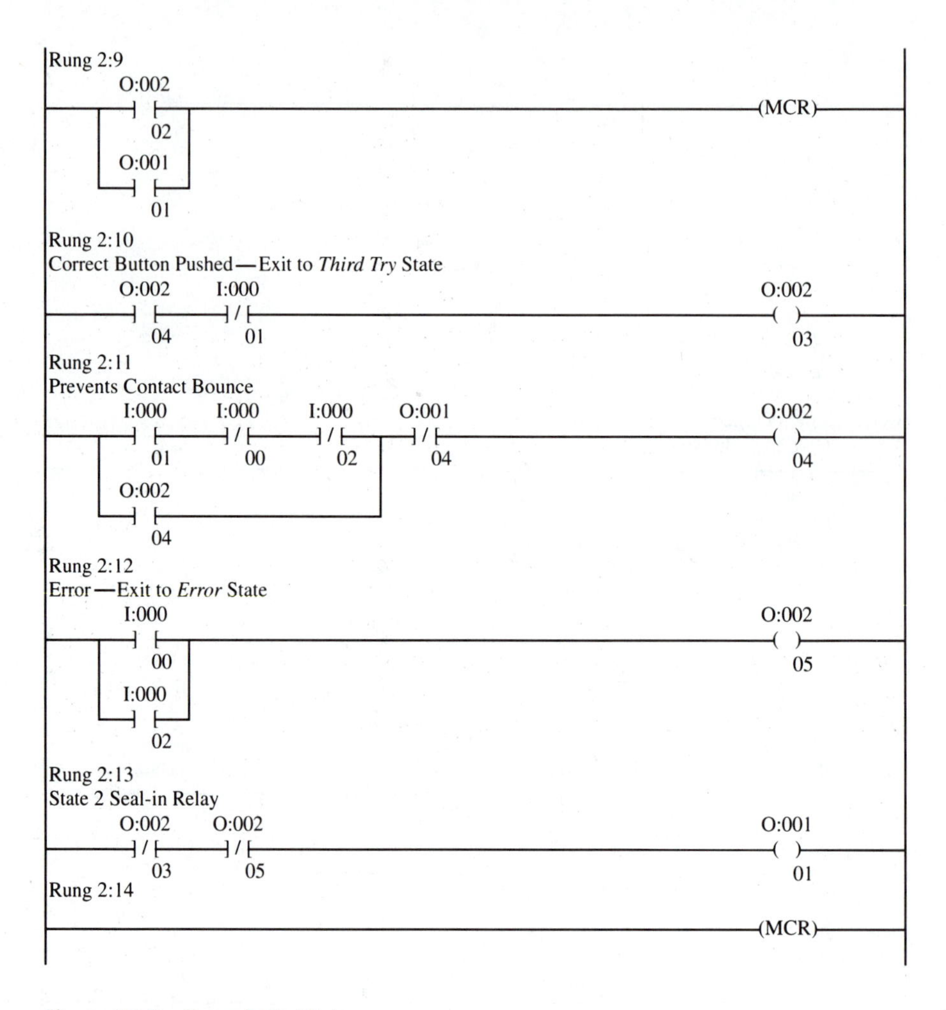

Figure 15-5 ***Second Try* State**

2. An initially full tank or a water level between full and low should keep both pumps turned off.

3. An initial condition below low but not empty should turn on one of the pumps until full is reached.

If the water level goes below empty while one pump is on, both pumps should come on and stay on until the tank is full. The pumps are to alternate during these fill cycles.

To alternate the pumps, we will set and reset a latching relay.

The state diagram for this control is given in Figure 15-9.

The following symbols are used in the state diagram: *E* = ''empty'' sensor; *L* = ''Half-full'' sensor; *F* = ''Full'' sensor; *Latch* = latching relay to alternate pumps; P1 = Pump 1; P2 = Pump 2.

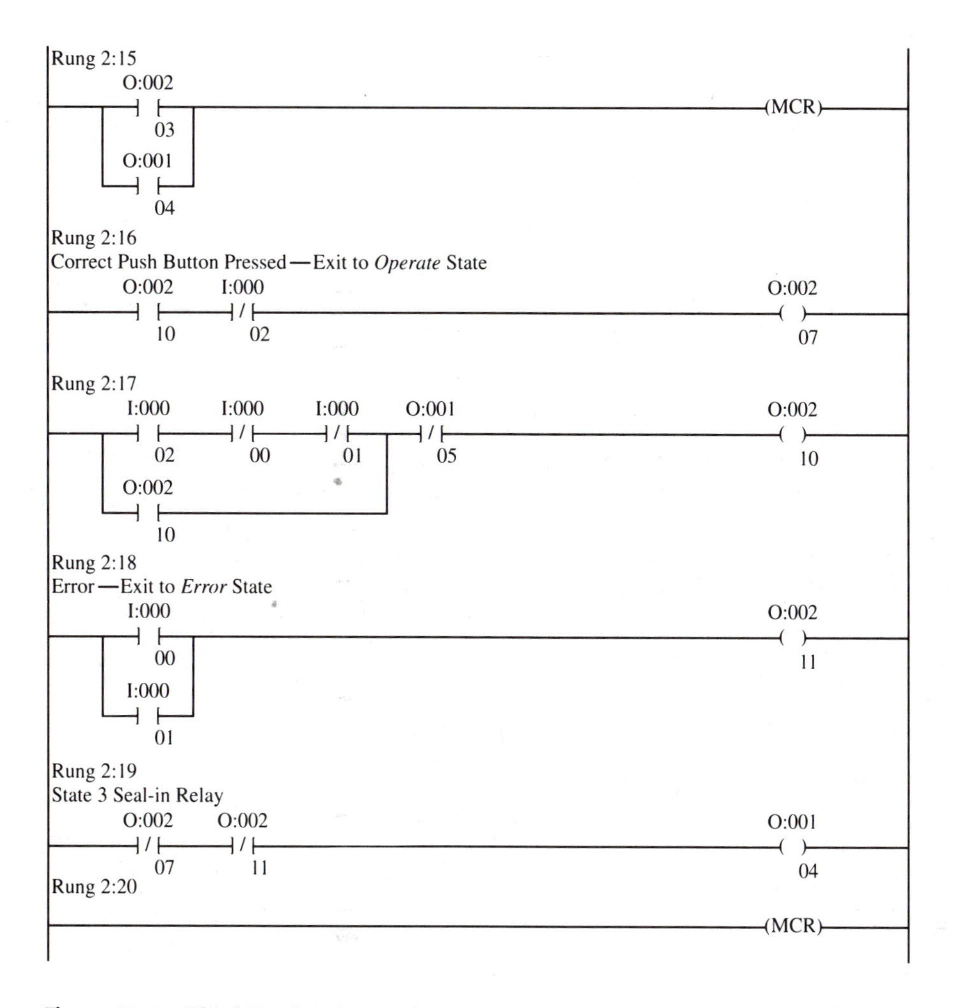

Figure 15-6 ***Third Try* State**

Sensors produce a contact closure when they touch the water; therefore, E means water is touching the sensor—contact is closed, and $\overline{E}$ means water is *not* touching the sensor—contact is open.

The method in Example 15-2 is applicable only if one state is active at a time. However, there are other methods available for handling parallel operations, such as the Petri network, which we will discuss next.

15-2 Petri Networks

Petri networks were first employed in computer programming as a structured way of handling programs in which parallel processes have to be accomplished

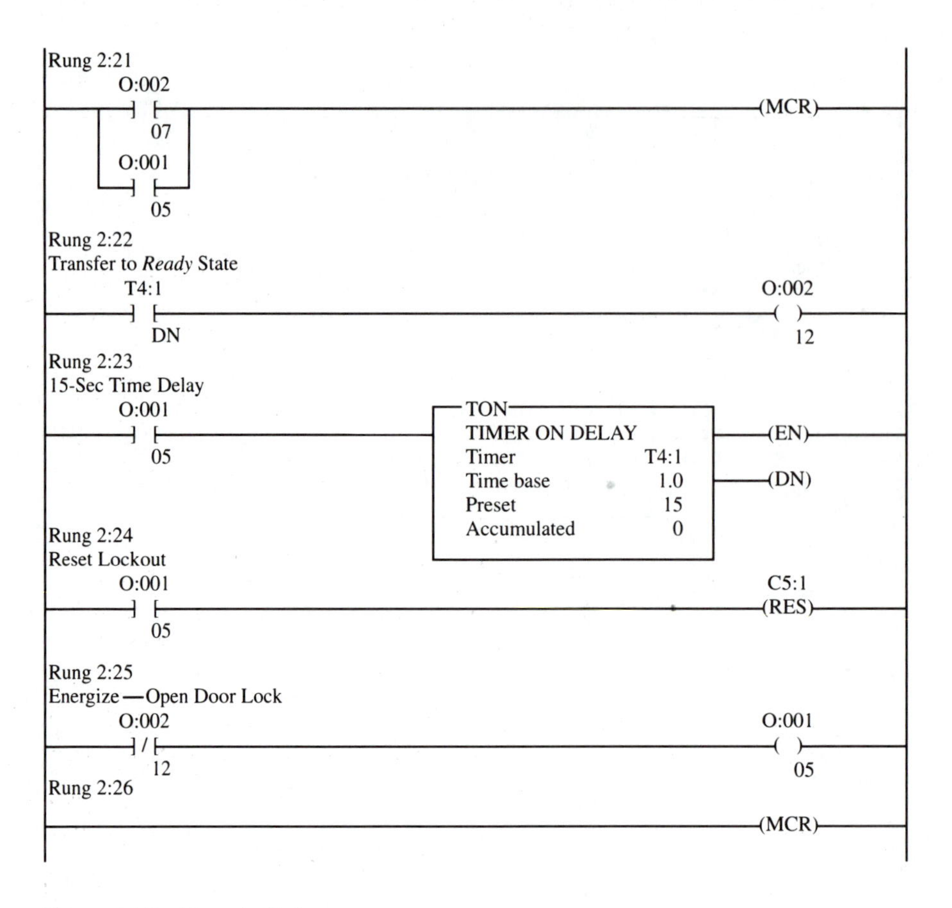

Figure 15-7 ***Operate*** **State**

simultaneously and each parallel process has to be completed in order to continue on to the next portion of the program. A **Petri diagram** is a pictorial representation of a control process that requires parallel branching and simultaneous processing. It shows the possible paths the process can take, the Boolean conditions necessary to go from one state to another, and where convergence is required to continue.

An example where multiple operations go on simultaneously is an automated assembly table that receives two different printed circuit boards that have to have parts mounted, leads soldered, and miscellaneous processes completed before they advance to a mother board to which the three individual boards are to be connected. Only after all three boards have been completed and connected can the mother board proceed to a new location where it will be automatically tested.

A Petri network of this process is shown in Figure 15-10. This can be implemented on any PLC by using a structured state diagram for each state, with the added requirement of passing a **token** (computer password) as we go from one state to another. Only a state that is currently active and holding the token may communicate with the processor.

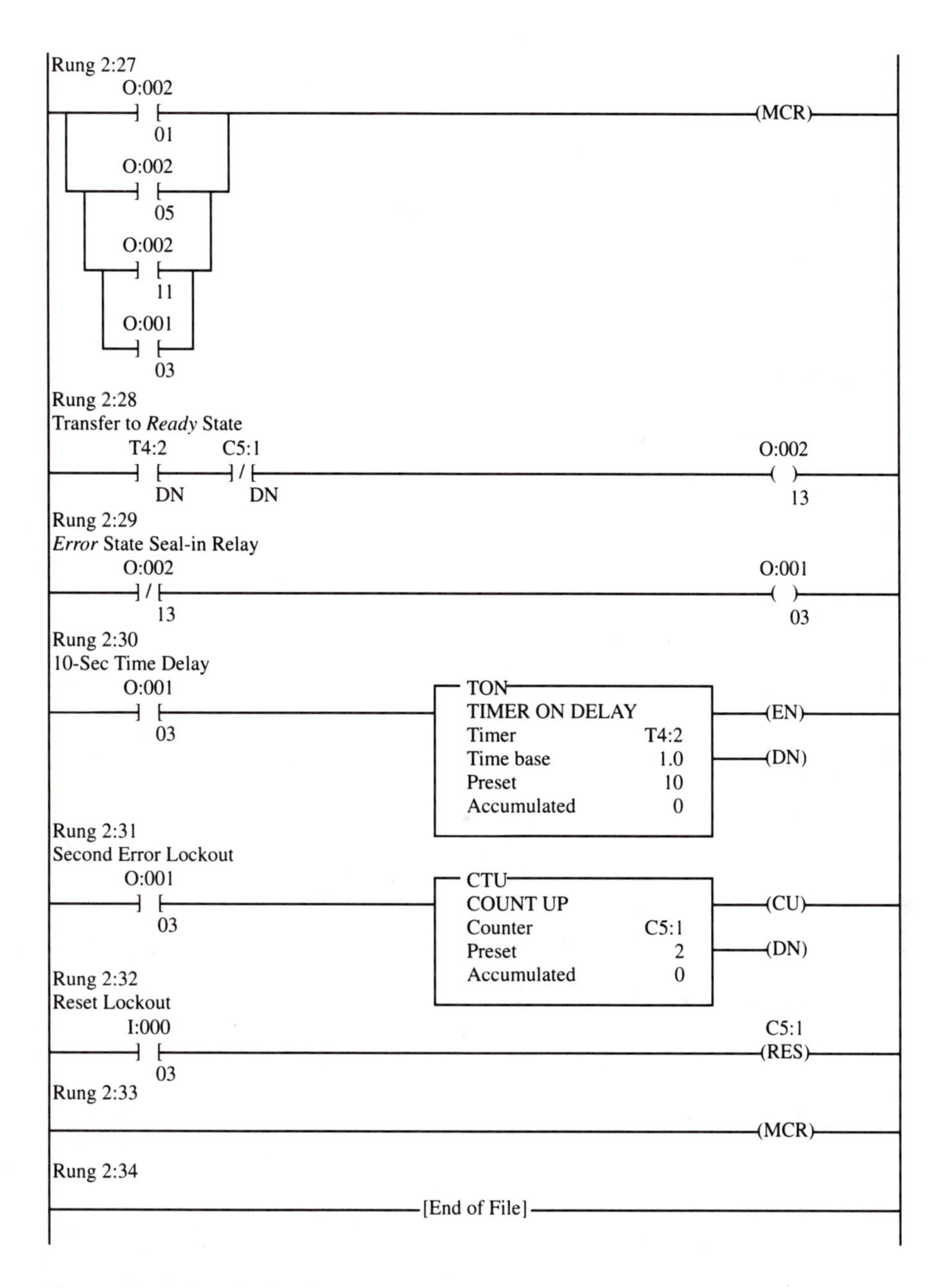

Figure 15-8 ***Error* State**

Note in figure 15-10 that a vertical line connects the states, and that this vertical line is crossed by a horizontal line with a notation such as PB1 adjacent to it. This indicates that PB1 must be true for a transition to the next state to take place. The token is passed automatically when a transition is made to a new state.

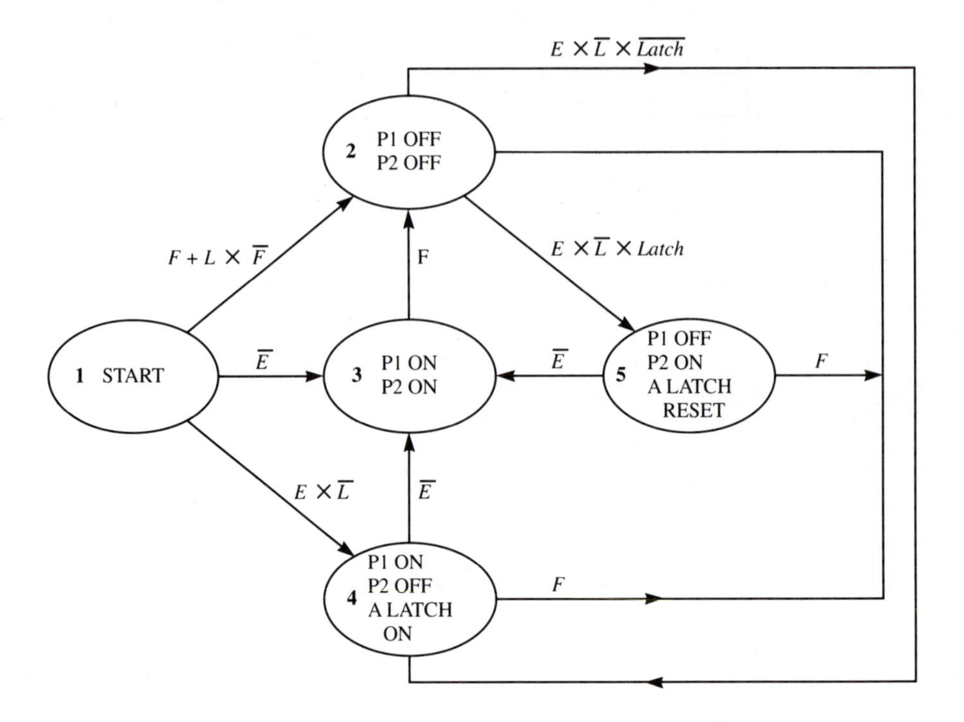

Legend
E, "Empty" sensor
L, "Half-full" sensor
F, "Full" sensor
Latch, Latch relay, to alternate pumps
P1, Pump 1
P2, Pump 2
Overbar (e.g., $\overline{E}$) indicates sensor contact is open.
No overbar (e.g., *E*) indicates sensor contact is closed.

Figure 15-9 State Diagram for Example 15-2 (Two-Pump Problem)

Where a double horizontal line appears on the diagram, all tokens must be present from all the parallel branches before the process can proceed.

15-3 Sequential Function Charts

The process in Figure 15-10 can be programmed on a PLC in a structured way by using a state diagram for each state and registers to hold and pass tokens. On a machine for which structured programing is unavailable, however, this takes a lot of time and programming. The Allen-Bradley PLC-5 has as an available option a special structured type of programming that allows you to do Petri networks using ladder logic. These **sequential function charts,** or SFC, are the trade name for Allen-Bradley's implementation of Petri networks. You do not have to worry

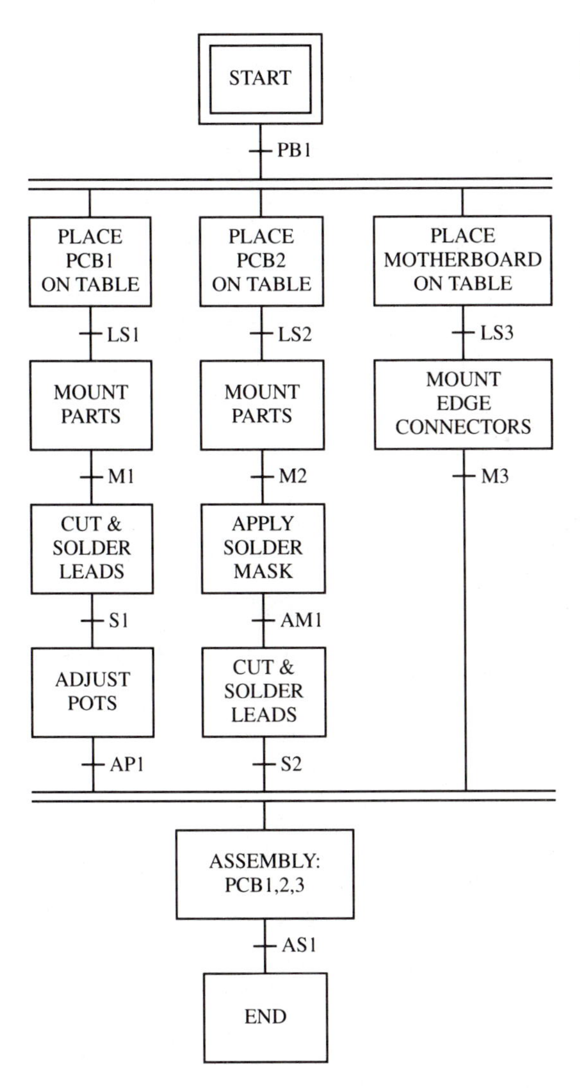

Figure 15-10 Example Petri Network

about token passing, since this is accomplished automatically by SFC. The program also automatically generates the necessary state diagram logic to implement a state, although you must put in the ladder logic needed for each state. All that is required for entry is the software that lets you enter the diagram of the process you want. Figure 15-10 would be implemented with such software, which would let you enter the state transition pairs, define their names, and then define the ladder logic for each state.

One advantage of SFC is that the scan becomes more efficient. Instead of scanning through all the logic for the entire control in SFC, only the logic for the states that are active is scanned and the rest are passed over. Let's say in our earlier example that we are mounting parts on PCB1 and soldering on PCB2 and that the mother board is finished and waiting for the other two boards. The only logic scanned is the logic for mounting parts on PCB1 and soldering on PCB2. The PLC skips over all other logic.

Figure 15-11 shows the Petri diagram as it would be implemented on the PLC-5 using sequential function charts. Note that each state and transition has a file number attached to it. Data files 0 and 1 are not used here, because they are reserved for the output and input image tables.

Let's now look at what has to be in the files for this particular SFC.

First let's look at data file 003. This is a transition point, and we have to tell the PLC the conditions required to move to the next state following this transition. The symbol for this transition in the SFC is:

003
PB1

We have to create a data file 003 and program it as follows:

I:001 (PB1)
00
(EOT)

This program is the relay ladder logic that will cause an *end of transition* (EOT) if PB1 is true. If input I:001/00 is energized, it will trigger the transition to the next state.

Each individual state has a symbol such as the following:

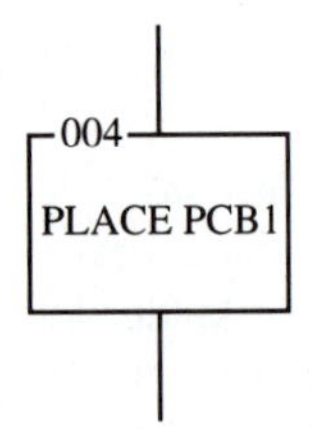

We must supply the necessary relay ladder logic programming for each state. This logic will tell what is to be done in that state. To place PCB1, data file 004 would have to be created and programmed. When this state is called on, a solenoid will be activated that will push one board onto the assembly table in the correct position. When the board slides into position, a limit switch will be closed, deactivat ing the solenoid. If the limit switch is not closed in 2 seconds, an alarm will go off, indicating a problem. The logic for this state is as follows:

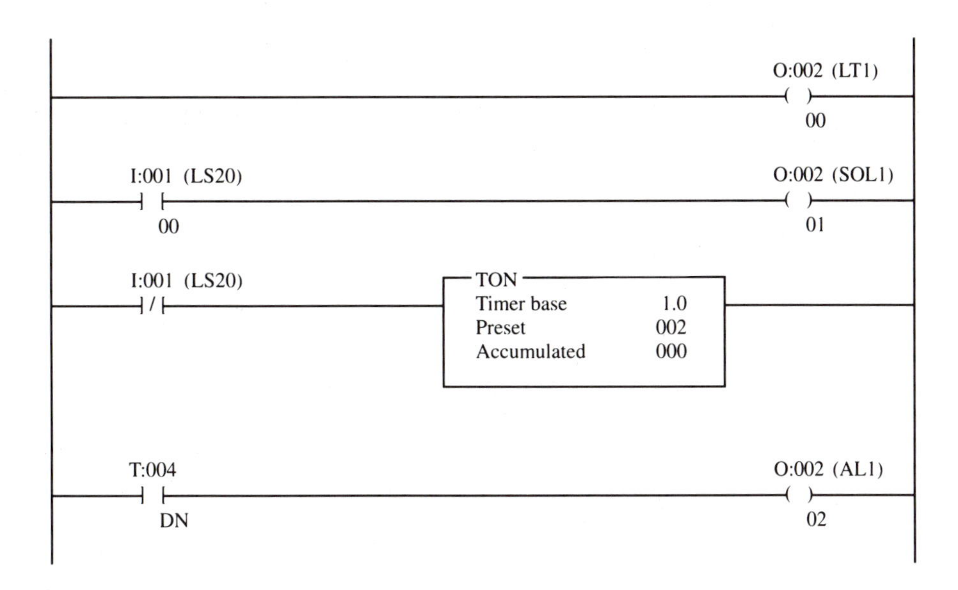

E X A M P L E 1 5 - 3
SEQUENTIAL LOCK DESIGN:
SFC IMPLEMENTATION

The three-push-button sequential lock earlier implemented by state diagrams is implemented here using sequential function charts (see Figures 15-12 through 15-17). This will show, in addition to the SFC, all the ladder logic required for each state and transition. Notice the similarity between the SFC ladder logic and the ladder logic used to implement the state diagram of Figure 15-1. The exit conditions have become the transition ladder logic in the sequential function charts, and the ladder logic between the exit conditions and the state relay in the state diagram implementation have become the SFC logic for the states. The SFC implementation requires fewer output relays and no zone control, and ultimately less ladder logic. Also, if parallel operations are required, then sequential function charts accomplish token passing automatically.

E X A M P L E 1 5 - 4
TWO-PUMP PROBLEM:
SFC IMPLEMENTATION

As another example of a sequential chart, the two-pump problem shown as a state diagram in Figure 15-9 is implemented in an SFC in Figure 15-18. Notice that states and transitions in this application are duplicated. When a branch is encountered, the program can diverge so the processor can select one of four parallel branches, depending on which transition is satisfied. Ultimately these four branches converge again. This is referred to as *selective branching*. The duplicated states and transitions only require the generation of one set of ladder logic.

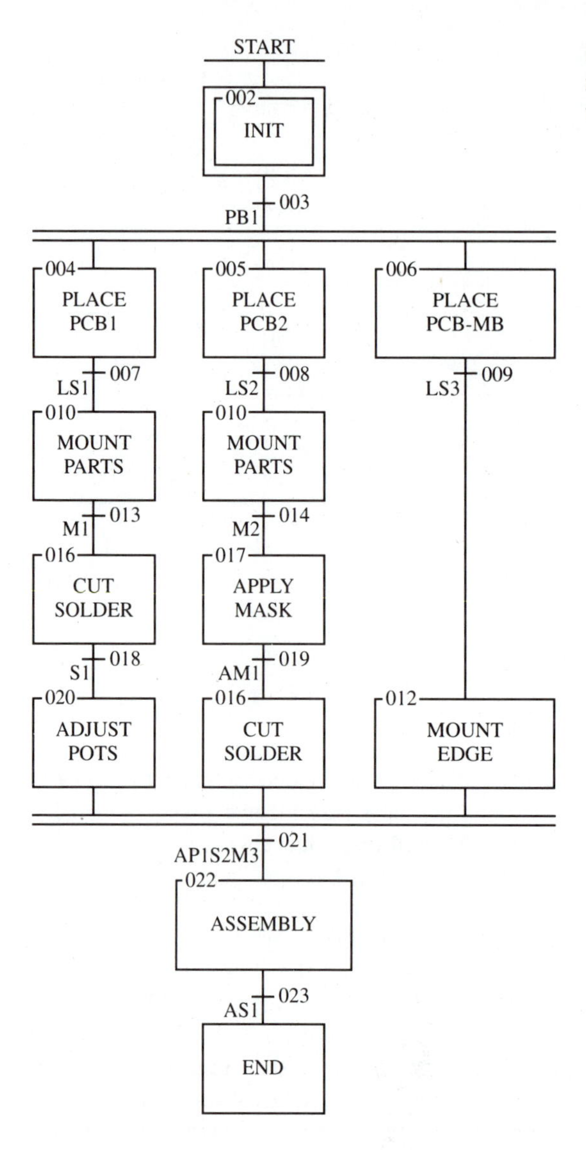

Figure 15-11 Petri Net Implemented Using Sequential Function Charts on the PLC-5

Parallel operation of states is referred to as *simultaneous branching*. Duplication of states and transitions are likely to occur under such conditions. Sequential function charts will function correctly with duplicate states and transitions.

Summary

There are so many advantages to having control programs structured that it is well worth the time it takes to learn one of the techniques. The two obvious advantages are (1) it breaks up complex control into small, manageable units, and (2) it

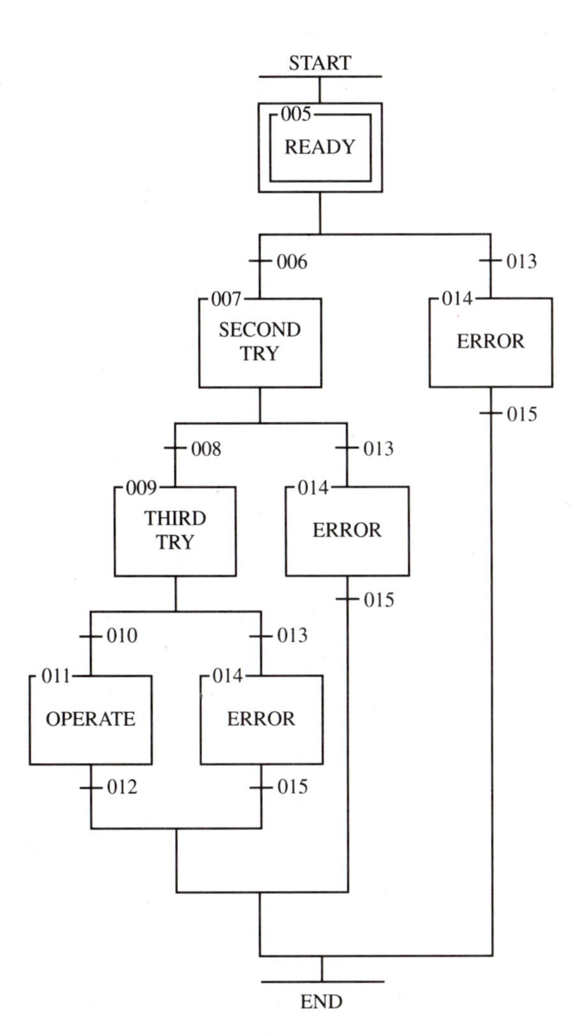

Figure 15-12 Three-Pushbutton Lock (Sequential Function Chart)

makes control much easier to troubleshoot. Another advantage is that it forces you to document what has been done so that others can understand the generated control. This avoids any chaos if you were to leave or become unavailable. Trial and error gets the job done, but it is often painful and wasteful, and if it isn't documented the next person must repeat the same time-consuming process.

Exercises

15-1. If Figure 15-2 were changed to Figure 15-19, what problems would result using a PLC?

15-2. Make a state diagram for the following control problem: A machine operator is required, for safety, to have both hands away from a machine while

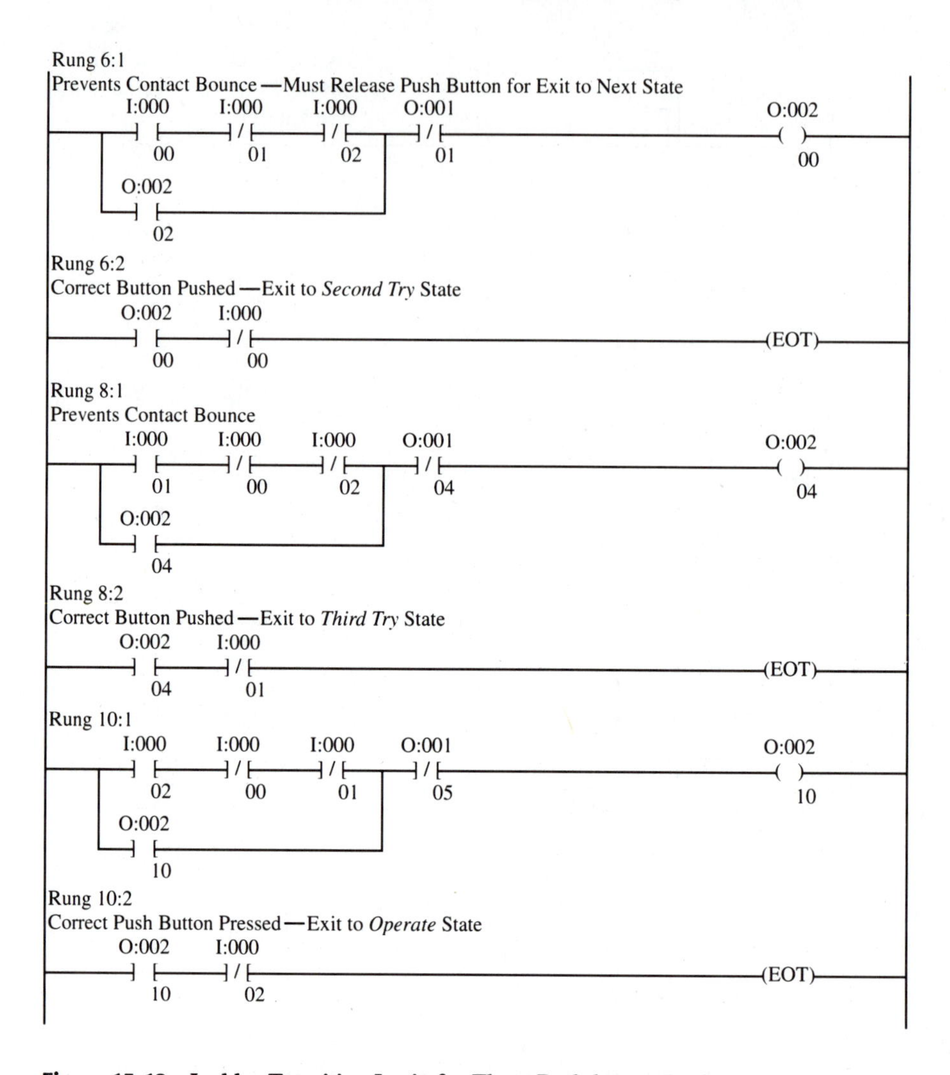

Figure 15-13 Ladder Transition Logic for Three-Push-button Lock

it is running. In order to ensure this, two push buttons must be pressed simultaneously to operate the machine, and they are far enough apart so this can't be done with one hand. After pressing both push buttons simultaneously, if one hand is removed from one of the buttons, the machine will shut off and lock out, and will not operate until both buttons are released and pushed simultaneously again.

15-3. For the state diagram you created for Exercise 15-2, draw the control required for an Allen-Bradley PLC-5.

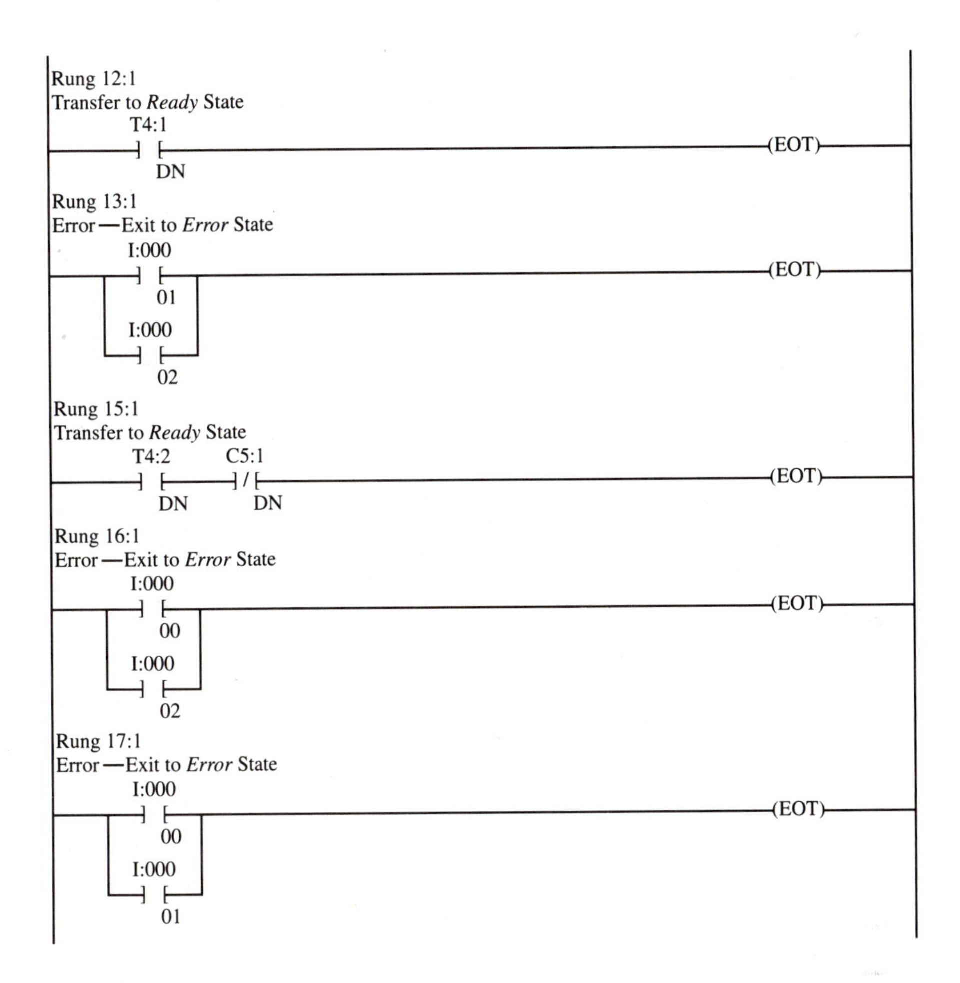

Figure 15-14 Ladder Transition Logic for Three-Push-button Lock (*Continued*)

15-4. Cite the advantages and disadvantages of using state diagrams for designing control.

15-5. What features of a PLC make it possible to implement control with state diagrams?

15-6. How could you use the state diagram to implement a control with mechanical relay logic rather than a PLC?

15-7. What are the main advantages of structured programming?

15-8. How do structured techniques of the state diagram and the Petri network make the scanning of the PLC more efficient?

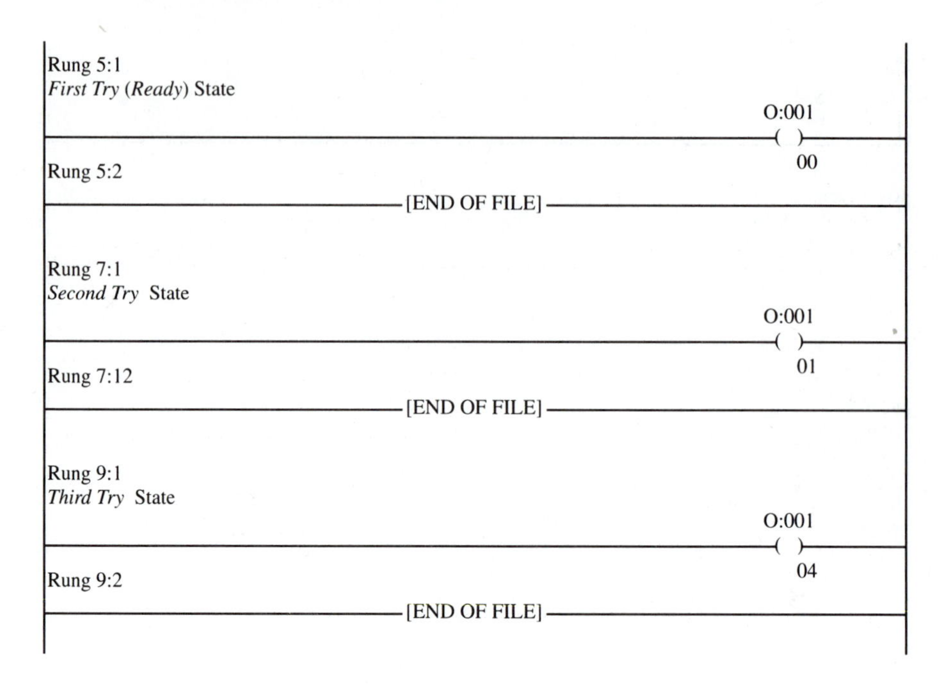

Figure 15-15 State Logic for Three-Push-button Lock

15-9. Give the Boolean equations for all the transitions in the sequential function chart in Figure 15-18.

15-10. Show the ladder logic for the SFC transition in Figure 15-18. Assume that devices are connected to modules per the following:

I:000/00	*E*, "Empty" sensor
I:000/01	*L*, "Half full" sensor
I:000/02	*F*, "Full" sensor
O:000/01	*Latch*, latching relay (to alternate pumps)

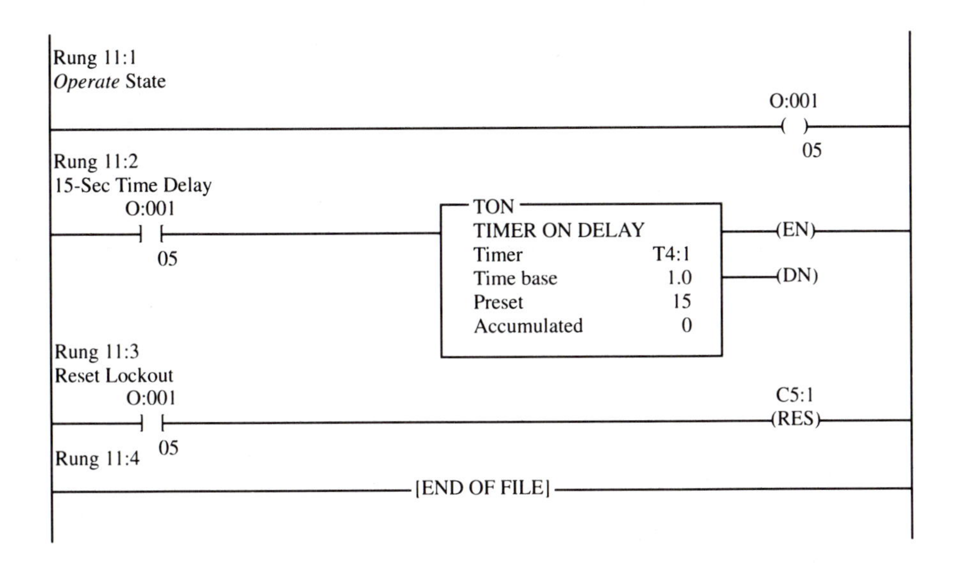

Figure 15-16 State Logic for Three-Push-button Lock (*Continued*)

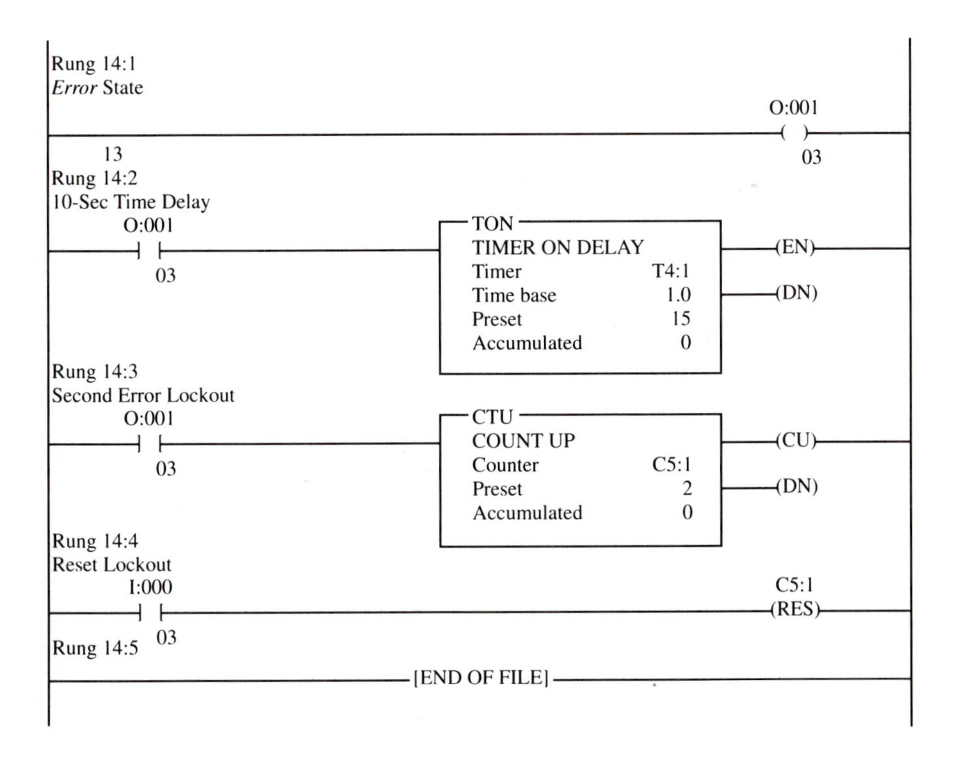

Figure 15-17 State Logic for Three-Push-button Lock (*Continued*)

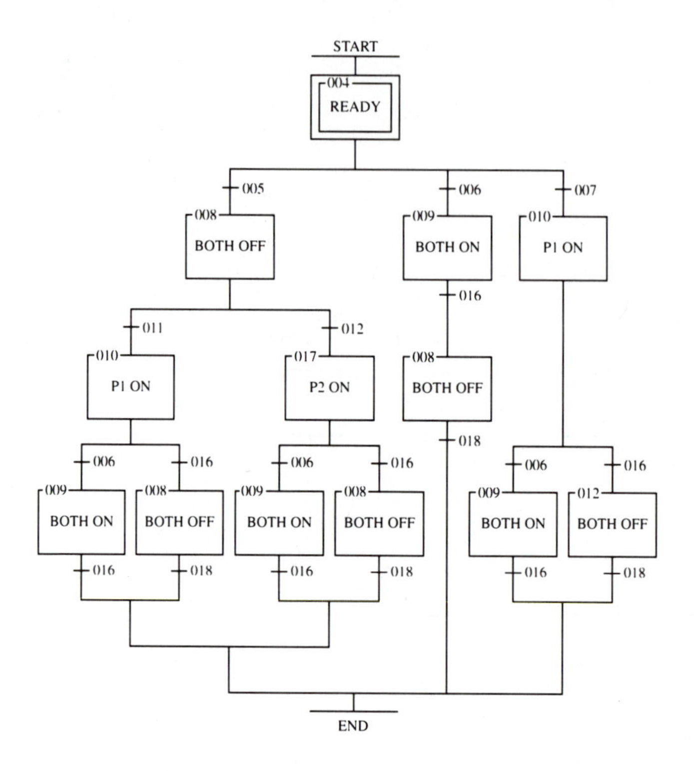

Figure 15-18 Sequential Function Chart for Example 15-4 (Two-Pump Problem)

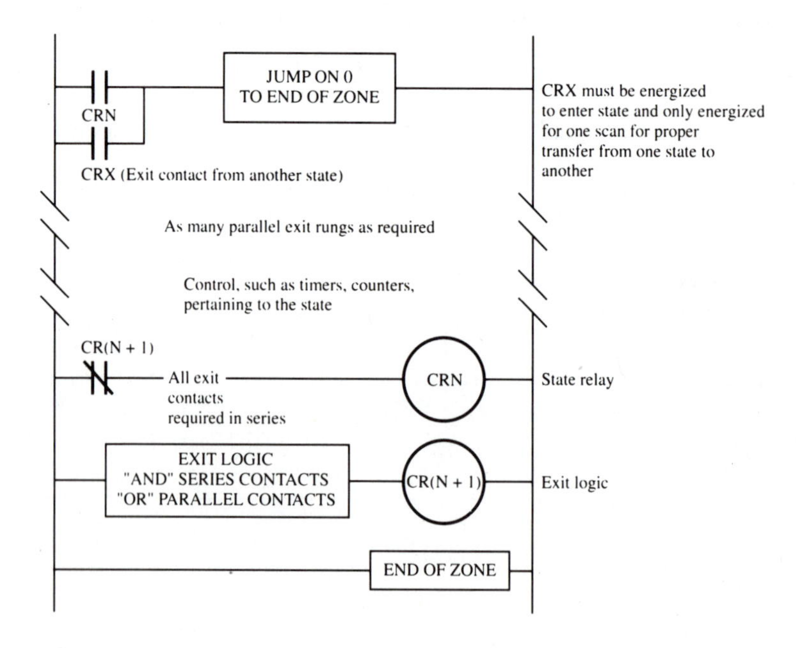

Figure 15-19

16

Communication with Other Programmable Controllers and Computers

OUTLINE

OBJECTIVES

Upon completion of this chapter, the student will be able to:

- Describe a local area network and its three most common configurations.
- Describe the Data Highway Plus network and its function.
- Discuss the hardware necessary to construct a Data Highway Plus network.
- Enter the proper parameters into the *message* instruction for communication between PLC-5's over a Data Highway Plus.

COMMUNICATION between programmable controllers or between programmable controllers and computers can be done over a data highway. The Allen-Bradley PLC-5 family of programmable controllers has a built-in port through which it communicates over the Allen-Bradley Data Highway Plus. The *Data Highway Plus* is a local area network (LAN) that allows peer-to-peer communications. There are basically three types of topology in LAN networks: the star, the ring, and the bus. Each configuration had its own characteristics. Communications between programmable controllers, or between programmable controllers and computers, has become a common application. It may be necessary to store information such as recipes, production requirements, and start-up sequences in the computers and then to download that information to the programmable controllers actually controlling the production. It also may be necessary for the programmable controller to perform production counts, diagnostics, and alarming and then to have this information transferred to computers for record keeping and data logging. In addition, program can be transferred between a computer and a programmable controller over such a network.

Because the protocol on local area networks is usually different, communication between different manufacturers' programmable controllers and even between different families of the same manufacturer's programmable controllers may involve additional hardware and programming.

16-1 Local Area Network (LAN)

A **local area network (LAN)** is a communications network with a data rate of up to 20 megabytes per second and a range of up to 20 kilometers. There are two basic means of transmission in an LAN: analog and digital.

Analog signals are either voltage-based or current-based, and information is transmitted by varying either the amplitude or the frequency of the waves. Varying the wave amplitude is called *amplitude modulation,* or AM. Varying the wave frequency is called *frequency modulation,* or FM. Cable television is an example of an analog system in which several frequencies are transmitted over the same cable. Because multiple frequencies may be transmitted over the same cable, this has become known as a **broadband system**. Broadband systems typically use coaxial cable for transmission, due to the wide bandwidth requirements.

Digital signals have either of two states: on and off. As previously mentioned, these states are called *bits*. Information is sent at a set rate, known as the *baud rate*, which is expressed in bits per second. Patterns of bits represent information. Computer networks commonly use this method of communication, also known as a **baseband system**. Baseband systems typically use a twisted, shielded pair of wires as transmission medium for most applications.

LAN's come in three basic *topologies* (that is, physical layout or configuration of the communication network): star, ring, and bus. The points where the devices connect to the transmission medium are known as **nodes** or **stations**.

In the **star** network, illustrated in Figure 16-1, a central control device is connected to a number of nodes. This allows for bidirectional communication between the central control device and each node. All transmission must be between the central control device and the nodes, since the central control device controls all communication. However, if the central control device fails, the total system is down.

The **bus** topology is illustrated in Figure 16-2. In a bus network, each node is connected to a central bus. When a node sends a message on the network, the

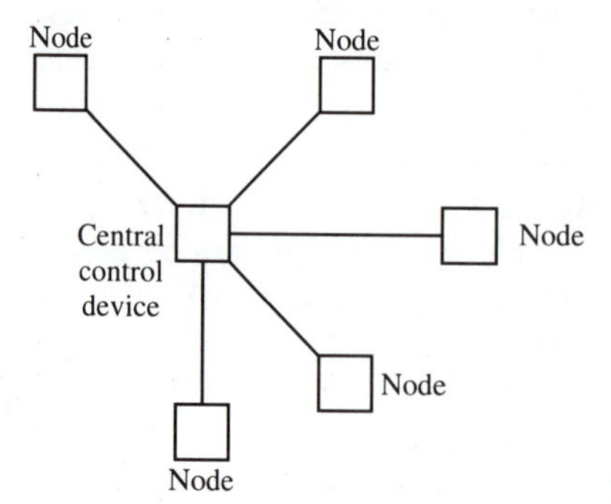

Figure 16-1 Star Network

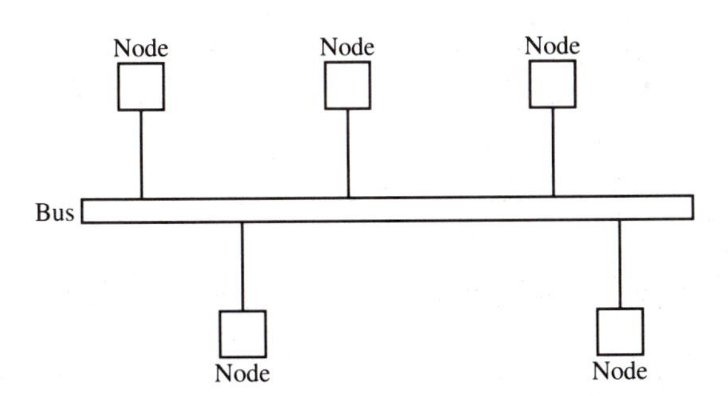

Figure 16-2 Bus Network

message is aimed at a particular station or node number. As the message moves along the total bus, each node is listening for its own node identification number and only accepts information sent to that number. Control can be either centralized or distributed among the nodes.

In the **ring** network, illustrated in Figure 16-3, each node is connected to another node, in ring fashion. There is no end or beginning to the network. Messages are aimed at a particular node or station number, with each node listening for its own identification number. Signals are passed around the ring, and are regenerated at each node. Control can be centralized or distributed.

Each network has its own advantages and disadvantages. The bus network requires the least amount of cable, but breaks in the cable are hard to detect. In the star topology, additions and deletions to the system can be made easily. But it requires the most cabling, and loss of the central control device shuts down the entire system. The ring network requires more cable than the bus network but less than the star network. Distances covered by the ring network can be greater, since the signal is regenerated at each node. However, modification of the network may be difficult.

16-2 Data Highway Plus

The PLC-5 family of programmable controllers has a built-in communications port that allows the PLC-5 to communicate via the Data Highway Plus. The **Data Highway Plus** is a token-passing, bus network that permits up to 64 stations on the network, octally numbered from 0 through 77. The token passing between stations gives each station a chance to send a message over the highway. A station

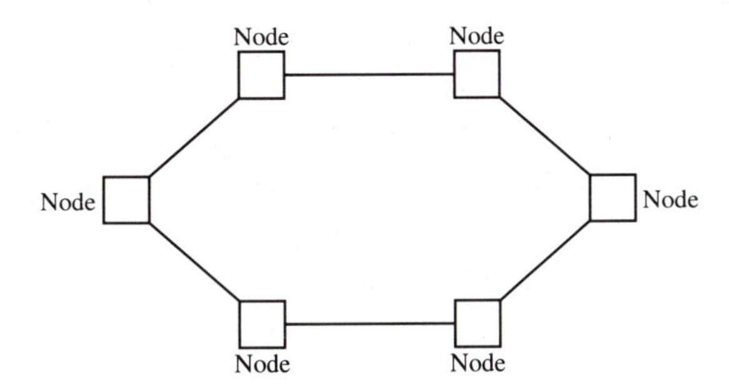

Figure 16-3 Ring Network

becomes a master on the network when it receives the token and sends a message to another station. This could be a *read* message from the other station or a *write* message to that station. When the message is completed, the master passes the token to the next higher station number. The maximum cable length on the highway, from one end to the other, is a trunk line of 10,000 cable feet, with drop lines down to each station from the trunk line of up to 100 feet. The baud rate, or communication rate, is 57.6 K baud (57,600 bits per second).

The Data Highway Plus also enables a computer, the programming device, to communicate with the programmable controllers on the network. The computer must have programming software that permits such communication and that enables it to edit and monitor the program in the programmable controller. The computer may also have a *driver program,* that is, a program allowing it to transfer data between itself and programmable controllers.

The Data Highway Plus is not intended for real-time control of data, since the transfer of information is relatively slow. This is because only two stations are communicating with each other at any one time, and there is communication back and forth between the two stations during the passing of data.

Hardware for Data Highway Plus

Figure 16-4 shows a typical Data Highway Plus network. Note the different modules (PLC-3 and PLC-2) needed to interface the Data Highway Plus with the computers. Only the PLC-5 has the built-in capability to communicate over the Data Highway Plus.

A number of different modules allow a computer to connect to the Data Highway Plus, for instance, the 1770-KF2 Series B, which is a desktop module; the 1785-KE module, which is a chassis-mount module that mounts in an I/O chassis and receives its power from the I/O chassis backplane; and the 1784-KT board, which plugs into a slot in an IBM or IBM-compatible computer. The 1785-KA3 module, which mounts in an I/O chassis and receives its power from the backplane, allows the PLC-2 family of processors to communicate over the Data Highway Plus. The 1775-S5 module mounts in the PLC-3 chassis and allows the PLC-3 to communicate over the Data Highway Plus.

Any of these modules enable the PLC-3, the PLC-2, the PLC-5, and computers to communicate over the same local network.

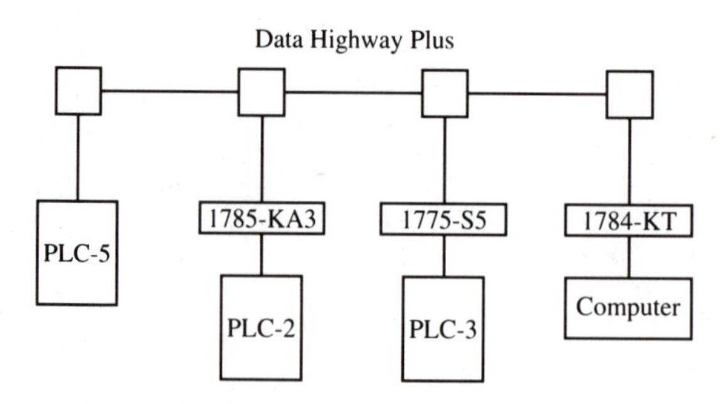

Figure 16-4 Typical Data Highway Plus Network

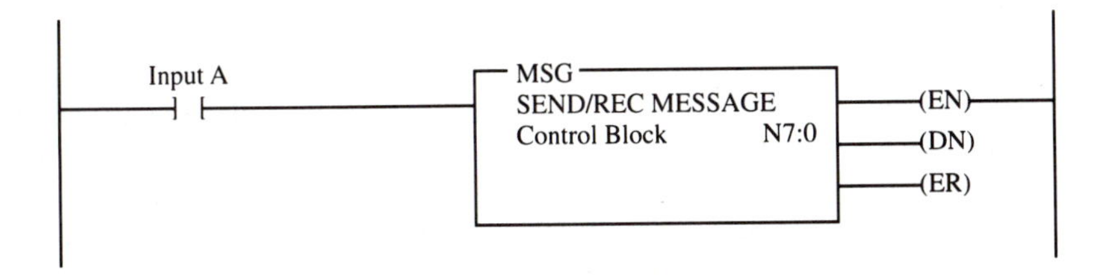

Figure 16-5 Example of the *Message* Instruction (MSG)

16-3 *Message* Instruction (MSG)

The *message* instruction (MSG) is an output instruction used by the PLC-5 programmable controller to initiate a message on the Data Highway Plus. This message will be either a *read* or a *write* of data to another programmable controller.

An example of the MSG instruction is shown in Figure 16-5. The only entry made directly into the message instruction is an integer-file address, which is the starting address of the control block. This is entered without the # symbol. The length of the control block ranges from 11 to 12 words for communication with a PLC-2 to 11 to 15 words for communication with a PLC-3 or PLC-5.

An example of the screen display for entering data into the control block of the *message* instruction is given in Figure 16-6. The first entry involves choosing *read* or *write*. The F2 control key toggles between *read* and *write*. The choice made depends on the programmable controller in which you are entering the message instruction: If this programmable controller is going to be doing a *write* to another programmable controller's data table, then toggle to *write;* if it is going to

MESSAGE INSTRUCTION DATA ENTRY FOR CONTROL BLOCK N7:0

Read/Write:	Write
PLC-5 Data-Table Address:	N7:100
Size, in Elements	10
Local/Remote:	Local
Remote Station:	N/A
Link ID:	N/A
Remote Link Type:	N/A
Local Node Address:	20
Processor Type:	PLC-5
Destination Data-Table Address:	N7:50

Block Size = 9 Words

Press a key to change parameter or <RETURN> to accept parameters.

Program Forces: None Edits: None PLC-5/15-File HWY

Read/ Write	PLC-5 Address	Size, in Elements	Local/ Remote	Remote Station	Link ID	Remote Link	Local Node	Proce Type	Destin Address
F1	F2	F3	F4	F5	F6	F7	F8	F9	F10

Figure 16-6 Typical Screen Display for Data Entry in MSG

a *read* from another programmable controller's data table, then toggle to *read*.

The F2 key allows you to enter the starting element address in the data table in the programmable controller in which you are entering the message instruction. If you are doing a *write*, this will be the address of the source file; and if you are doing a *read*, it will be the address of the destination file.

The F3 key enables you to enter the length of the file that was selected via the F2 key, in number of elements (1 through 999). This is the length of the file that will be transferred by the message instruction.

The F4 key toggles between *local* and *remote*. *Local* is for connection on the Data Highway Plus. *Remote* is for connection to Data Highway through an interface module. (Data Highway is another Allen-Bradley local area network. We cover only the Data Highway Plus here, since that is the LAN capability built into the PLC-5.)

Keys F5, F6, and F7 do not allow entries when *local* is choosen with the F4 key.

The F8 key allows you to enter the local station address, which is the target PLC's address (the address to which the message is being sent), 0 through 77 octal.

The F9 key lets you choose between communicating with the PLC-2, the PLC-3, or the PLC-5 programmable controller.

The F10 key allows you to enter the starting address of the target programmable controller's data table. This is the destination-file address if *write* was chosen in the first entry, and the source file if *read* was chosen.

The information entered in the control block as shown in Figure 16-6 would write data from integer-file address N7:100, ten words long, to the PLC-5, station address 20, to its integer file N7:50. Figure 16-7 illustrates this example.

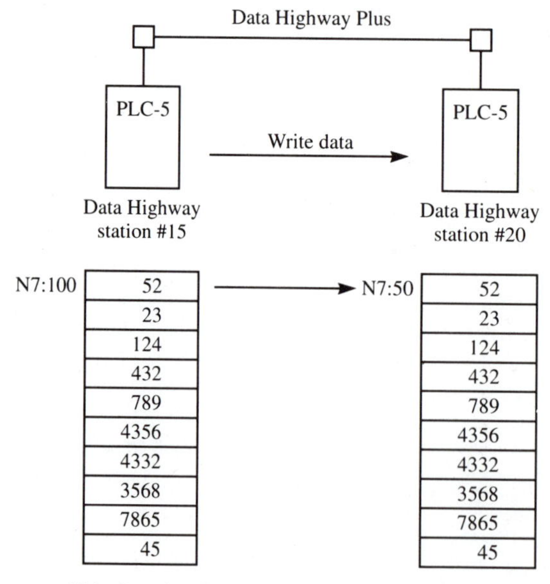

Figure 16-7 Data Highway Plus Message for Figure 16-6

Summary

A local area network (LAN) is a communication network with a data rate of up to 20 megabytes per second and a range of up to 20 kilometers. The Data Highway Plus is an LAN that allows for the transfer of data between programmable controllers or the transfer of data or program between programmable controllers and computers. It is built into the Allen-Bradley PLC-5, with additional communication modules allowing communication with other Allen-Bradley programmable controllers. The Data Highway Plus may be up to 10,000 cable-feet in length, and operates at a baud rate of 57.6 k baud. The transfer of data between PLC-5's is accomplished with the *message* instruction, which may be used to write or read up to 1000 elements between the programmable controllers.

Exercises

16-1. What are the three different topologies for LAN's?

16-2. Which topology is used by the Data Highway Plus?

16-3. What is the maximum length over which two programmable controllers may communicate over the Data Highway Plus?

16-4. At what baud rate does communication take place over the Data Highway Plus?

16-5. What is the maximum number of stations allowed on the Data Highway Plus? How are they numbered?

16-6. What is the maximum length of data that may be transferred by the *message* instruction?

16-7. What information would have to be entered into the *message* instruction to read 25 elements of data from a PLC-5's data file, starting with address N7:25 and copying that data into the PLC-5, where you are entering the *message* instruction, data-file address starting with N7:40? You are reading the data from station number 10 of the PLC-5. The control block for the message instruction is N7:100.

17 PLC Programming Languages

OUTLINE

OBJECTIVES

Upon completion of this chapter, the student will be able to:

- Explain the difference between a machine-language program and a high-level language.
- Explain the purpose of an assembler or interpreter.
- Give the advantages and disadvantages of Ladder, Boolean, and BASIC programming languages.
- Analyze a simple control program in Ladder, Boolean, or BASIC.

THE manufacturers of programmable controllers went to great lengths to make the early PLC's user-friendly to the people involved with control. This meant developing a special language that could be used immediately without special training. The first such language, ladder logic, is based on the use of relay ladder logic symbols for entering a program. For those involved in the area of digital logic, ladder logic was not so friendly, so machines were designed that would accept Boolean statements. Finally, for those who were implementing control via computers programmed in a high-level language, it made sense to add BASIC to PLC's. Thus, you can see that the PLC has continually evolved and is becoming more flexible. We can expect this evolution to continue, with manufacturers adding features that are in demand. This chapter will briefly cover some of the most common languages available.

17-1 PLC's CPU, Machine Code, and Languages

Presentday programmable controllers have one or more microcomputer chips at their heart, and with this comes the potential for having available any high-level language. We can see how this is possible by understanding how the microcomputer chips are programmed. The specific processor used by various manufacturers varies but will probably have been made by one of the two principal manufacturers of this technology, Intel and Motorola. The central processor unit, the CPU, is the heart of the microcomputer, and it gets its instructions from reading patterns of 1's and 0's from a program stored in memory. If we have a processor with an eight-bit data bus, it can read eight bits of data at a time. This means it can receive a set of commands ranging from all eight bits being 0's, or 00000000, to all bits being 1's, or 11111111. There are actually 256 possible combinations of 0's and 1's on an eight-bit machine. Each eight-bit pattern means something very specific to the CPU. Let's use Intel's popular 8085 CPU as an example.

Machine Code Pattern	Meaning
00111010	Read an eight-bit pattern from memory
00110010	Store an eight-bit pattern in memory
10000110	Add two eight-bit patterns
00101111	Change an eight-bit pattern so the 0's are 1's and the 1's are 0's

Machine language uses binary patterns to program a microcomputer. Every PLC manufacturer must write a program with these codes to make it function. Writing machine code requires the programmer to have an intimate knowledge of the particular CPU and demands special training and skill. To make this job easier, the codes are first written in assembly language. That is, instead of writing the binary pattern, an assembly program is written using lettered codes called *mnemonics* that the manufacturer creates and documents in an instruction set. The machine codes used in the preceding example have the following mnemonics:

Mnemonic	Machine Code	Meaning
LDA	00111010	Read an eight-bit pattern
STA	00110010	Store an eight-bit pattern in memory
ADD	10000110	Add two eight-bit patterns
CMP	00101111	Change an eight-bit pattern so the 0's are 1's and the 1's are 0's

It is much easier for humans to write programs via mnemonic codes than with binary patterns.

Once the mnemonic codes for a program have been written in a correct and logical order, they could be changed by hand to their equivalent machine code. However, this is rather tedious and error-prone because of the ease of miscopying the machine codes. Instead, a computer can run a program called an **assembler** that converts an assembly-language program into machine code.

PLC users do not need to know or write machine code, because the PLC's have machine codes stored in memory that allow programming via a keyboard. An **interpreter** is a machine-language program written so that certain keystrokes or groups of key strokes, called *commands*, can be understood by the microcomputer. The commands may be part of any high-level language. To use BASIC on a PLC, you would need a BASIC module with a BASIC interpreter; to use C or FORTH, you would have to have those respective interpreters. In essence, the interpreter translates the high-level commands into machine language as the commands are entered. The processor in the PLC can understand only machine language.

Relay ladder logic programming of a PLC requires that a relay logic interpreter be stored permanently in memory so that these keys and commands can be changed to machine code. The only language the CPU understands is 1's and 0's.

Manufacturers can make available an interpreter for any high-level language. Whether this is done depends on the demand for that language. **High-level languages** accept normally English words or groups of words and convert them into machine-language routines that enable those commands to be executed by the CPU.

For each of the languages discussed in this chapter, we will use the same control problem and implement it on an Allen-Bradley PLC-5. This is the control problem introduced as an example in Chapter 3. It is repeated here:

Control Problem: Design a control scheme that requires a person to press two push buttons, one at a time in the proper sequence, to operate a particular machine. Once the two buttons have been pushed correctly, the machine will stay on until a reset button is pushed. The correct sequence is PB1 followed by PB2. A relay will detect any wrong sequence and light an error light, which will stay on until the control is reset. Once an error has occurred, no operation of the push buttons can turn on the machine. The relay control for this is shown in Figure 17-1.

17-2 Ladder Logic

Ladder logic was the first and most popular language available on the PLC, and it is still popular. Since it is covered thoroughly in other chapters of this book, we will not go into much detail here. The ladder logic interpreter stored in read-only memory enables control to be programmed via keys with symbols of relay contacts, relay coils, timers, counters, etc.—the devices and symbols used in relay control. Many control engineers know relay logic because they have used it for years. The symbols have similar functions to the relay devices themselves. For example, when the symbol for a coil is pressed and entered in a program, a routine is available in memory to simulate a coil's operation. When programs are printed out for review, they look very similar to standard relay logic and can be interpreted basically the same way. Shown in Figure 17-2 is the two-push-button lock implemented via ladder logic on the Allen-Bradley PLC-5.

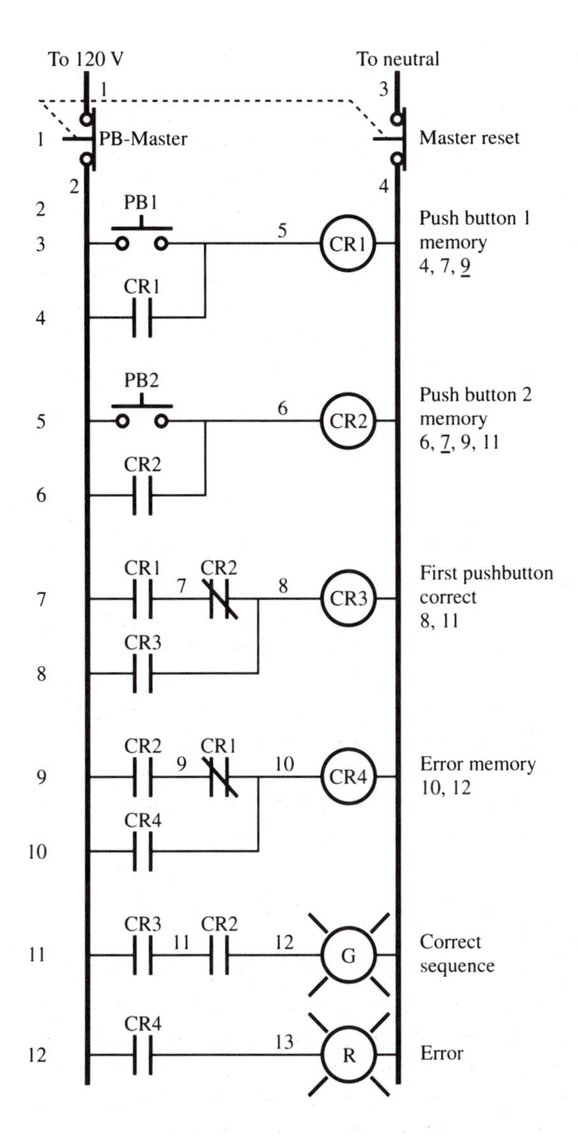

Figure 17-1 Two-Push-Button Lock Implemented with Relays

An alternative to standard relay logic is a structured approach using sequential function charts, also available on the PLC-5. Figure 17-3 shows the control problem implemented this way. Files 02 through 14 would have to be set up. (Review Chapter 15 for the procedures for setting up a sequential function chart, if necessary.)

17-3 Boolean

For engineers who used digital gates in the form of integrated circuits to create control and who were unfamiliar with relay logic, it made sense to make Boolean logic available on PLC's. Chapter 4 introduced Boolean algebra and the gates used, such as AND, OR, INV, EXOR. When programming in Boolean algebra,

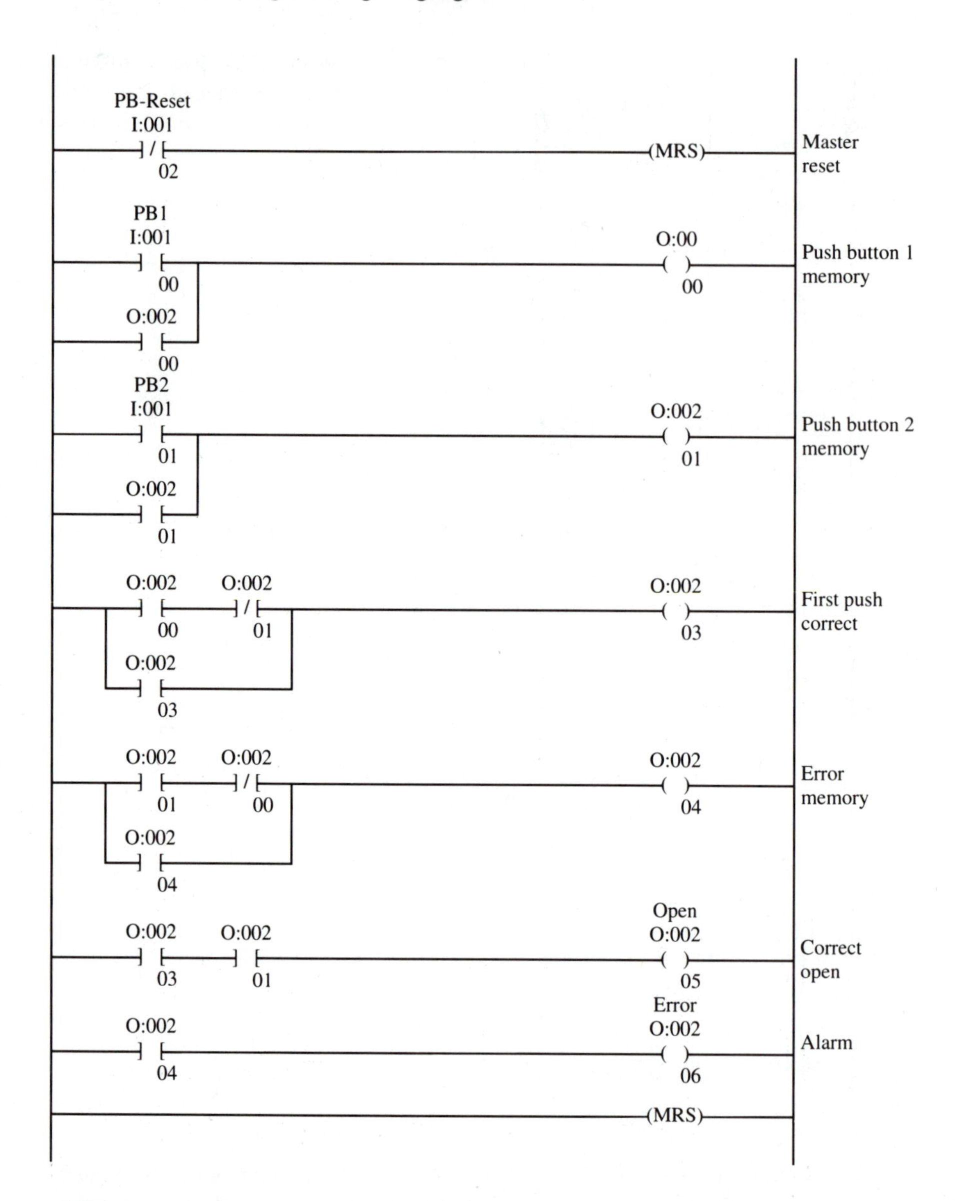

Figure 17-2 Two-Push-Button Lock Implemented on the PLC-5 with Ladder Logic

equations can be written to duplicate the logic. (For example, $A + B = C$ represents an OR gate with A and B as inputs and C as an output.

The PLC-5 doesn't support direct entry of these equations, but it does have the logic functions available. The gates and logic devices can be represented by selecting the function you want via the keyboard and assigning addresses for the inputs and outputs as required. The PLC simply needs an interpreter to understand and convert these statements into subroutines so the machine can execute these logic

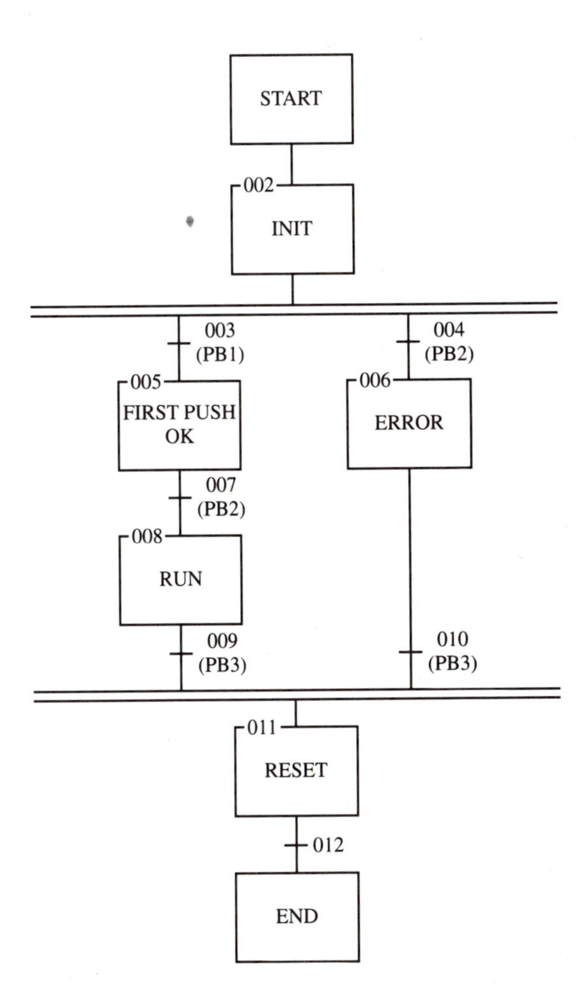

Figure 17-3 Two-Push-Button Lock Implemented on the PLC-5 with Sequential Function Charts

functions. Instead of placing the symbols for logic gates, you define the function you want and assign addressing. Since the PLC will be constantly scanning the inputs and updating outputs for the correct operation, the logic function must be placed in the correct order. Shown in Figure 17-4 is the two-push-button lock implemented via standard IC logic gates. Figures 17-5a, b, c, and d present the two-push button lock implemented on the PLC-5 using the logic functions available.

17-4 BASIC

BASIC (*B*eginner's *A*ll-purpose *S*ymbolic *I*nstruction *C*ode) is a very popular programming language. It uses English words and statements instead of symbols, and consequently it is very easy to use once a few rules have been learned. A BASIC interpreter to convert the English commands into machine language must be available. Different computers need different Basic interpreters. For example, the IBM personal computer and the Apple Macintosh computer each require a different inter-

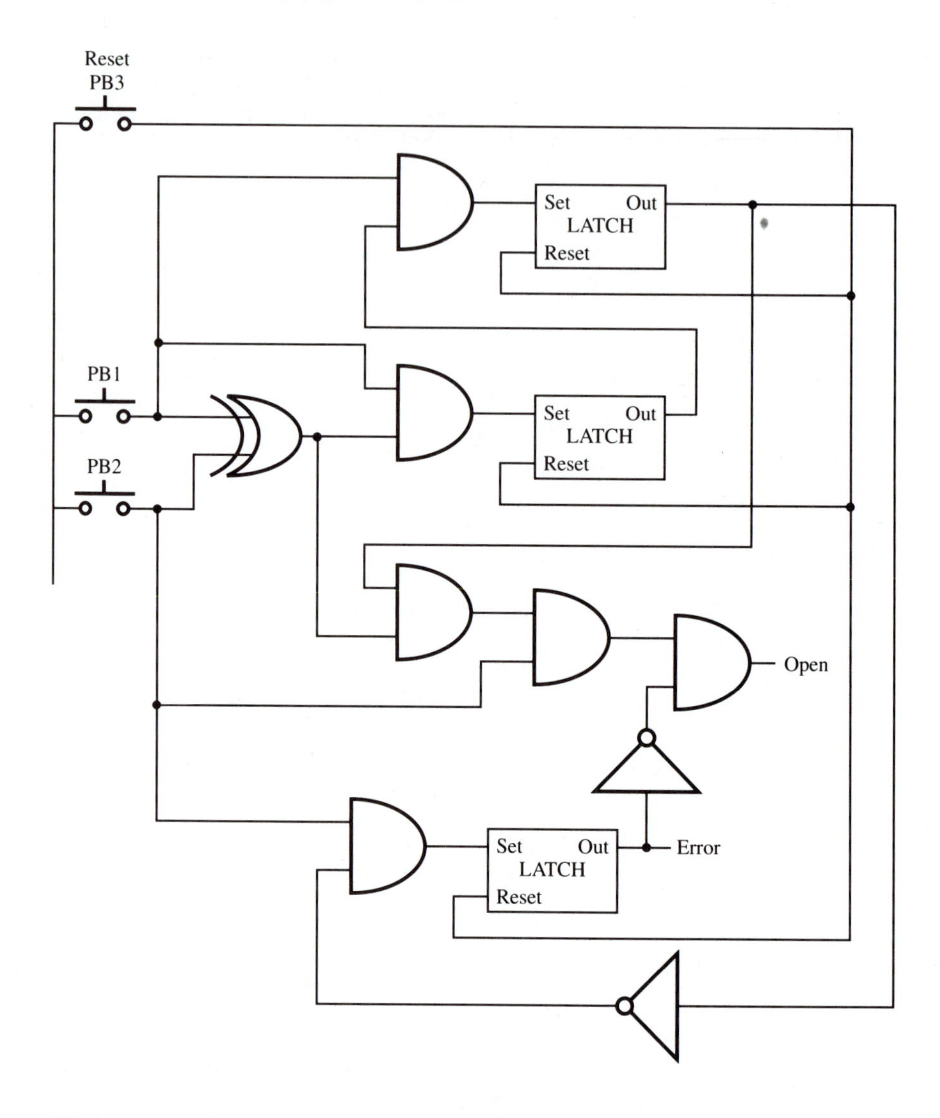

Figure 17-4 Digital Logic for the Two-Push-Button Lock

preter. Only BASIC will be discussed in this chapter. Other high-level languages, such as PASCAL, FORTH, and C, are not at present usable with the PLC-5.

A module is available for the PLC-5 that can execute BASIC programs. This module, which resides in the PLC-5 chassis, contains a BASIC interpreter for Intel BASIC-52 programming as well as communications circuitry that enables the BASIC module to communicate with the PLC-5 processor. This communication is accomplished via bidirectional block transfers of data. The ladder logic program that must be written for this is shown in Figure 17-6. This program alternately lets the BASIC module read the input table and modify the output table. The input table in the processor is read via the *block transfer write* (BTW), and

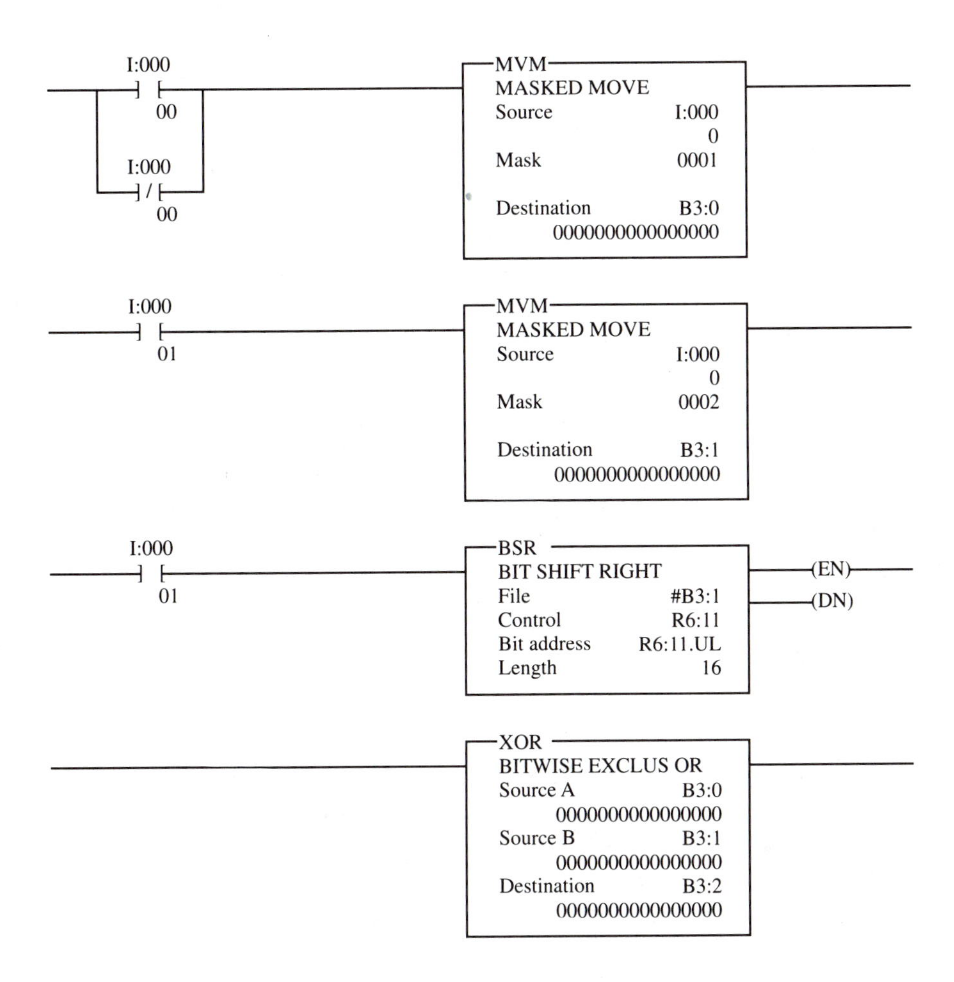

Figure 17-5a Digital Logic for the Two-Push-Button Lock Implemented on the PLC-5

the output table is modified via the *block transfer read* (BTR). Only one *block transfer* instruction is executed per scan, and this ladder logic causes the block transfers to alternate between BTW and BTR. How the output table is modified depends on a BASIC program, which must be written by the user.

We will shortly show the two-push-button lock as implemented in BASIC on the PLC-5. The inputs for this example are from push button 1, which is fed into the module for I/O group 0, bit 0, and push button 2, which is fed into the module for I/O group 0, bit 1. The reset push button is fed into the module for I/O group 0, bit 2. Two-slot addressing is used with two 8-input modules in group 0 and two 8-output modules in group 1. The BASIC module is in group 2, slot 1.

Understanding the program requires a knowledge of BASIC. Many good books are available on programming in BASIC, so that will not be covered here.

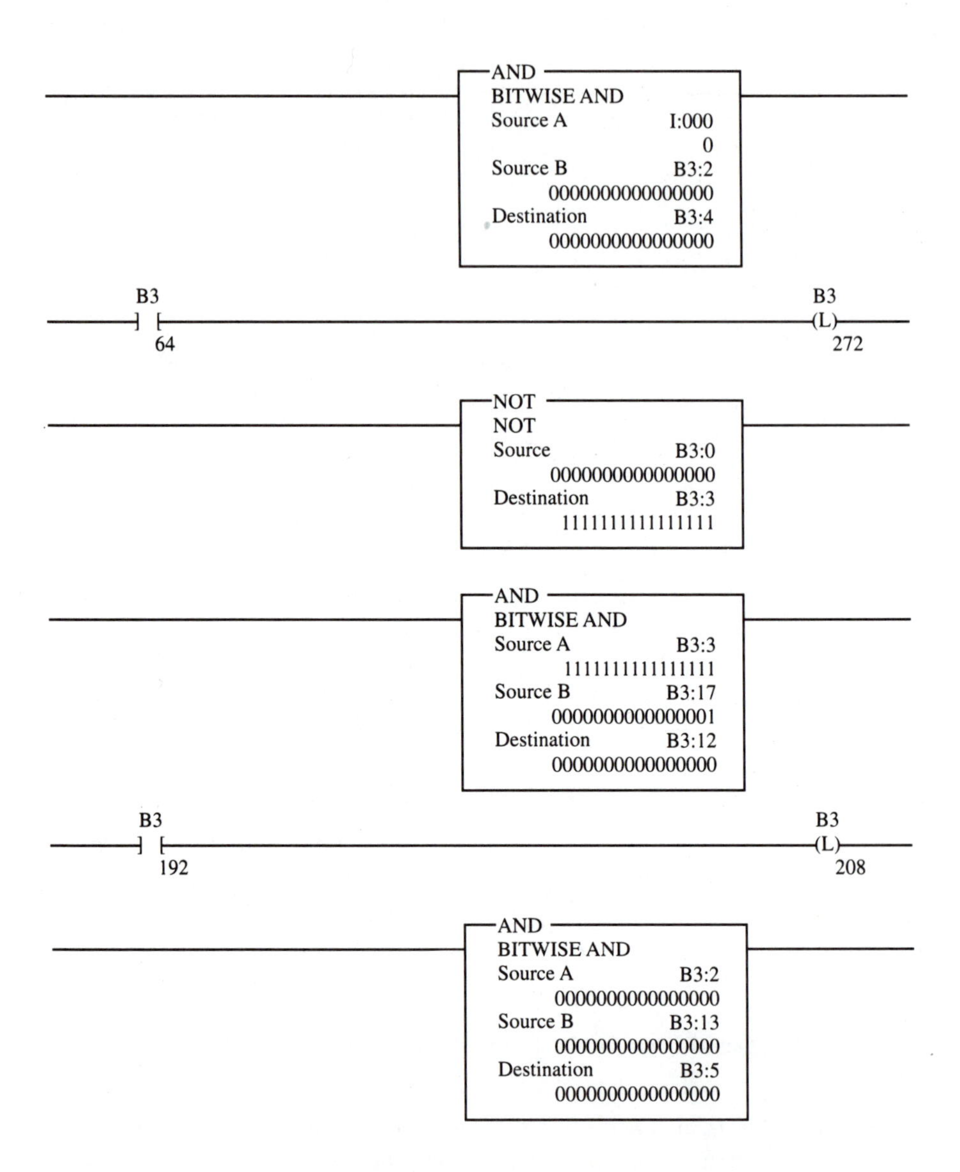

Figure 17-5b Digital Logic for the Two-Push-Button Lock Implemented on the PLC-5 (*Continued*)

The CALLs in this BASIC program are user subroutines in ROM and come with Allen-Bradley's BASIC module. Here is a summary of the CALLs used and the tasks they accomplish.

Subroutine	Purpose
CALL 2	Transfers the BTR buffer to the auxiliary processor on the BASIC module

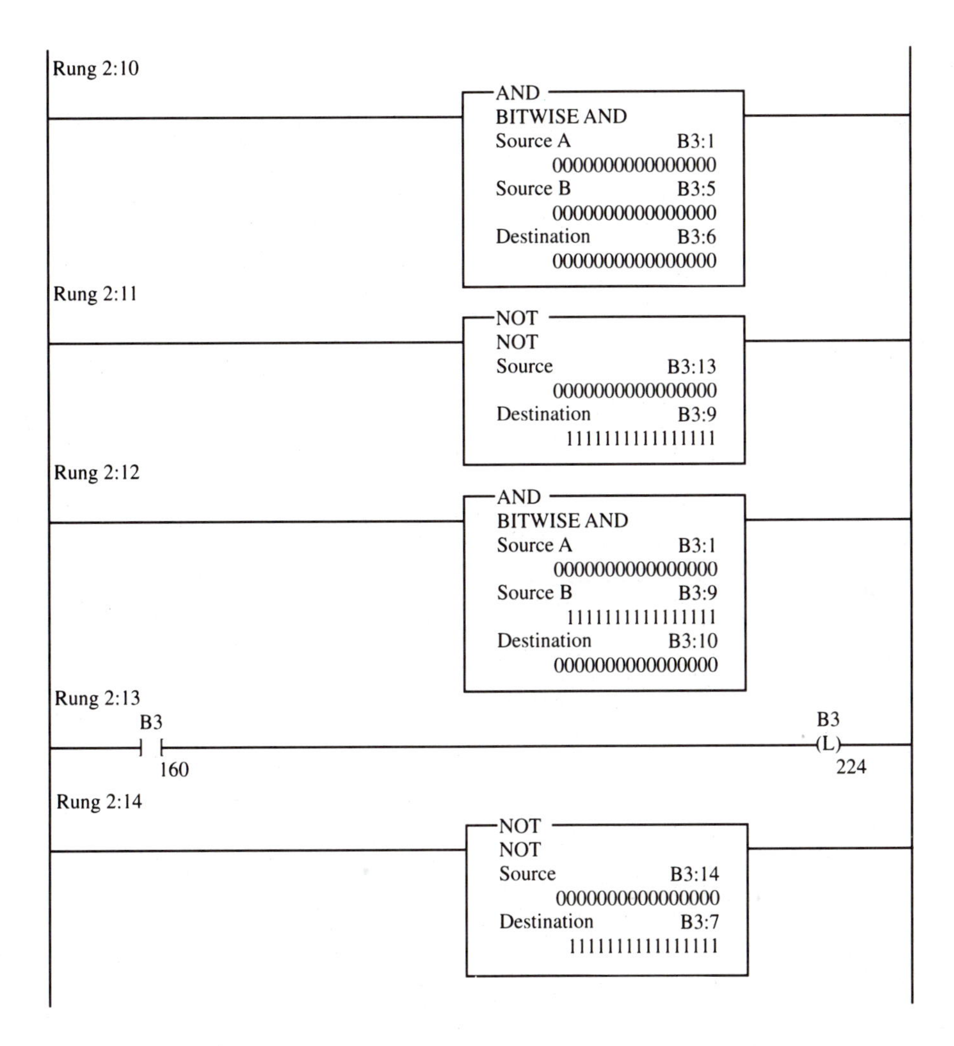

Figure 17-5c Digital Logic for the Two-Push-Button Lock Implemented on the PLC-5 (*Continued*)

CALL 3	Transfers the BTW buffer to the auxiliary processor on the BASIC module
CALL 4	Sets the number of words for BTW to be transferred between the BASIC module and the processor
CALL 5	Sets the number of words for the BTR to be transferred between the BASIC module and the processor
CALL 6	Transfers the BTW buffer of the auxiliary processor on the BASIC module to the BASIC BTW buffer
CALL 7	Transfers the BTW buffer of the auxiliary processor on the BASIC module to the BASIC BTW buffer
CALL 11	Changes 16-bit binary to internal floating point
CALL 21	Changes internal floating point to 16-bit binary

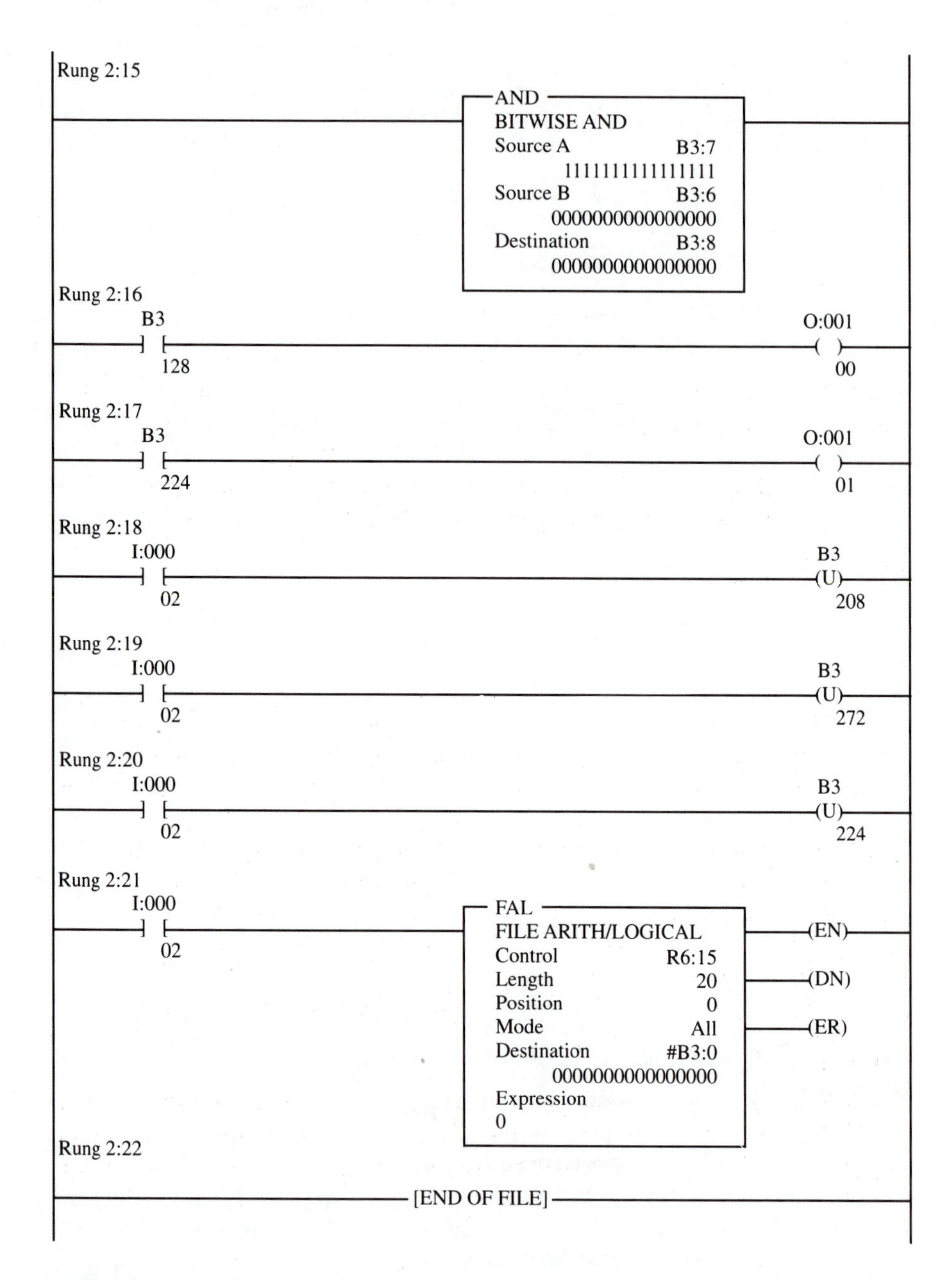

Figure 17-5d Digital Logic for the Two-Push-Button Lock Implemented on the PLC-5 (*Continued*)

```
10 REM: This program is written to implement a two-
   push-button sequential lock. A reset operation
```

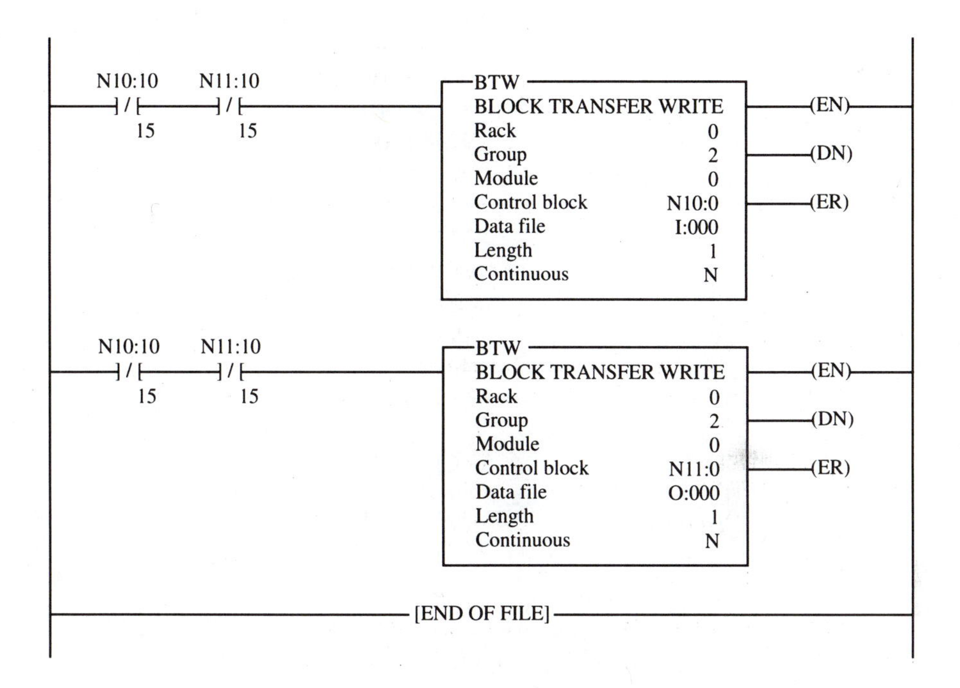

Figure 17-6 Ladder Logic to Implement Data Exchange with BASIC Module on the PLC-5

```
     is also provided and is initiated by a reset
     push button.
20   PUSH 10:CALL4:REM: Sets BTW length to 1 word
30   PUSH 10:CALL2:REM: Sets BTR length to 1 word
40   GOSUB 200
50   IF x = 1 THEN GOSUB 300
60   IF x = 2 THEN GOSUB 400
70   IF x > or = 4 THEN GOSUB 700
80   GOTO 40
90   END

200  REM: This subroutine reads the BTW buffer and
      sets the 16-bit word to a floating-point vari-
      able in an array A.
210  CALL6:REM: Read the BTW buffer
220  PUSH (I): CALL 11: POP x
230  RETURN
300  REM: This subroutine sets a flag C indicating
      the first push button pressed was OK
310  LET C = 1
320  RETURN
```

```
400  REM: This subroutine determines if it is OK to
     operate, or if an error has occurred when the
     second push button is pushed.
410  IF C = 1 THEN Y = 1: REM: Correct sequence, op-
     erate output.
420  GOSUB 600
430  IF C = 0 THEN Y = 2: REM: Error has occurred, ac-
     tivate error output.
440  GOSUB 600
450  Return

600  REM: This subroutine writes to the BTR buffer
     so that the outputs can be changed.
610  PUSH Y:PUSH 1: CALL 21
620  REM: Write to the BTR buffer
630  CALL7
640  RETURN

700  REM: This subroutine resets the outputs to
     zero when the reset push button is pressed.
710  Let Y = 0
720  GOSUB 600
730  RETURN
```

BASIC is more often used for number crunching, data manipulation, and report writing than for implementing control programs. However, as this example demonstrates, it can be used for control if desired.

Summary

The programmable controller is designed to be very flexible and to accommodate the needs of engineers and technicians in control so that they have several choices. Each type of implementation has something to offer. Relay logic is easy to teach and graphically lets the user see how the control is implemented. If the control doesn't get too complicated, it is not hard to troubleshoot. Digital logic is easy to teach and graphically lets the user see how the control is implemented. It is a bit harder to troubleshoot, but also is more powerful than relay logic. BASIC, the most sophisticated implementation, lets you take full advantage of the powerful computing capability of the PLC. Learning a computer language requires more effort and training than do the first two implementations, but the payoff is the ability to handle more complicated control problems. We have introduced only BASIC in this chapter, but other languages are sure to be made available as PLC manufacturers respond to the needs of the market.

Exercises

17-1 What are the six different ways the two-push-button control problem was implemented in this chapter?

17-2 What is the function of an interpreter program? What type of memory does it require?

17-3 Explain what machine language is and why manufacturers usually don't require users to program in this language.

17-4 What is meant by the term *high-level language*? Give one example.

17-5 What must a PLC manufacturer do in order to make a new high-level language available on a PLC?

17-6 What is the function of an assembler? Discuss what type of programs it works with and what type of programs it creates.

18 Selecting a PLC

OUTLINE

OBJECTIVES

Upon completion of this chapter, the student will be able to:

- List what a written specification for purchasing a PLC should contain.
- Explain the difference between single-ended, multitask, and control-management applications for a PLC, and describe how these applications affect selection of a PLC.
- Evaluate approximate memory size needed when purchasing a PLC.
- Discuss the advantages and disadvantages of fixed vs. modular I/O.
- Name some of the available I/O modules and special modules.
- Describe and show a diagram of the following networks: daisy chain, star, loop, and multidrop.
- List several communication protocols for PLC's.
- Name the languages presently available for programming PLC's.
- Compare the main peripherals available for programming PLC's.
- Name several environmental conditions that necessitate special protection for a PLC.
- Discuss how support, stability, cost, compatibility, and reliability enter into choosing acceptable manufacturers.
- Discuss the importance of matching PLC to level of employee skill.

SELECTING a PLC can be a daunting task if you start by surveying all the equipment available. With the many companies making PLC's and the number of different models available, there are well over a hundred machines from which to choose. We could fill half this book with the various specifications for these machines. Looking through all the specifications can be confusing, which could make selection very difficult.

The approach we will take in this chapter is to analyze the particular application requirements and then write a specification that details what is needed, for the actual application makes some features important and others trivial. Once the specification is complete, it can be submitted to PLC manufacturers for bidding.

18-1 Specification

When buying a product from another company, it is wise to generate a document called a **specification**. By describing in detail the requirements of the product the seller is to provide, the specification narrows the risk of getting something you don't want. If the seller agrees to the specification, the document becomes a legal contract to which the manufacturer must adhere.

Specifications can be simple, such as a specification for a washer, or quite extensive and voluminous, such as a specification for an airplane. There is no limit to what the buyer can put into a specification, but when well written it is succinct and contains only items important to the buyer. Here are some typical topics one might find in a specification:

Description of item(s) purchaser wants
Billing procedures
Change procedures
Extras
Service engineer or technician
Correspondence procedures
Bill of materials
Acceptable suppliers
Warranty
Shipping
Inspection rights
Types of wire used
Testing required
Codes to be met
Applicable standards
Types of acceptable material
Finishes
Wire-pulling procedures
Correct tools
Test reports
Date required
Penalty clause (what happens if delivery date is missed)

The list could go on endlessly.

This chapter will discuss the criteria necessary to correctly specify a PLC.

18-2 Application

Your particular application is the most important criterion, because it defines the category of PLC you need. Keep in mind that the PLC is just a tool for implementing the control you desire. The tool should not dictate how you proceed or limit what can be done. When you buy a car, you probably first decide if you need a luxury, standard, or compact model. Similarly, with a PLC the specification categories are a function of the PLC's use.

There are three major types of application: single-ended, multitask, and control management. See Table 18-1.

Table 18-1 Application Categories

Category	Meaning
Single-ended	One PLC controlling one process
Multitask	One PLC controlling several processes
Control management	One PLC controlling several PLC's

Single-ended This type of application may indicate a small PLC or a micro, although it could be complicated as well as simple. The PLC for this kind of application would be a stand-alone unit and would not be used for communicating with other computers or PLC's. The size and sophistication of the process being controlled is obviously a factor in determining which PLC to select. Single-ended applications can be as simple as controlling the motors for a car wash or as complex as running a complicated robot that requires feedback and continual adjusting to operate properly. The applications could dictate a large processor, but usually this category requires a small PLC.

Multitask The multitask application probably means a medium PLC, since one PLC is programmed to perform many tasks. The number of I/O points becomes significant. If the PLC will be a subsystem of a larger process and have to communicate with a central PLC, you will need the accessories required to communicate with the computers or PLC upstream.

Control Management This type of application usually points to a large, sophisticated machine that will be supervising several PLC's. It probably will be downloading programs that tell the other PLC's what has to be done. It may need to be capable of being connected to all the PLC's so it can communicate with any one it wishes, by proper addressing. This kind of application requires a machine designed to communicate with other PLC's, and possibly with a computer. The computer would be used for processing data the control-management PLC collects from the other PLC.

18-3 Memory Need

It is important to determine the size of memory required, because otherwise you might buy a PLC with insufficient memory for your application or one unable to have its memory expanded to the level needed. Not anticipating your memory need could mean the costly purchase of a larger PLC at a later date.

Table 18-2 Factors Affecting Memory Size Needed

Number of I/O points
Size of control program
Use of structured programming
Data-collecting requirements
Supervisory function

Calculating the memory size required can be tricky, because different machines require different amounts of memory to accomplish the same task. The safest way to get what is needed is to include in the specification the items listed in Table 18-2. The *number of I/O points* is simple. For the *size of the control program*, you could specify the number and type of each instruction needed to implement the control, or you could include the actual program. If *structured programming* is involved, include the state diagram or Petri network you will be using. To help determine how much memory *data collection* will require, specify the number of devices and the amount of data. The *supervisory function* should be shown in a diagram indicating how many machines the supervisory machine will be controlling.

Once the size of memory needed is decided, it would be wise to add 50% more, to ensure sufficient memory is available and to provide for future expansion.

User memory (RAM) can usually be added to a machine in 1K increments (1K = 1024 words of memory). The length of a word is dependent on the machine you buy. Some machines may use eight-bit words; others may use 16-bit words. As the technology continues to develop, 32-bit or larger words are possible. If you'll need the capability to expand memory at a later date, this should be included in your specs.

If you require a nonvolatile memory (for instance, for a control program so essential that you do not want to rely on battery backup for its integrity), then most PLC's will accept a PROM, EPROM, or EEPROM module that can hold critical programs during a total loss of power.

18-4 Fixed vs. Modular I/O

There are two ways in which I/O is incorporated into the PLC: fixed and modular.

Fixed I/O is typical of small machines, particularly the micros that come all in one package with no separate, removable units. The processor and I/O are packaged together, and the I/O terminals are available but cannot be changed. The obvious advantage of this type of packaging is lower cost (as little as a few hundred dollars). The number of available I/O points varies and usually can be expanded by buying additional units of fixed I/O. Such add-on units come in the same type of physical package, except more I/O is available per package because the processor is not needed. The disadvantage of fixed I/O is its lack of flexibility; you have to take what you can get in the quantities dictated by the packaging. Also, if any part in the unit fails, the whole unit has to be replaced.

Modular I/O is provided by compartments into which separate modules can be plugged. This feature greatly increases the options and flexibility. You can choose from the modules available from the manufacturer and mix them any way you desire. The increased flexibility makes it easy to switch the PLC from one process to another, which may be impossible with fixed I/O. Troubleshooting becomes much simpler, and problems with one module can be remedied by simply changing the module. The PLC can also be operated with the module removed if the module isn't critical to the process. The obvious disadvantage is higher cost (up to thousands of dollars).

I/O Modules and Special Modules

Chapter 7 describes in detail the possible input, output, and special modules available from manufacturers. An alphabetical list of these follows.

Analog Input Module converts analog input signals to digital signals the processor can read.

Analog Output Module converts digital signals from the processor to isolated analog signals that can be used to drive output devices.

ASCII Input Module converts ASCII-code input information from an external peripheral to alphanumeric information the PLC can understand.

ASCII Output Module converts alphanumeric information from the PLC to ASCII code to be sent to an external peripheral via one of the standard communication interfaces, such as an RS-232 or RS422.

BCD Input Module allows the processor to accept four-bit BCD digital codes.

BCD Output Module enables a PLC to operate devices that require BCD-coded signals to operate.

Communication Module allows the user to connect the PLC to high-speed local networks that may differ from the network communication provided with the PLC.

Discrete Input Module lets the user make two-state signals available to the PLC's for use in the control program.

Discrete Output Module enables the PLC's processor to control output devices by changing a digital-level signal to the level required by the devices being controlled.

Dry Contact Output Module enables the PLC's processor to control output devices by providing a contact that is isolated electrically from any power source.

Encoder Counter Module enables continual monitoring of an incremental or absolute encoder. (Encoders keep track of the angular position of shafts or axis.)

Grey Encoder Module converts the grey-code signal from an input device into straight binary.

Isolated Input Module enables the PLC's processor to receive dry contacts as inputs.

Language Module enables the user to write programs in a high-level language. The languages available differ from one manufacturer to another. BASIC is the most popular. Other language modules available are C, FORTH, and PASCAL.

PID Module proportional integral-derivative closed-loop control lets the user hold a process variable at a desired set point.

Servo Module closed-loop control is accomplished via feedback from the device.

Stepper Motor Module provides pulse trains to a stepper motor translator that enables control of a stepper motor.

Thumb-Wheel Module enables the use of thumb-wheel switches to feed information in parallel to the PLC to be used in the control program.

TTL Input Module enables devices that produce TTL-level signals to communicate with the PLC's processor.

TTL Output Module enables a PLC to operate devices that require TTL-level signals to operate.

This is just a sample of the modules available. New modules are constantly being introduced to the market. The problem for the buyer of a PLC is that these modules are not all available from one manufacturer. If you require a particular module for an application, you must check that that module is available. For instance, if you wanted to use FORTH to program a PLC, you would need to find out which manufacturers support this high-level language. At present, only a few support FORTH. However, if demand increases for this language, manufacturers will respond. If you like a particular PLC that is not supporting what you want, see if that function will become available in the near future.

Another problem is compatibility between the module and the application device. For example, if you purchase a manufacturer's PID module, will it be compatible with the device with which it will be used? Check with the manufacturer to see which devices it supports. If a special module is needed, this may limit the manufacturers you can consider. Include in the specification any special requirements.

18-5 Communication

Communication may be required either for remote I/O or for networking PLC's.

Remote I/O is I/O under the control of the PLC but not in the main PLC chassis. Such I/O is located at a different location, which could be miles away. Each location has its own chassis or racks that may contain many I/O points. Communication between the remote I/O is via a twisted pair, coaxial cable, or a fiber-optic link. The advantage of remote I/O is the very substantial savings in wiring materials and labor, because the wiring between the device being operated and the I/O module can be greatly reduced (remote I/O usually means one large PLC controlling a high number of I/O points). Another advantage is that remote location can be taken out of service to be worked on while the rest of the system is up and running. A serial interface adapter or module is required in the main PLC unit, and a remote serial interface adapter or module is required in the remote chassis or rack.

Networking PLC's require special consideration, and we could devote an entire book to this subject alone. First, you must decide on the type of communication link that will be used for the network. The usual choices are twisted pair, coax cable, telephone, radio link, microwave link, and optic fiber. Second, you must choose the configuration for the network: daisy chain, star, loop, or multidrop. Third, computer communication protocols must be chosen. The present choices are RS-232C, RS-449, current loop, and IEEE 488. The machine you pick must be compatible with the network you are going to use. Networking will require a large, sophisticated machine for the server, which could be a computer or a PLC. The size of the PLC's connected to the network will be determined largely by their application. Data collection and transmission will require that the machines connected to the network be compatible.

18-6 Languages

One big difference between a personal computer and a PLC is that the PLC comes with a program language built into its permanent memory, whereas a personal computer comes with a disk operating system. Which PLC you buy can limit you to the language it comes with, unless it is a modular type that enables you to plug in a language module. PC's have large user memory. Several megabytes is normal, which enables you to load and use software stored on a disk. Any language capability can be made available via this software, which can be loaded into the memory via these disks. PLC's are not designed with this kind of flexibility but are meant to be specialized computers for control, and the features available enhance interfacing and control with external devices.

When choosing the language you are going to use for programming, it is important to consider the personnel who are going to be programming and maintaining the PLC. What would happen if the person you are presently relying on leaves? If you pick a language only one person understands, you are taking a considerable risk. It may be the right choice, but you will need to educate personnel to ensure proper use and maintenance of the PLC. Be prepared to educate others so that you can replace personnel if needed.

Ladder Logic is the most popular language for PLC's, for the reasons discussed in Chapter 17. It is the oldest method of implementing control, so programmers know how to use it. It is also easy to understand, because the resulting ladder logic shows graphically what the control is going to do.

Boolean is the next most popular language. It is used for developing control with discrete logic and is also widely known. PLC's have been provided with Boolean because of the demand for it.

High-level Computer Languages are available and probably will become more popular. Examples are BASIC, PASCAL, and FORTH. These are the main control language on some machines, or they can be added by purchasing a plug-in module that has the necessary interpreter. If you already know one of these languages and are fluent enough to program control, they are well worth looking into. If you don't know the language, you must to be willing to learn it. This takes a considerable amount of time, with a correlation between how powerful a language is and how long it takes to master it. Use of any of these languages will limit the choice of manufacturers and machine. If your application involves data collection and number crunching, then a high-level language is indicated. Control languages aren't really very good for these because they weren't designed for these functions.

18-7 Peripherals

You will need some means to program your PLC. The following three options are usually available: hand-held programmer, portable programmer, and programming software plus an interface between a personal computer and the PLC.

Hand-held programmer This is the least expensive option, but it has limitations. It enables you to place the control program into the PLC, although you must put the instructions in one step at a time. You cannot easily troubleshoot your program because it doesn't let you see several rungs of control at once. It is intended as simply a download device and is not useful or practical for developing and troubleshooting control.

Portable Programmer This option usually lets you review and monitor your program by means of an LCD screen. It will have editing capability, on-line and off-line programming, and monitoring while control is running on the PLC. These features allow you to develop and test control programs readily. Portable programmers are more expensive than the hand-held programmer and less expensive than using a software package on a PLC. You can get a printout of your control for documentation and troubleshooting. If one of your requirements is to develop a limited amount of control, this is a wise choice.

Programming Software Plus Interface Between PC and PLC This is generally more expensive than the other two options, but it has advantages nonetheless. It requires the purchase of the software and the necessary interface card for communication between the PC and the PLC. The software and the interface give you a powerful development system for generating control programs. You can generally see larger portions of the control at one time, and the editing features are more powerful and make development of control easier and efficient; for example, cut-and-paste capability, merging of several small programs into one, and structured programming such as state diagrams and Petri networks. Better and easier documentation is also available. If one of your requirements is to develop a large amount of control, this is a good choice.

18-8 Environment

The environment into which you will be placing the PLC may require some special consideration. There are several types of hostile environments that could result in serious problems. Table 18-3 lists examples of environmental situations requiring special care.

Table 18-3 Environments Requiring Special Care

Excessive ambient electrical noise
Corrosive atmosphere
Dusty atmosphere
Tropical atmosphere
Excessive moisture or water
Vibration
Hazardous atmosphere
Excessive ambient heat or cold

Any of these will cause the PLC to fail if no protection is provided. The solution in most of these cases is to place the PLC in a special protective enclosure. *Electrical noise* will require filtering and surge protection. *Vibration* will require special mounting. *Hazardous atmosphere* will require special expensive enclosures.

18-9 Manufacturer

Choosing acceptable manufacturers depends on a number of factors: support, stability, cost, compatibility, and reliability.

Support basically means having: good technical assistance; short required lead times for delivery of equipment; quick repair and turnaround; and readily available replacement parts. Part of technical assistance is a sales staff able to show how their PLC can help solve your control problems and get technical questions answered quickly and accurately. After the sale, training becomes important. The

PLC is just a high-tech tool, and you must know how to use the tool properly to be effective, which requires training. Thus, the company you buy a PLC from needs to have training available. Alternatively, if you are going to teach your own personnel, then good reference manuals and documentation are essential. It is a good idea to include in your specification a requirement for reference manuals and training documentation plus a clause about training personnel.

Stability determines whether the company will be there when you need them. Nothing is more disheartening than buying a product from a manufacturer who later goes out of business, leaving you with no technical assistance or parts for maintenance and limited chance of future expansion based on that product. In the extreme, it can be very costly. The number of PLC manufacturers exploded in the 1980s to somewhere around forty. The 1990s will surely see the number of vendors reduced, probably to a dozen. Unless you want to look into each company and study its prospects for survival, it is best to go with a company that has been in the field for a while and has a substantial share of the market. This may not be where the real bargain is to be found, but the risk of being left holding the bag is diminished. One way of evaluating this is to look at the company's position in the PLC market, information that is readily available in PLC trade magazines.

Cost If cost is no object, you can ignore this paragraph, but seldom is that the case. As just pointed out, for stability, you should go with well-established companies; however, well-established companies can dictate prices, and often their prices are the highest. Good bargains can be found in emerging companies that can be lean and mean because they are smaller and very competitive. A percentage of these companies will grow into major contenders. To limit your risk, do your homework and research the company.

Compatibility Will the equipment you buy now be compatible with equipment you already have or will purchase later? Answering this will take a good deal of research, but can be very costly if overlooked. Also, when a company brings out newer, more advanced machines that completely change a product, you could be left out in the cold as the earlier versions are phased out. Most companies recognize this and adjust the design of their new machines to be compatible.

Reliability The reliability of a manufacturer's product should be checked. Many companies keep repair records so they can monitor equipment reliability. Ask the manufacturer's sales department to supply you the names of users of the particular PLC. Then contact these companies for their comments on satisfaction and copies of performance reports, if available. Also, check trade magazines, which often carry articles evaluating reliability and customer satisfaction.

18-10 Matching PLC to Employee Skill Level

It is important to take into account the skill level of the people who are going to be using the equipment. The complexity of PLC's varies. The higher-end models are quite sophisticated and will require highly skilled people to take advantage of their capability. Generally, the more powerful the PLC, the harder it will be to use. Training will be necessary to bring staff up to the level required by the machine. Personnel probably will have to be sent to manufacturer training programs. And when they return, time will have to be invested to master the skill. A match must be found between the level of skill required by the machine and people's ability to learn those skills. Otherwise, expensive equipment will be underutilized or possibly not used at all.

Summary

Selecting a PLC can be simple or a complex process, depending on the application. You should first define the task to be done and then select the machine that can best fill your needs. Selecting a PLC first and then trying to make it fit your requirements can result in costly problems and delays. Get independent outside help if you don't have the knowledge necessary to make a good selection. The various manufacturers will be glad to give you this help, but they have a vested interest in their product. The time spent at the beginning in defining the task and researching which PLC's can do the job properly will be well worth it. This chapter touched on some of the common requirements you will need to consider. A well-written specification will aid in the purchase of a PLC that meets your requirements and will ensure a proper contract between you and the manufacturer.

Exercises

The following exercises will require you to do some library research. This type of research is similar to what you will have to do when selecting a PLC. Consult the bibliography of this book for a start on good sources.

18-1. Find two manufacturers of PLC's that have structured programming available. List the manufacturer and the name of the structured programming.

18-2. List fifteen companies presently making PLC's.

18-3. Pick a manufacturer and then a model of PLC that has a modular chassis. List a sample of fifteen modules available.

18-4. Explain and show a diagram for one of the following networks:

Daisy chain

Star

Loop

Multidrop

18-5. Explain how communication is accomplished with one of the following protocols:

RS-232C

RS-449

Current loop

IEEE 488

18-6. Explain how you would protect a PLC from each of the following hostile environments:

Excessive ambient heat or cold

Excessive ambient electrical noise

Corrosive atmosphere

Dusty atmosphere

Tropical atmosphere

Excessive moisture or water

Vibration

Hazardous atmosphere

18-7. List three manufacturers and the models with software available to enable their PLC to be programmed with a personal computer.

18-8. Find a manufacturer and model of PLC that can be programmed in each of the following languages:

Boolean

BASIC

PASCAL

FORTH

APPENDIX A

Graphic Symbols for Electrical Control Diagrams

NOTE: These symbols, which represent only some of those used, were taken from the National Machine Tool Builders Assoc. (NMTBA) and Joint International Congress (JIC) standards. The choice of standards is not uniform throughout the industry. Other graphic standards used include the American National Standards Institute (ANSI) standards.

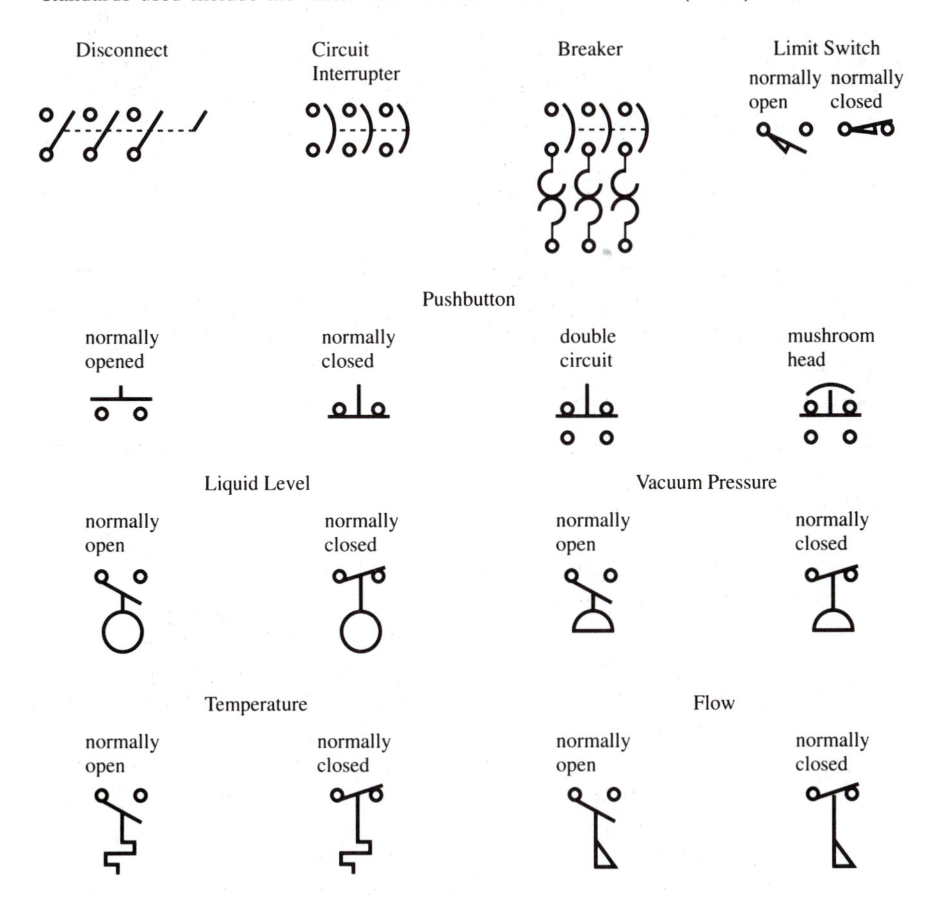

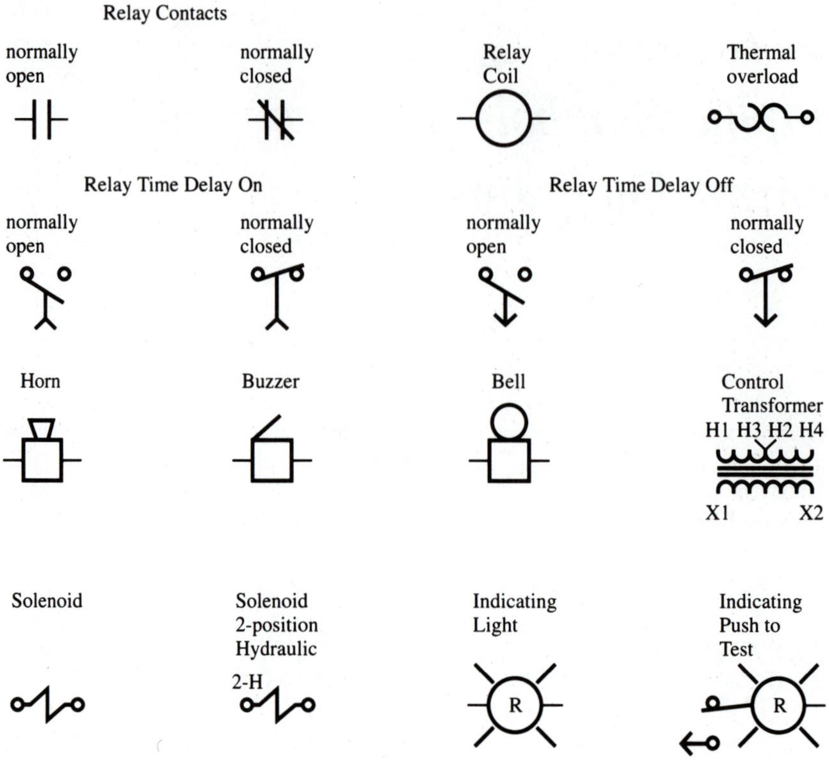

Relay Contacts

normally open

normally closed

Relay Coil

Thermal overload

Relay Time Delay On

normally open

normally closed

Relay Time Delay Off

normally open

normally closed

Horn

Buzzer

Bell

Control Transformer

H1 H3 H2 H4

X1 X2

Solenoid

Solenoid 2-position Hydraulic

2-H

Indicating Light

R

Indicating Push to Test

R

APPENDIX B

Projects for Programmable Controllers

Outline

THE laboratory setup for the programmable controller and its suggested I/O configuration will be covered in this appendix. It will include a recommended hardware setup along with suggested laboratory functions that can be implemented with this hardware.

B-1 Hardware Recommendations

Hardware can be set up in a number of configurations. We will discuss the standalone, the scanner-adapter configuration, and the Data Highway Plus setup.

The minimum processor and I/O recommended *per station* is as follows:

1 Allen-Bradley PLC-5 processor
1 12-slot I/O chassis
2 16-DC input cards
2 16-DC output cards
1 8-amp slot-mounted power supply
6 normally open push buttons
2 normally closed push buttons (*stop* buttons)
8 two-position selector switches
1 four-digit binary-coded-decimal LED display
1 four-digit binary-coded-decimal thumb-wheel

(optional, but recommended)

1 12-bit-resolution analog-input card

1 12-bit-resolution analog-output card

4 0–10-volt potentiometers

2 0–10-volt voltmeters

2 two-position selector switches to select analog output channels to voltmeters

The PLC-5 processor can function in either the adapter or the scanner mode and has built-in Data Highway Plus capabilities. The sixteen input switches and sixteen pilot lights are discrete-input and discrete-output devices that allow for the simulation of on–off control. They will enable you to explore the relay-type instructions and to simulate real-world input and output devices, including limit switches, photo eyes, pressure switches, and other on–off input devices, as well as motor starter coils, solenoid valves, and other on–off output devices. The sixteen switches and sixteen pilot lights also match the sixteen-point input and output I/O cards.

The four-digit BCD thumbwheels are inputs connected to the sixteen-point input card, and allow data to be entered into the program from the thumbwheels. Thus you can enter data into timer and counter presets and enter data into data files for manipulation by word and file instructions. The four-digit BCD LED's are outputs fed to the sixteen-point output card and allow for the display of timer and counter accumulated values and serve as destination for data obtained in word- and file-manipulation instructions.

The optional analog-input and analog-output cards allow you to bring analog data in and out of the programmable controller, which handles digital data. The analog-input card handles sixteen single-ended or eight differential inputs, and converts an analog signal brought into the card into a digital signal that can be transferred to the processor. The analog-output card handles four analog outputs and converts the digital signal from the processor into an analog signal, which is the output from the analog card.

The open slots can be used for additional modules, for instance, input cards allowing for the connection of input devices such as limit switches, proximity switches, photo eyes, and other inputs from lab devices. Output cards could also be inserted, which would allow for control of motor starters, solenoid valves, and other output devices.

The analog cards could also interface with analog-input and analog-output signals, either voltage or current, for positioning control, temperature control, and other analog control.

A sample lab configuration for the recommended hardware is presented in Figure B-1.

B-2 Programmable Controller Applications

Originally, programmable controllers acted to replace relays. They were used for basic sequential control, where they were applied to the control of conveyors,

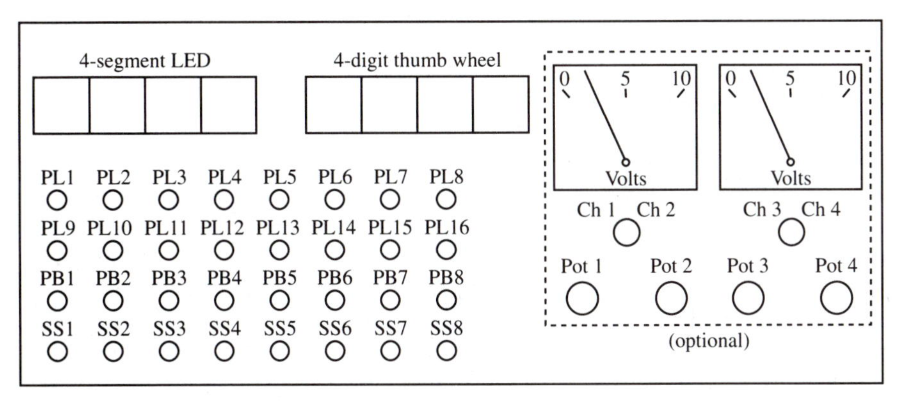

Control Panel

PLC-5	16-pt input module	16-pt input module	16-pt output module	16-pt output module	Open slot	8-amp power supply	Open slot	Analog input	Analog output	Open slot	Open slot	Open slot

I/O Chassis Layout

Figure B-1 Sample Lab Configuration for Recommended Hardware

basic material-handling equipment, sequencing of pumps, and other basic controls. They also served timing and counting functions. The first applications were in the automotive industry. Today, their use has expanded to many different applications in various industrial and nonindustrial situations.

Let's first look at applications where they still function as relay replacers: basic sequential control, starting and stopping motors, shifting solenoid valves that control hydraulic cylinders, and other basic on–off control. In these basic applications, the programmable controllers also serve timing and counting functions, such as controlling the amount of time a part stays in a position and counting the number of parts being made. Examples of such applications include moving parts on an assembly line, controlling a car wash, and controlling a small saw mill.

When the programmable controller can manipulate data, its capabilities are greatly enhanced. With the basic word-level instructions, data can be moved and manipulated. Timer and counter preset values can be changed via thumb-wheels, meaning an operator can change such values without employing a programming device. Math calculations and comparisons can be done, and then the programmable controller can make decisions based on such calculations and comparisons.

Percentages of good parts to total parts made during a shift can be readily determined by monitoring the information coming from the line and performing the math in the programmable controller.

Operating on file data gives the programmable controller true computer capabilities. Here, recipes can be stored in the programmable controllers, and operators can change the whole operation with just the push of a button. Material flow can be tracked with bit-shift registers or FIFO instructions. The program can be made more compact by using the sequencer instructions to control outputs and to make comparisons with input data. Diagnostics can be accomplished using the file logic instructions and information fed to screen displays to aid in troubleshooting. Groups of data can be analyzed to determine averages and do flow totalization.

Intelligent I/O modules can be added to the system in order to achieve analog-to-digital and digital-to-analog data conversions. Positioning modules can control linear positioning. Servo control can be accomplished via a servo module. Process control can be achieved through loop control. There are many general and specific applications that utilize the intelligent I/O modules.

Communication between the programmable controllers and programmable controllers and computers also greatly enhance their capabilities. Production reports can be generated promptly from information monitored by the programmable controller and sent to computers anywhere in the world. Production scheduling can be downloaded from computers to programmable controllers in the same manner.

Programmable controllers started off doing on–off control in the automotive industry and the expansion from there has been tremendous. Applications are no longer limited to the industrial environment, but have reached into many other areas.

When you take a ride on the latest amusement park roller coaster, there is a good chance the cars are being controlled by programmable controllers. When you ride a tram from one end of a hotel to the other, it too may be controlled by a programmable controller. Theme park rides and trams and moving monsters are probably controlled by programmable controllers. The elaborate scene changes on today's theater stages are likely to be controlled by programmable controllers. These are all applications that you might overlook when thinking of programmable controllers.

Another area where the use of programmable controllers has multiplied at an enormous rate is the food industry. When you drink a can of soda or beer, you can be sure programmable controllers controlled the making of the can lid, the can itself, the filling of the can, the packaging, and the sending of it to the warehouse. The production schedule was downloaded from a computer to the programmable controllers in order to determine how much and what types were to be made. These represent just a few of the many possible applications.

B-3 Suggested Projects

Many of this text's chapter exercises can be implemented via a demonstration panel such as that illustrated in Figure B-1. The rungs of logic or programs can be entered and their operation verified by using the demonstration panels. Numerous different student projects also can be implemented via the demonstration panels. The switches can simulate any on–off field-input device, such as a limit switch, photo eye, proximity switch, or pressure switch. The pilot lights can simulate motor starters, control valves, and other output devices. Also, additional input and output cards can be added that can be connected to such devices.

Student projects can start with basic ones, such as sequencing the starting of a series of motors on a time or an event basis. Another good example for sequencing is to construct a traffic light program. This can be easily simulated on the light panel in two different ways, first by using the basic bit instructions and timers, and then by using the sequencer output instruction and a timer. This is a good example for demonstrating the efficiency of programming a sequential operation via the sequencer output instruction.

Another good project is an automatic car wash sequence that moves the car from various stations, such as rinse, wash, rinse, and dry. The motion of the conveyor and the timing of the valve control can be easily simulated with the control panel.

Loading a tank with different quantities of ingredients, mixing them, heating them to a certain temperature, and pumping them out of the tank also can be easily simulated. Here, different recipes can be implemented by changing the ingredient amounts, and the temperature and the bake time can be varied by the operator. All of this can be programmed via the processor and the simulator panel.

An analog-output card controlling a variable-speed drive that uses an analog voltage or current signal to control the drive speed is a good demonstration of the programmable controller's analog capabilities. Input readings may be taken into the processor, and the digital information needed to control the speed can be determined by the program and transferred to the analog card, to be converted into analog form and sent to the drive.

Another project is to control the temperature in a tank or an oven. Temperature input is brought into the processor, either from a thermocouple card or an analog-input card. The processor then can decide what output value will maintain the temperature at a desired setpoint, and control the output accordingly. The PLC-5 processor can accomplish this via loop control.

Summary

The applications for programmable controllers in industrial and commercial applications are varied. The instruction set contained in the processor makes the programmer versatile and gives it the tools for developing a program that is both efficient and productive. There is an application for programmable controllers practically anywhere we are controlling mechanical motion, temperature, flow, or pressure.

Answers to Odd-numbered Exercises

Answers for Chapter 1 Exercises

1-1. Some of the advantages of programmable controllers over relays:

Control applications can be changed without having to change hard-wiring, by simply modifying programs stored in memory.

A PLC has a computer to do many functions not possible with relay control.

Since the same PLC can be used for many applications, one type of PLC can be mass produced, thus making PLC's more economical than one-of-a-kind relay controls.

PLC's can communicate easily with one another because they are computers.

PLC's take up less space than relay controls.

1-3. *Advantages:*

Rugged and able to take electrical surges and noise.

Simple to troubleshoot.

Good isolation between contacts and control.

Can safely predict the states of contacts when control circuits are deenergized.

Disadvantages:

Expensive to change control because of required rewiring.

Takes up more space than a PLC.

Requires more power for control.

1-5. When a new product is introduced, some of the major factors that will determine its acceptance are: cost, reliability, availability, technical support, ease of use, and flexibility.

1-7. The difference between a personal computer you can buy today and a programmable controller stems from the programmable controller's following special features:

Hardened to work in an industrial environment.

Equipped with special I/O that can handle the voltages common in industrial control.

Has a special keyboard and a language developed specifically for industrial control.

1-9. Many manufacturing processes could be cited. A good example of a sequential process we use often is the automatic, drive-through car wash. A programmable controller could be used effectively for controlling the wetting, the soaping, the rinsing, the application of hot wax, the blow-drying, the moving of the car, etc.

1-11. Some problems the solid-state industry had when introducing new solid-state controls into control where high-voltage power was used are:

High voltage wasn't properly isolated from the solid-state devices. Such voltage easily damaged the solid-state devices and caused failure. (If you pass 120 volts through the base of a bipolar transistor, it is almost certainly going to be damaged.)

High-voltage surges coupled to the solid-state devices via electrical fields set up by these surges or through magnetic coupling. Such surges cause failures in solid-state equipment.

Answers for Chapter 2 Exercises

2-1. Contacts are shown in the deenergized state in control schemes to standardize electrical schematics. It is similar to deciding whether to drive on the right side or left side of the road—it is much easier if everyone makes the same choice. When the standards are followed, schematics from various sources can be understood without problems; if no standard is present, confusion results.

2-3. The difference between a motor starter and a contactor is that the starter has an overload device in series with the contacts, whereas a contactor does not. The overload relay is designed to match the overload characteristics of the motor to which the starter is connected. The contactor is really just a heavy-duty relay designed to pick up loads the general-purpose relay cannot handle. A starter is a contactor and overload relay wired together into a single unit.

2-5. Some consequences of exceeding the voltage rating on a device are: loss of device's life expectancy, electrical failure, shock hazard, and overheating.

2-7. Isolation is achieved between the source of power operating the coil of the relay and the source of power connected to the contacts of the relay because the coil is wound with insulated wire which physically insulates it from the contacts.

Answers for Chapter 3 Exercises

3-1. Nodes on a schematic are labeled in order to identify nodes so point-to-point wiring diagrams can be created. Once the control is hard-wired with labeled wires, the labels serve in conjunction with the schematic as an aid to troubleshooting and testing.

3-3. The characteristics of a ladder diagram that differentiate from other types of schematics are:

It looks like a ladder with rails and rungs.

Contacts and inputs are on the right and outputs are on the left.

Each control line is numbered, and contact locations of coils are designated.

Nodes are labeled for wiring diagrams and testing.

3-5. See accompanying figure.

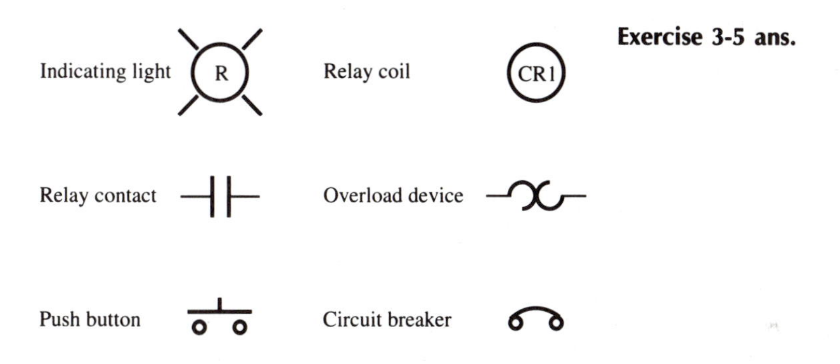

Exercise 3-5 ans.

3-7. Node 2 goes to the following addresses: B2, D2, E1, E2, E3, F2, F3, TB3.

3-9. See accompanying figure.

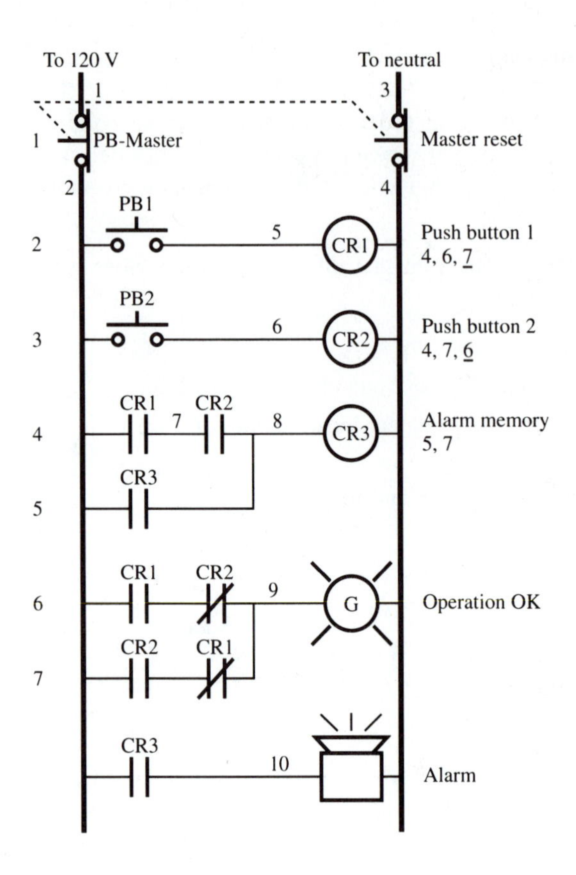

Exercise 3-9 ans.

Answers for Chapter 4 Exercises

4-1.

A	B	C	D	Y
0	0	0	0	0
0	0	0	1	0
0	0	1	0	0
0	0	1	1	1
0	1	0	0	0
0	1	0	1	1
0	1	1	0	1
0	1	1	1	1
1	0	0	0	0
1	0	0	1	1
1	0	1	0	1
1	0	1	1	1
1	1	0	0	1
1	1	0	1	1
1	1	1	0	1
1	1	1	1	1

4-3. ORing the fundamental products $Y = \overline{A}\overline{B}\overline{C} + \overline{A}B\overline{C}$

Factoring $= \overline{A}\overline{C}(\overline{B} + B)$

Reducing $= \overline{A}\overline{C}$

4-5. $Y = \overline{A}\overline{B} + \overline{C}D + BD$ See accompanying figure.

	$\overline{C}\overline{D}$	$\overline{C}D$	CD	$C\overline{D}$
$\overline{A}\overline{B}$	1	1	1	1
$\overline{A}B$		1	1	
AB		1	1	
$A\overline{B}$		1		

Exercise 4-5 ans.

4-7. The truth table is:

A	B	C	Y
0	0	0	0
0	0	1	0
0	1	0	1
0	1	1	1
1	0	0	1
1	0	1	1
1	1	0	1
1	1	1	1

The Karnaugh map derived from this truth table is shown in the accompanying figure. There are two groups of four, and the map reduces to $Y = A + B$.

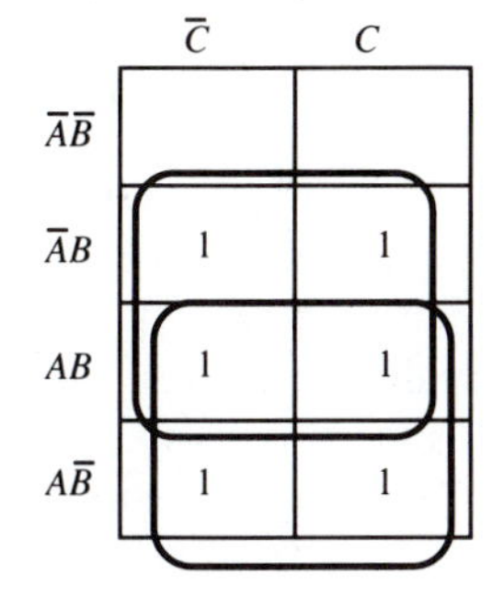

Exercise 4-7 ans.

4-9. $Y = \overline{A}\overline{B}(\overline{C} + B\overline{C}) + A\overline{C}$

Multiplying through $Y = \overline{A}\overline{B}\overline{C} + \overline{A}\overline{B}BC + A\overline{C}$

Using $B \times \overline{B} = 0$ $Y = \overline{A}\overline{B}\overline{C} + A\overline{C}$

4-11. First simplify the equation:

$$Y = (AB + ACD)C = ABC + ACD = AC(B + D)$$

Then implement per the accompanying figure.

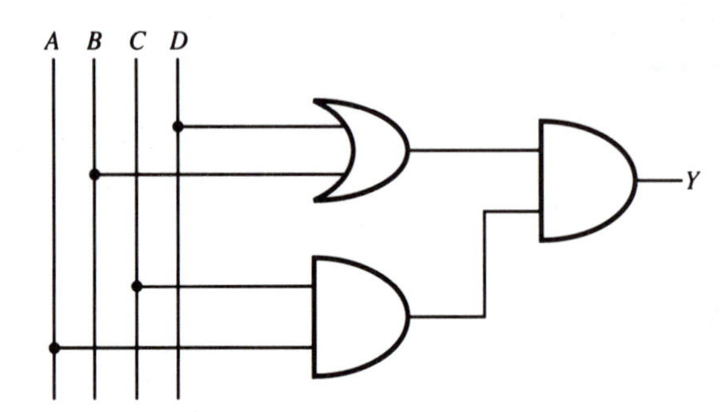

Exercise 4-11 ans.

Answers for Chapter 5 Exercises

5-1. The three reasons for grounding are:

(1) *Safety.* An example is grounding the metal housing of a washing machine in case of insulation failure.

(2) *Prevent interference.* An example is grounding a coaxial cable shield that completely encloses the signal-carrying conductor. The shield shorts out any interference from outside sources.

(3) *Common reference.* An example is using the chassis of a TV as the return conductor.

5-3. The shield wire on a coaxial cable acts to enclose the signal conductor with a metal or conducting housing that can be grounded to prevent interference from outside sources by shorting them to ground.

5-5. Ground loops are a problem for low-level signals in the millivolt range.

5-7. An electrostatic charge is built up by rubbing two unlike materials together. For example, walking across a carpet with rubber-sole shoes would cause a voltage build-up on your body that could easily be hundreds of volts. When you pick up a device containing IC's, this build-up may discharge to ground via a conductor through an IC and damage the IC. A good way to prevent this is to ground yourself on a good ground before handling such devices. Even better would be to attach a ground to your body when handling any sensitive devices.

5-9. An example in which setting the error detection to leave the outputs in the last state could create a problem is a cement mixing plant in which a PLC was in the process of loading sand, gravel, and cement mix into a truck. The PLC could have opened all the gates from the bins for the three materials, then if the error detection left them in that state, the truck and driver could be buried. Obviously, you want these gates closed or shut off if an error occurs.

5-11. Five factors which would affect the location of a PLC are:

Clear space to work on the PLC
Excessive heat environment
Dust or corrosive atmospheres
Excessive vibration
Need to control PLC's closeness to equipment

Answers for Chapter 6 Exercises

6-1. (a) $18.75 = 10010.11_2$ (b) $497 = 111110001_2$

6-3. (a) $532_{10} = 214_{16}$ (b) $356.25_{10} = 164.4_{16}$

6-5. (a) $555_{10} = 1053_8$ (b) $786_{10} = 1422_8$

6-7. (a) $000110010101_{BCD} = 195_{10}$
(b) $010110010101_{BCD} = 595_{10}$
(c) $110010101_2 = 101011111_{grey}$
(d) $10010101_{grey} = 11100110_2$

6-9. The most common language for programming programmable controllers at present is relay ladder language. This may change as PLC's become more sophisticated, since the relay ladder language is limited in scope.

6-11. *BCD* means "binary-coded decimal." Its most common use is for easy conversion from decimal digits to binary. Each decimal digit is converted to a binary code. The most popular code is the 8421, or natural, BCD: a four-bit code the binary equivalent of the value of the digit. There are hundreds of other codes that involve various numbers of binary digits for each decimal digit. Some are meant to make error detection easy when communicating, others to simplify mathematical operations.

6-13. Every programmable controller needs a bootstrap program that enables the PLC to get started when it is first turned on. This program has to be there all the time. A PLC without a bootstrap program could do nothing when turned on, and could not even communicate with you via a keyboard.

6-15. **(1)** A prototype unit will be made, and the programs to run the PLC will be in developmental stages and will need constant changing. The easiest and most economical way to make such changes is through the use of an EPROM or EEPROM. Even though these are the most expensive memory chips, the quantities required at this stage are very small.
(2) After the prototype unit has proved itself, a limited number of PLC's can be produced and sent to selected customers to be beta-tested.

These customers may find problems the prototype testing did not find, problems that will be corrected before full production. PROMs, which are less expensive than EPROMs or EEPROMs, are a good choice for this stage, for they can be easily programmed and sent to the beta-test customers as needed. The hundreds of memory chips needed at this stage are still not enough to have a special ROM chip designed and manufactured.

(3) A full production run next. Now the manufacturer is confident of the design of the PLC, and read-only memory chips are required in large quantities. This makes the ROM the only economical choice, because any other memory chips would be too expensive.

Answers for Chapter 7 Exercises

7-1. The red light will be off for 0–5 seconds and then come on; the yellow light will be on for 0–5 seconds and then go off.

7-3. The white light will go on as soon as the push button is closed, and will stay that way until 2 seconds after the push button is released. The green light will be on initially until PB2 is pushed. Then it will go off and stay that way until 2 seconds after the push button is released, after which it will come on and stay on.

7-5. The input/outputs cannot be directly connected to the internal data bus on a PLC because the data bus is shared between memory and the CPU as well as I/O. The data bus changes constantly while the PLC is operating because information is going to various places via the data bus. Inputs would conflict with these changes, causing shorts when they were different, and outputs would be turning off and on randomly. Special modules must be designed so that signals are sent and received under control of the CPU.

7-7. A *latch* is a relay that will be mechanically held once it is picked up electrically. The mechanical latch can be reset either by hand or with a special reset coil. Latches have two separate coils—one to operate the relay, and one to reset the relay.

7-9. An input module functions to convert the voltage/current signal to the proper voltage for the PLC data bus and to pass this voltage to the data bus. When an output module receives a command from the PLC, it reads and latches the digital signal, then converts the digital signal to the proper output voltage for output.

Answers to Chapter 8 Exercises

8-1. The three parts of the processor memory are: the program files, the data files, and unused memory.

8-3. Subroutines are assigned to program files 3 through 999.

8-5. Decimal and ASCII.

8-7. O:014/14

8-9. Three words make up the timer element: the control word, the preset word, and the accumulated word.

Answers to Chapter 9 Exercises

9-1. On-line

9-3. Zero

9-5. See accompanying figure.

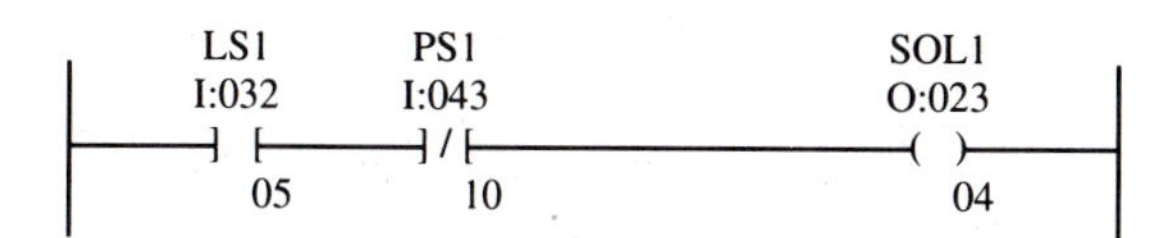

Exercise 9-5 ans.

9-7. See accompanying figure.

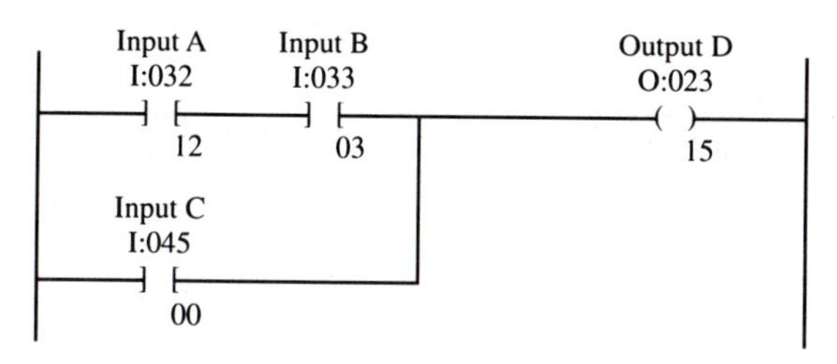

Exercise 9-7 ans.

9-9. See accompanying figure.

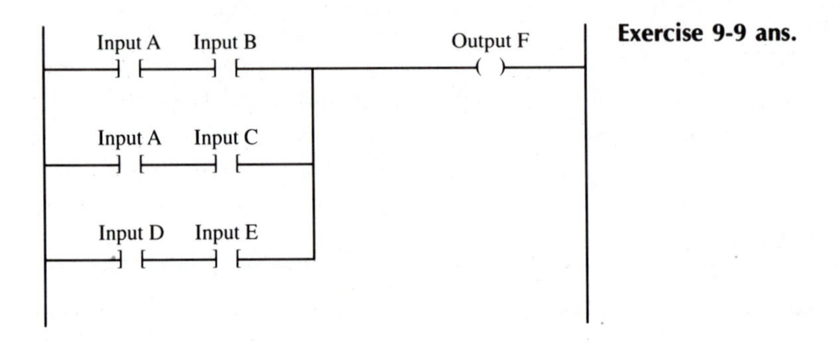

Exercise 9-9 ans.

Answers to Chapter 10 Exercises

10-1. Three data-table words: the control word, the accumulated word, and the preset word.

10-3. 32,767 seconds

10-5. Off-delay timer (TOF)

10-7. See accompanying figure.

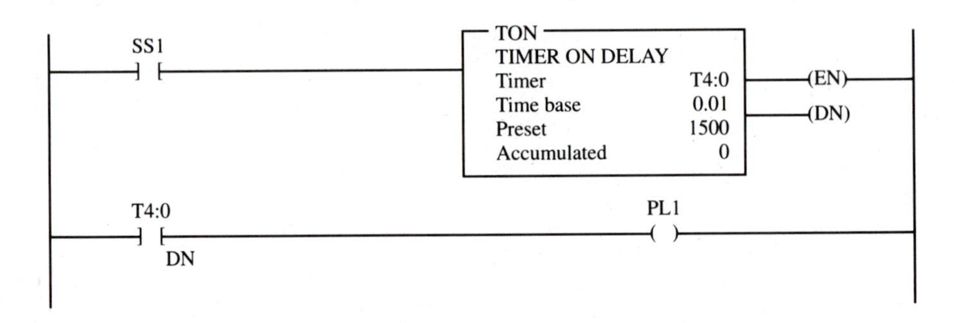

Exercise 10-7 ans.

10-9. See accompanying figure.

PB1
CTU
COUNT-UP COUNTER
Counter C5:0
Preset 10
Accumulated 0
(CU)
(DN)
C5:0
DN
PL3
( )
PB2
C5:0
(RES)

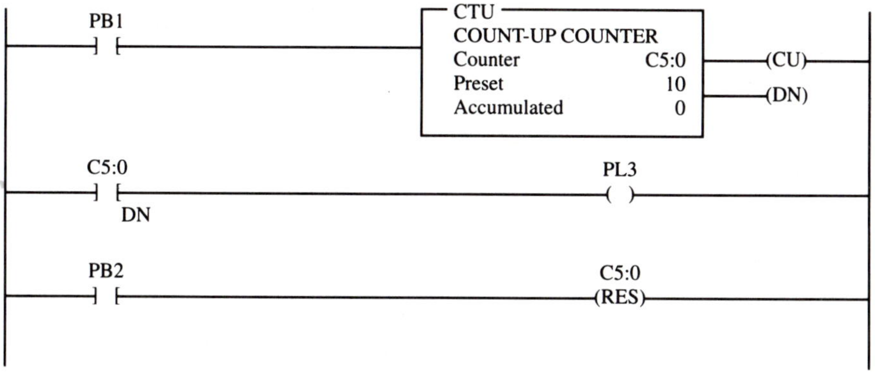

Exercise 10-9 ans.

10-11. The counter minimum preset is −32,768; its maximum preset is +32,767.

10-13. See accompanying figure.

PB1
CTU
COUNT-UP COUNTER
Counter C5:6
Preset 10
Accumulated 0
(CU)
(DN)
PB2
CTD
COUNT-DOWN COUNTER
Counter C5:6
Preset 10
Accumulated 0
(CD)
(DN)
C5:6
DN
PL2
( )
PB3
C5:6
(RES)

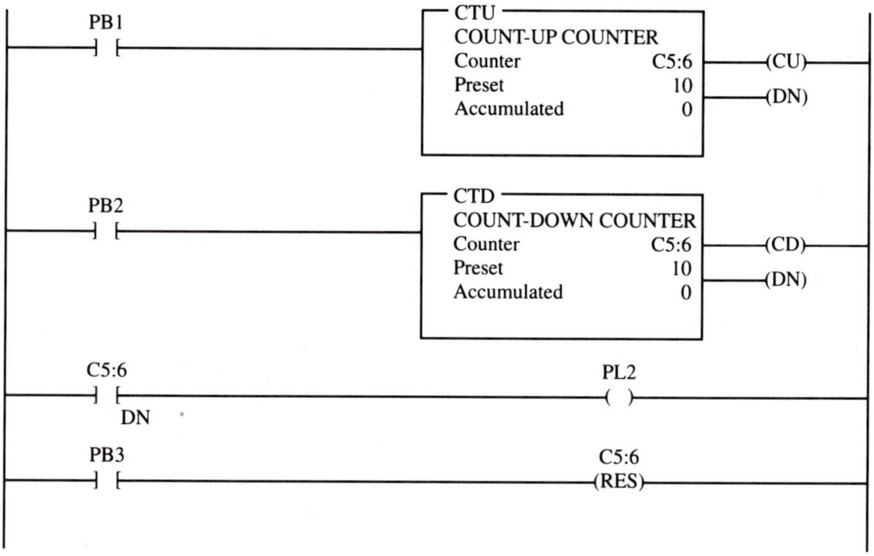

Exercise 10-13 ans.

10-15. See accompanying figure.

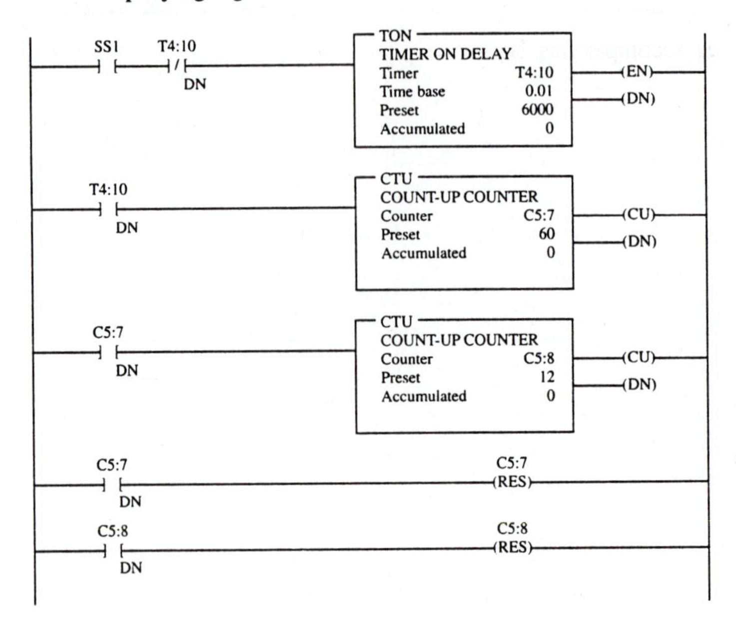

Exercise 10-15 ans.

10-17. See accompanying figure.

SN2 T4:10
DN
TON
TIMER ON DELAY
Timer T4:10
Time base 0.01
Preset 2000
Accumulated 0
(EN)
(DN)
T4:10
DN
CTU
COUNT-UP COUNTER
Counter C5:7
Preset 3
Accumulated 0
(CU)
(DN)
T4:10
DN
PL5
(L)
C5:7
DN
PL5
(U)
C5:7
DN
C5:7
(RES)

Exercise 10-17 ans.

Answers to Chapter 11 Exercises

11-1. See accompanying figure.

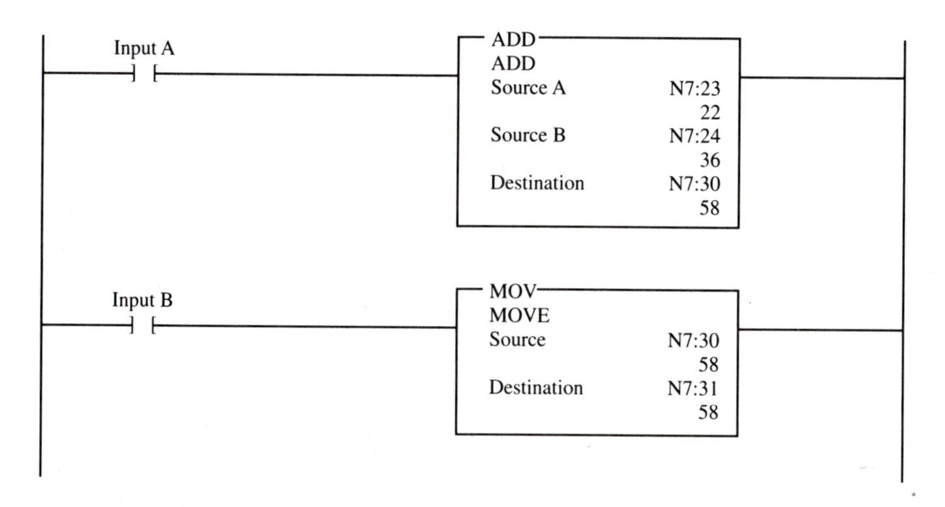

Exercise 11-1 ans.

11-3. See accompanying figure.

Input C
DIV
DIVIDE
Source A N7:48
333
Source B 50
Destination F8:0
6.66

Exercise 11-3 ans.

11-5. See accompanying figure. *Note:* For an explanation of the values stored in the TOD and FRD instructions, refer to p. 147 in the text.

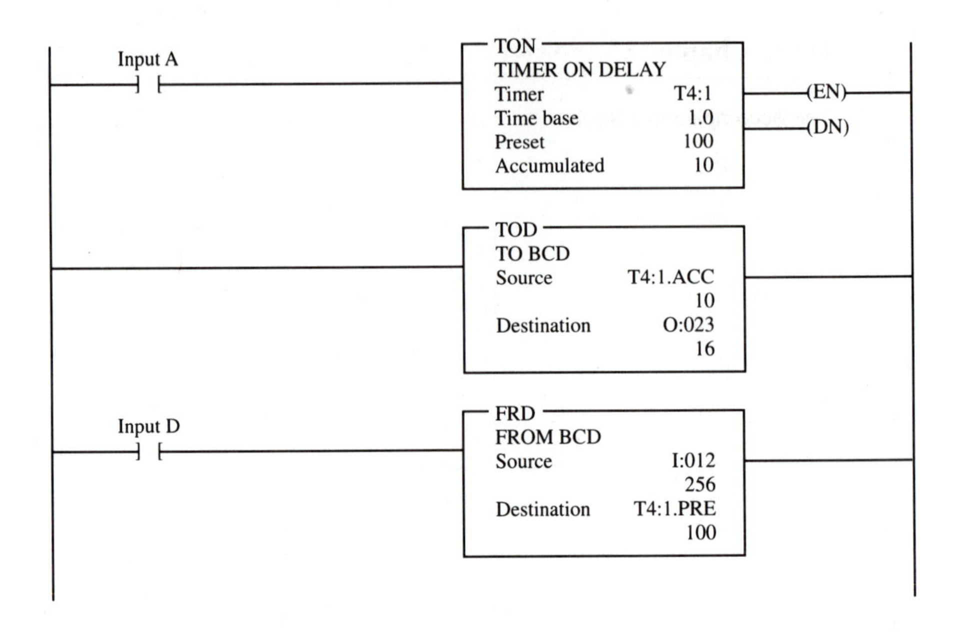

Exercise 11-5 ans.

11-7. No, the instruction is not true. In order for the instruction to be true, the result stored at the compare address, B3:34, would have to be: 1100XXXX1100XXXX. The X locations can have either a 1 or a 0, since those bit locations are being masked out.

11-9. See accompanying figure.

Input H

XOR
BITWISE EXCLUS OR
Source A B3:21
1100110011001100
Source B B3:15
0111000001110101
Destination B3:10
1011110010111001

Exercise 11-9 ans.

Answers to Chapter 12 Exercises

12-1. See accompanying figure.

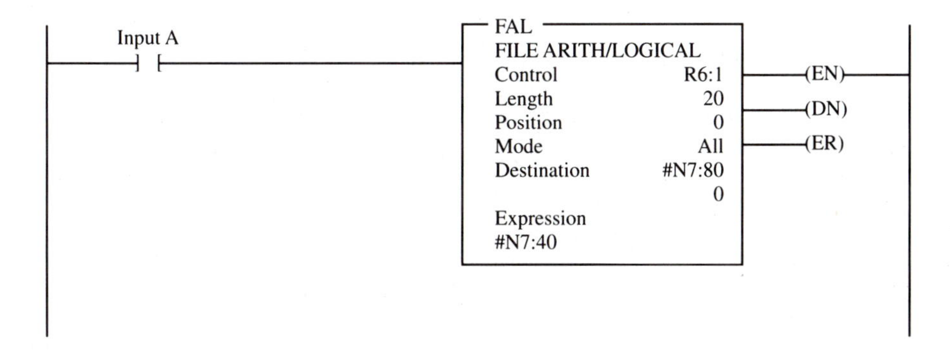

Exercise 12-1 ans.

12-3. See accompanying figure.

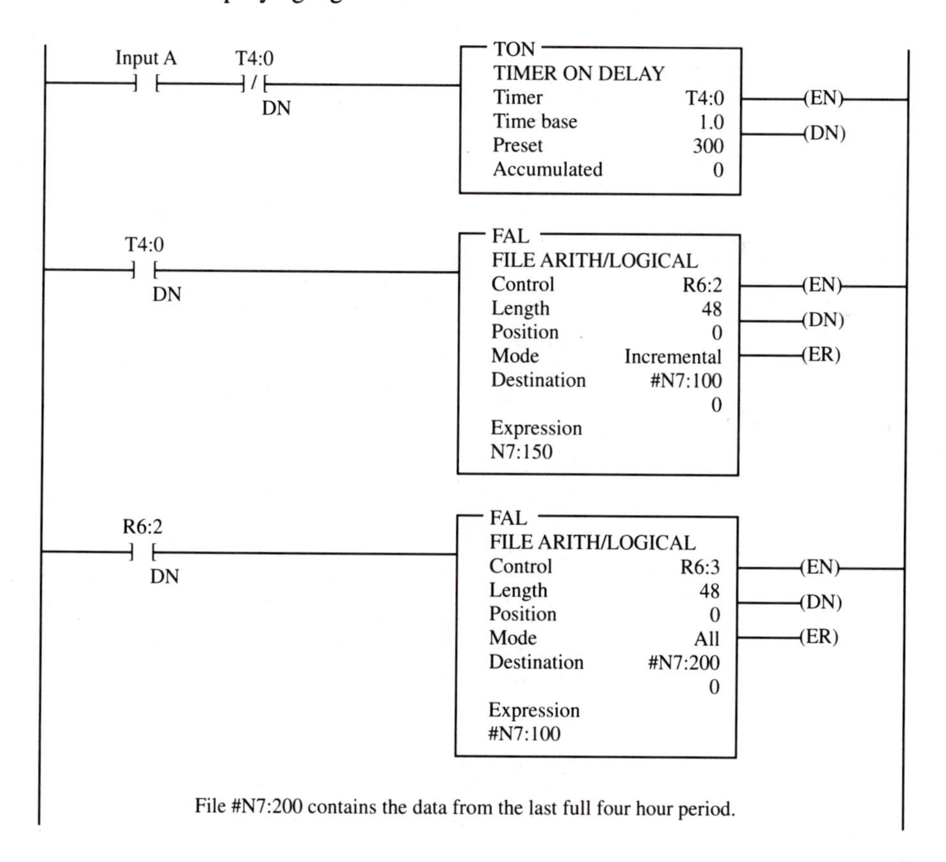

Exercise 12-3 ans.

12-5. 1000.

12-7. Have the instruction take less time to operate per scan, and spread the operation out over more than one scan. This will shorten the time of each program scan.

12-9. See accompanying figure.

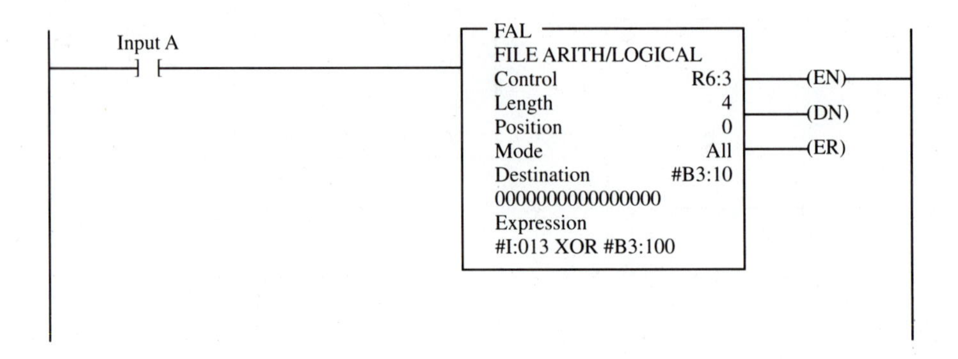

Exercise 12-9 ans.

12-11. See accompanying figure.

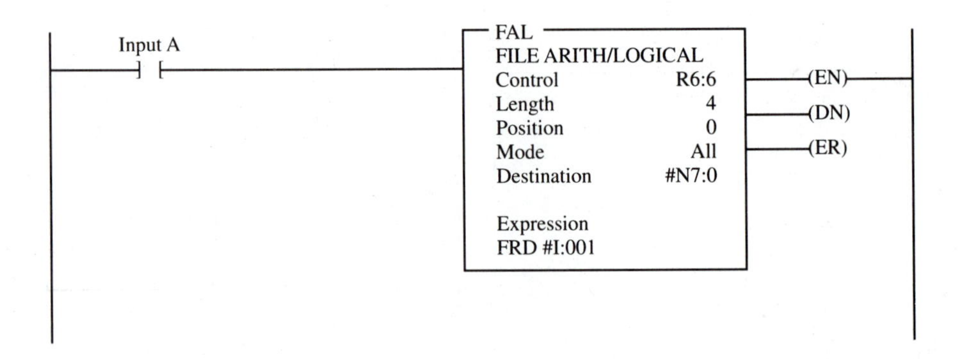

Exercise 12-11 ans.

Answers to Chapter 13 Exercises

13-1. Bits

13-3. See accompanying figure.

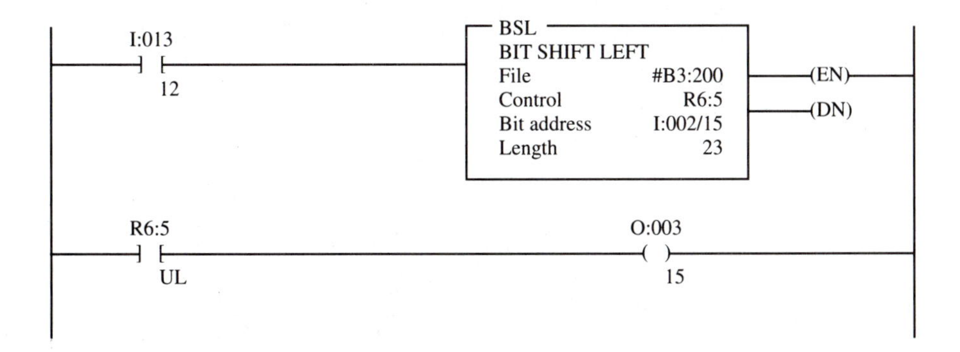

Exercise 13-3 ans.

13-5. The same control address.

13-7. See accompanying figure.

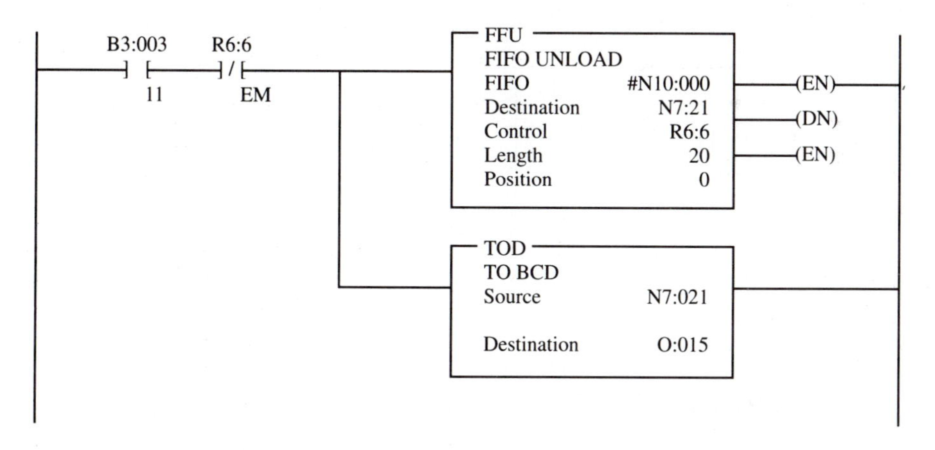

Exercise 13-7 ans.

13-9. See accompanying figure.

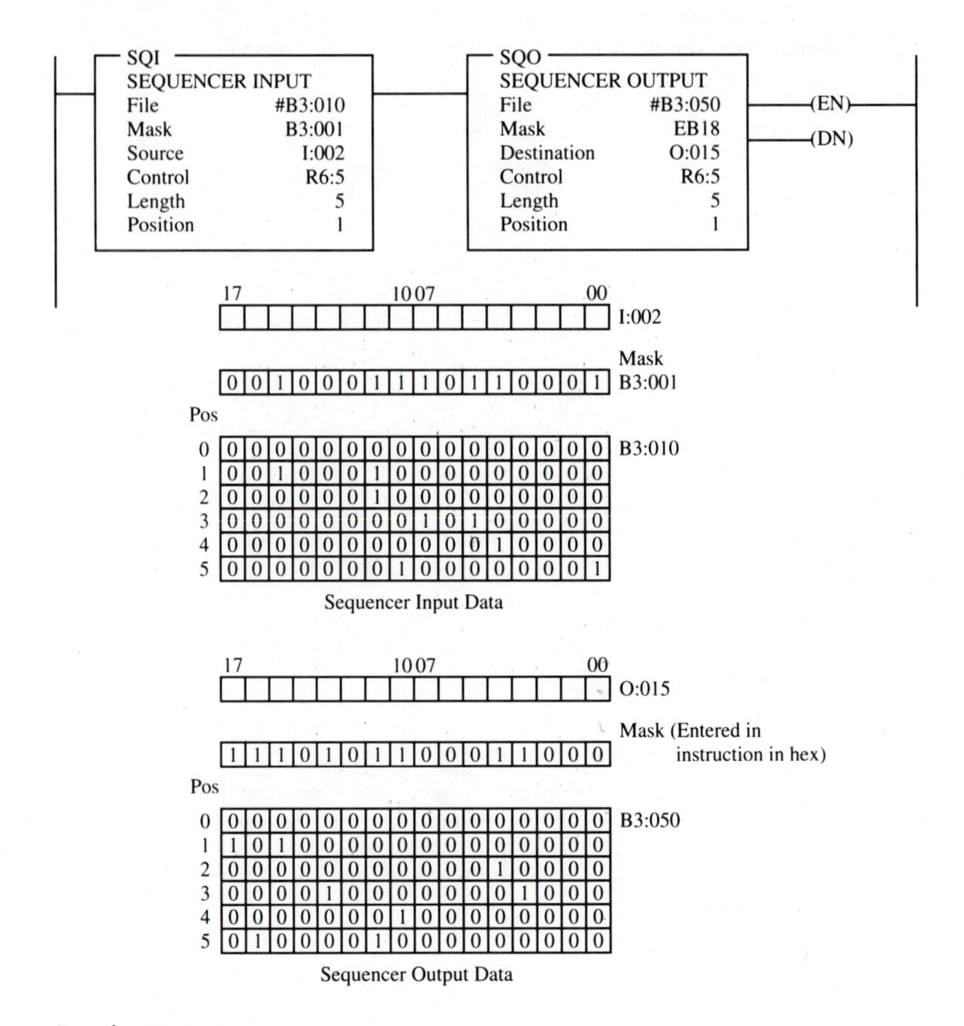

Exercise 13-9 ans.

Answers to Chapter 14 Exercises

14-1. An MCR zone is created by a conditional start fence and an unconditional end fence. MCR zones may not be overlapped or nested.

14-3. The number of MCR zones permitted is limited only by the processor memory.

14-5. The number of JMP instructions is limited only by the processor memory, whereas each program file can have 32 LBL instructions, numbered from 0 through 31.

14-7. It creates a loop, and there is a danger that the watchdog timer may time out if the processor scan is not completed in the allocated time. A scan counter could be used inside the loop to restrict the number of times the scan is allowed to loop.

14-9. See accompanying figure.

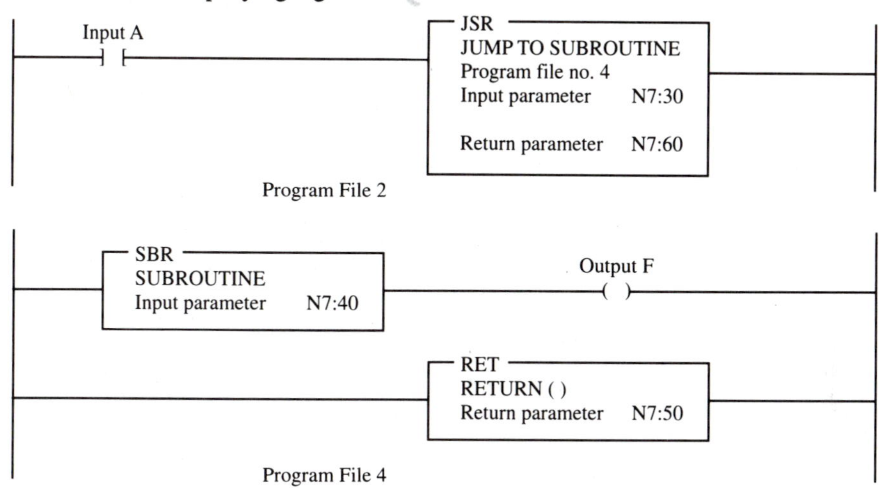

Exercise 14-9 ans.

14-11. The fault routine allows for either an orderly shutdown in case of a fault or a chance to continue if the fault routine clears the major fault word. The program file that is the fault routine is determined by the value stored in status word 29. A 0 stored in word 29 disables the fault routine.

14-13. The *immediate output* instruction should come after the output instruction controlling it, so that its current status is transferred to the real-world output.

14-15. Rack and I/O group

Answers to Chapter 15 Exercises

15-1. If Figure 15-2 were changed per figure 15-19, the resulting problem would be that two states would be active at one time for one scan, which could cause serious damage or a hazardous condition. For example, if one state were a machining operation and the next state were moving the piece to a new location, then moving the piece before the machining is finished would obviously damage both the piece and the machining device.

15-3. The control required for an Allen-Bradley PLC-5 is shown in the accompanying five figures.

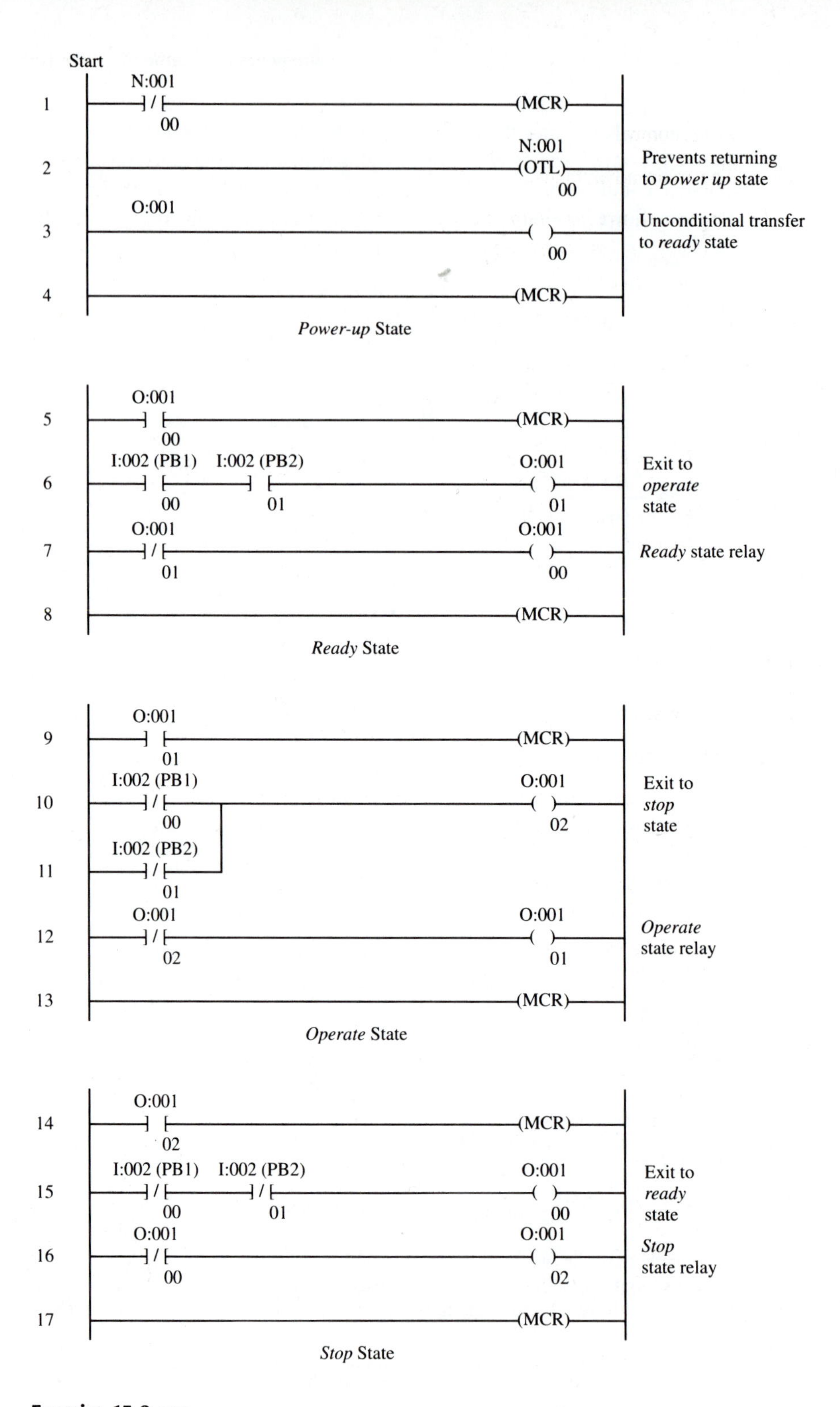

Exercise 15-3 ans.

15-5. Zone control.

15-7. The main advantages:

Organizes the problem design so that solutions are obtained more efficiently, in less time, and with fewer errors.

Easier to troubleshoot.

Easier for a person to follow.

15-9.

Transition	*Boolean Equation*
5	$F + L * \overline{F}$
6	$\overline{E}$
7	$E * \overline{L}$
11	$E * \overline{L} * \overline{\text{Latch}}$
12	$E * \overline{L} * \text{Latch}$
16	F
18	1 (unconditional transition)

Answers to Chapter 16 Exercises

16-1. Star, bus, and ring

16-3. 10,000 cable-feet

16-5. Sixty-four stations, octally numbered 0 through 77

16-7. See accompanying figure.

Message Instruction Data Entry for Control Block N7:100

Read/Write	Read
PLC-5 data-table address	N7:40
Size, in elements	25
Local/Remote	Local
Remote station	N/A
Link ID	N/A
Remote link type	N/A
Local node address	10
Processor type	PLC-5
Destination data-table address	N7:25

Block size = 9 words

Press a key to change parameter or <RETURN> to accept parameters.

Program Forces: None Edits: None PLC-5/15-File HWY

Read/ Write	PLC-5 Address	Size, in Elements	Local/ Remote	Remote Station	Link ID	Remote Link	Local Node	Proce Type	Destin Address
F1	F2	F3	F4	F5	F6	F7	F8	F9	F10

Screen Display for Data Entry in MSG

Exercise 16-7 ans.

Answers to Chapter 17 Exercises

17-1. The six different ways the two-push-button control problem was implemented in this chapter are: mechanical relay logic; relay logic on PLC-5; sequential function charts on PLC-5; digital logic gates; Boolean on PLC-5; BASIC on PLC-5.

17-3. *Machine language* is the binary code that the CPU can work with and interpret. The CPU can only work with 1's and 0's. The user is not required to program in machine language because it is very difficult (user unfriendly), quite tedious, and error-prone for people to use.

17-5. The PLC manufacturer must provide an interpreter in order to make a new high-level language available on a PLC.

Answers to Chapter 18 Exercises

18-1. Two manufacturers of PLC's who have structured programming available are: (1) Allen-Bradley—sequential function charts; (2) Telemecanique—Grafset charts.

18-3. The Allen-Bradley PLC-5 has a modular chassis. The following is a sample of the available modules.

1771-ASB	Remote I/O Adapter Module
1771-DA	ASCII I/O Module
1771-DB	Basic Module
1771-DC	Real-Time Clock Module
1771-DCM	Direct Communication Module
1771-DE	Absolute Encoder Module
1771-DL	Grey Encoder 12–24-V Input Module
1771-DR	I/O Logic Controller Module
1771-DS	Latching Input Module, 10–27-V DC, 8 inputs
1771-DW	Wire Fault Module, 12–24-V DC, 7 inputs
1771-IAD	120-V AC Input Module, 16 inputs
1771-1EO1	Analog Input Module, 8 Bits, 8 Sgl-End 1–5-V
1771-IJ	Encoder/Counter Module TTL, 5-V DC
1771-1R	RTD Input Module, 6 3-Wire RTD's
1771-1S	5-V DC Multiplexer Input Module, 72 inputs

18-5. Example answer for RS-232C:

It would be impossible to connect different types of equipment together without some standard methods of communication. When two devices want to talk to each other, some handshaking needs to take place. For example, when information is going to be transmitted, the receiving device must be on and ready, otherwise the information would be lost. Handshaking was developed to remedy these types of problems. One device can determine if another is ready by checking the status of certain pins, with a high voltage meaning a yes and a low voltage no.

Another example is the downloading of information to a printer: Information can be sent to a printer much faster than the printer can print. Most printers have an RAM storage area (a *buffer*) used to hold information waiting to be printed. Transmitted information that overflows this buffer is lost. The remedy is a signal-on pin, which tells when the buffer is full so that transmission can be stopped, and another pin that tells when it is okay to proceed.

RS-232C is one of the standard protocols for handling the necessary handshaking, and is also one of the most common communication protocols in use today. The normal standard is a 25-pin D-type connector. Here are nine of the most commonly used pin numbers, signal designations, and functions:

Pin No.	Signal Designation	Function
1	Ground	Equipment ground
2	Transmit data	Transmit line
3	Receive data	Receive line
4	Request to send	Enter transmit mode
5	Clear to send	Ready to transmit
6	Data set ready	Establish communication
7	Signal ground	Common ground reference for communication
8	Carrier detect	Carrier being received
20	Data terminal ready	Connect, power up, ready

The RS232C protocol basically serves for serial communication, because information is transmitted and received on a single conductor. The receive and transmit lines each have a signal conductor of their own, so transmitting and receiving can take place simultaneously. Operating in this mode is referred to as *full duplex*. *Simplex* involves communication in one direction only. *Half duplex* involves communication in both directions, but only in one direction at a time.

The number of conductors between the two devices is a function of how much handshaking is required. The simplest setup requires only three conductors, whereas a fully developed RS-232 interface might need 22.

The RS-232C Protocol has a special bit format for communicating that requires a start bit, parity, and a stop bit as part of a word of information. Successful communication also depends on the devices' being set at the same baud rate of transmission.

18-7. Three manufacturers and the models with available software that enable their PLC to be programmed with personal computers are:

Allen-Bradley model SLC-500

Eaton Corp. Cutler-Hammer model D500CPU25

Siemens model S5-100U/CPUI03

Bibliography

Ball, Kenneth E., "An Update on the Micros," *Programmable Controls—The User Magazine,* Sept.–Oct. 1988.

Basic Module: User's Manual. Publication 1771–6.5.34. Allen-Bradley, 1987.

Goody, Roy W. *The Intelligent Microcomputer.* Chicago: Science Research Associates, 1982.

Jones, C. T., and L. A. Bryan. *Programmable Controllers Concepts and Applications.* Atlanta: IPC/ASTEC, 1983.

Kissell, Thomas E. *Understanding and Using Programmable Controllers.* Englewood Cliffs, N.J.: Prentice Hall, 1986.

Otter, Job Den. *Programmable Logic Controllers: Operation, Interfacing, and Programming.* Englewood Cliffs, N.J.: Prentice Hall, 1988.

PLC-5 Family Programmable Controllers: Assembly and Installation Manual. Publication 1785-6.6.1. Allen-Bradley, 1989.

PLC-5 Family Programmable Controllers: Processor Manual. Publication 1785-6.8.2. Allen-Bradley, 1987.

PLC-5 Programming Software: User's Manual. Publication ICCG-5.1. Allen-Bradley, 1988.

Romero, Stephen G. "A Comprehensive Evaluation of PLC: Micro, Mini, and Large," *Programmable Controls—The User Magazine,* May–June 1987.

Tocci, Ronald J. *Digital Systems: Principles and Applications.* 4th ed. Englewood Cliffs, N.J.: Prentice Hall, 1988.

Warnock, Ian G. *Programmable Controllers: Operation and Application.* Hertfordshire, U.K.: Prentice Hall International, 1988.

Web, John W. *Programmable Controllers: Principles and Applications.* Englewood Cliffs, N.J.: Prentice Hall, 1988.

Weingarten, Warren, Robert F. Filer, Paul Lewis, and Wayne Weingarten. "Improving PLC Programming Techniques," *Programmable Controls—The User Magazine,* May–June 1988.

Wilhelm, Robert E. *Programmable Controller Handbook.* Hasbrouck Heights, N.J.: Hayden Book Co., 1984.

Glossary

analog-input module Converts analog signals to digital signals. Isolates the input analog signal and changes it to digital form and the correct level so the changed signals are compatible with the PLC's data bus.

analog-output module Converts digital signals from the processor into isolated analog signals that can be used to drive output devices.

AND gate A logic device requiring all its inputs to be satisfied before activating its output.

ANSI/IEEE Stands for "American National Standards Institute/Institute of Electrical and Electronic Engineers," the group that develops standards for the manufacturing and testing of electrical products.

ASCII-input module Converts ASCII-code input information from an external peripheral into alphanumeric information a PLC can understand. Uses one of the standard communication interfaces, such as an RS-232 or RS-422.

ASCII-output module Converts alphanumeric information from the PLC into ASCII code to be sent to an external peripheral via one of the standard communication interfaces such as an RS-232 or RS-422.

assembler A computer software program that converts an assembly-language program into machine code.

automatic control A process in which the output is kept at a desired level by using feedback from the output to control the input.

BCD-input module Allows the processor to accept four-bit BCD digital codes.

BCD-output module Enables a PLC to operate devices that require BCD-coded signals to operate.

binary A number system with two digits: 0 and 1.

binary-coded decimal Usually referred to as BCD; a group of binary digits that represent the decimal numbers 0 through 9. The most popular form uses a four-bit binary number equal to the decimal digits when converted into decimal.

bit A binary digit.

central processing unit (CPU) The part of a computer integrated-circuit chip that decodes and executes instructions, performs arithmetic and logic functions, controls data and address buses, and sends necessary control signals to memory and I/O.

communication module Allows the user to connect the PLC to high-speed local networks that may differ from the network communication provided with the PLC.

contactor A special type of relay designed to handle heavy power loads that are beyond the capability of control relays.

control relay A low-power electromechanical device with a wound coil; can be activated and deactivated by applying voltage to the coil. Associated with this coil are sets of normally opened and normally closed contacts that change with the activating and deactivating of the coil.

data files The area of processor memory in which are stored status and data values that can be accessed by the ladder logic program.

discrete control A process that allows inputs and outputs to be in only one or the other of two states, for example, off or on.

discrete-input module Lets the user make two-state signals available to the PLC's for use in the control program. Isolates the input signal and changes the form and the level so the changed signals are compatible with the PLC's data bus.

discrete-output module Enables the PLC's processor to control output devices by changing a digital-level signal to the level required by the devices being controlled.

dry-contact-output module Enables a PLC's processor to control output devices by providing a contact that is isolated electrically from any power source.

EEPROM Stands for "electrically erasable, programmable read-only memory." An electronic device on which the program may be stored as a backup to the program in the processor. It is electrically erasable.

element Uses either one, two, or three words in the data file. Timer, counter, and control elements each consist of three words; floating-point elements consist of two words; all other data types consist of one word per element.

encoder-counter module Enables continual monitoring of an incremental or absolute encoder.

EPROM Stands for "erasable, programmable read-only memory." An electronic device on which the program may be stored as a backup to the program in the processor. It may be erased by exposing it to ultraviolet light.

EXclusive OR gate A logic device requiring one or the other but not both of its inputs to be satisfied before activating its output.

fault-routine file A special subroutine that, if assigned, executes when the processor has a major fault. Unless the fault word is cleared, it is scanned a single time before the processor shuts down. The subroutine may be numbered from 3 to 999, and is assigned by placing the file number in the appropriate word in the status data file.

general-purpose control relay A simple, low-power electrical device with a coil that, when energized or deenergized, causes the states of its associated contacts to change from closed to opened or vice versa. The contacts usually are rated below 10 amps.

grey code A special binary code developed for decoding angular displacement with a resolution equal to the least significant bit.

grey-encoder module Converts the grey-code signal from an input device into straight binary.

hexadecimal A number system with sixteen digits: 0 through 9 and A through F.

high-level languages Accepts words or groups of words, normally English, and converts these commands into machine-language routines that allows these commands to be executed by the CPU.

industrial control relay Similar to a general-purpose relay, except that it is constructed more stoutly and has contacts generally rated to withstand more voltage and current. Being designed to operate reliably in industrial environments, it is physically bigger and tougher.

industrial terminal The device used to enter and monitor the program in a PLC.

instruction The part of ladder logic that tells the processor which function to perform.

interpreter A machine-language program written so certain keystrokes or groups of keystrokes (called *commands*) can be understood by the microcomputer.

INVERTER A logic device that causes the opposite condition of its input to occur at the output.

I/O group A logical-addressed unit consisting of sixteen input points and sixteen output points. Eight I/O groups make up one rack.

isolated-input module Enables a PLC's processor to receive dry contacts as inputs. Monitors these contacts and produces two-state digital signals the processor can read.

JIC (Joint International Congress) standards Standards used in drafting relay logic.

Karnaugh maps Used to obtain the simplified logic expression for a given truth table by analyzing a special array of the fundamental products.

language module Enables the user to write programs in a high-level language. BASIC is the most popular language module. Other language modules available include C, FORTH, and PASCAL.

machine language Employs binary patterns to program a microcomputer.

motor starter A special relay designed to provide power to motors; has both a contactor relay and an overload relay connected in series and prewired so that if the overload operates, the contactor is deenergized.

nested branches A branch that begins or ends within another branch.

nonvolatile RAM Memory that will not be lost if the power turns off.

octal A number system with eight digits: 0 through 7.

off-line programming The ability to write program and store it in the personal computer without being connected to a PLC.

on-line programming Entering program directly into a PLC.

OR gate A logic device requiring one or the other or both of its inputs to be satisfied before activating its output.

Petri diagram A pictorial representation of a control process that requires parallel branching and simultaneous processing. Shows the possible paths the process can take, the necessary Boolean conditions to go from one state to another, and where convergence is required to continue.

PID module Proportional integral-derivative closed-loop control lets the user hold a process variable at a desired setpoint.

postulate A self-evident truth.

program files The area of processor memory where the ladder logic programming is stored.

programmable controller A computer that has been hardened to work in an industrial environment and is equipped with special I/O and a control programming language.

PROM Stands for "programmable read-only memory."

rack An addressable unit consisting of 128 input points and 128 output points. Since an I/O group has sixteen input points and sixteen output points, a rack consists of eight I/O groups, numbered from 0 through 7.

RAM Stands for "random-access memory." A read/write memory.

rated current The value of current a device can carry continuously without damage.

rated voltage The value of continuously applied voltage at which a device will operate properly and without any problems.

register A collection of memory cells used for temporary storage of binary information.

relay logic Discrete control implemented through the use of relay coils and contacts.

ROM Stands for "read-only memory." Such memory is permanently written and cannot be altered.

selectable timed-interrupt file A special subroutine file that can be assigned to any program file from 3 through 999. Will be executed on a time basis rather than on an event basis. The file is set by entering in the status section of the data files the file number and the interval at which it is to be executed.

sequential control A process that dictates the correct order of events and allows one event to occur only after the completion of another.

sequential function charts Allen-Bradley's software available on their PLC that implements state diagrams and Petri networks using relay logic.

servo module The device whose feedback is used to accomplish closed-loop control. Though programmed through a PLC, once programmed it can independently control a device without interfering with the PLC's normal operation.

solenoid A simple electromechanical device used to get mechanical displacement from an electrical signal.

state diagram A pictorial representation of a sequential control process that shows the possible paths the process can take and the Boolean conditions needed to go from one state to another.

stepper-motor module Provides pulse trains to a stepper-motor translator that enables control of a stepper motor.

theorem A statement that can be proven.

thumb-wheel module Enables thumb-wheel switches to feed information in parallel to a PLC to be used in the control program.

token A computer password that is passed from state to state.

truth table A chart showing all the possible input variables in all possible combinations and the desired output for each of these combinations.

TTL-input module Enables devices that produce TTL-level signals to communicate with a PLC's processor.

TTL-output module Enables a PLC to operate devices requiring TTL-level signals to operate.

unused memory The area of processor memory that is unassigned and can be allocated to either data files or program files.

volatile RAM Memory that will be lost if power to the computer is turned off.

wiring diagram A drawing that shows the way the electrical conductors are connected to devices to implement the desired control shown on the schematic.

word Sixteen bits.

Index